Global Perspectives on Health Geography

Series Editor
Valorie Crooks, Department of Geography
Simon Fraser University
Burnaby, BC, Canada

Global Perspectives on Health Geography showcases cutting-edge health geography research that addresses pressing, contemporary aspects of the health-place interface. The bi-directional influence between health and place has been acknowledged for centuries, and understanding traditional and contemporary aspects of this connection is at the core of the discipline of health geography. Health geographers, for example, have: shown the complex ways in which places influence and directly impact our health; documented how and why we seek specific spaces to improve our wellbeing; and revealed how policies and practices across multiple scales affect health care delivery and receipt.

The series publishes a comprehensive portfolio of monographs and edited volumes that document the latest research in this important discipline. Proposals are accepted across a broad and ever-developing swath of topics as diverse as the discipline of health geography itself, including transnational health mobilities, experiential accounts of health and wellbeing, global-local health policies and practices, mHealth, environmental health (in)equity, theoretical approaches, and emerging spatial technologies as they relate to health and health services. Volumes in this series draw forth new methods, ways of thinking, and approaches to examining spatial and place-based aspects of health and health care across scales. They also weave together connections between health geography and other health and social science disciplines, and in doing so highlight the importance of spatial thinking.

Dr. Valorie Crooks (Simon Fraser University, crooks@sfu.ca) is the Series Editor of Global Perspectives on Health Geography. An author/editor questionnaire and book proposal form can be obtained from Publishing Editor Zachary Romano (zachary.romano@springer.com).

Melinda Laituri
Robert B. Richardson • Junghwan Kim
Editors

The Geographies of COVID-19

Geospatial Stories of a Global Pandemic

Editors
Melinda Laituri
Colorado State University
Fort Collins, CO, USA

Robert B. Richardson
Michigan State University
East Lansing, MI, USA

Junghwan Kim
Virginia Tech
Blacksburg, VA, USA

ISSN 2522-8005 ISSN 2522-8013 (electronic)
Global Perspectives on Health Geography
ISBN 978-3-031-11777-0 ISBN 978-3-031-11775-6 (eBook)
https://doi.org/10.1007/978-3-031-11775-6

This Springer imprint is published by the registered company Springer Nature Switzerland AG
The registered company address is: Gewerbestrasse 11, 6330 Cham, Switzerland

This book is dedicated to the scientists and researchers around the world who have contributed to our understanding of the global pandemic. And to the lives lost – 6,000,000 and counting.

Foreword

I count myself fortunate to have had the opportunity to work with Dr. Melinda Laituri and Dr. Robby Richardson during their tenures as Jefferson Science Fellows in the U.S. Department of State's Office of the Geographer and Global Issues, which I direct in my role as the Department's Geographer. This office conducts research and analysis with a geographer's lens on a broad range of issues that includes, among others, food security, water security, climate change, migration, infectious disease, and human rights. The COVID-19 global pandemic impacts each of these issues, regionally and differentially. And I can think of few scientists better positioned than the editors of this book to provide expert guidance in stitching together this collection of geospatial "stories" the pandemic has spawned.

During the course of their fellowships in the Office of the Geographer, Dr. Laituri and Dr. Richardson helped my office spearhead participatory mapping projects in more than a dozen "secondary cities" around the world. These projects were designed to tap local knowledge to create geospatial data and build human geography content and knowledge in order to explore rapidly growing urban areas that were often underexamined, poorly mapped, and ill understood by scientific inquiry. A spin-off of one of those projects, the "COVID-19 Cities' Mapping and Mitigation" initiative, introduced me to Junghwan Kim, who is a postdoctoral fellow at Harvard University's Center for Geographic Analysis and the third editor of this book. The combination of expertise these three editors bring to this endeavor are both exceptional and complementary. Dr. Laituri is a geographer who specializes in participatory mapping and geospatial tools and techniques; Dr. Richardson is an ecological economist who studies the contribution of ecosystem services to socioeconomic well-being; and Dr. Kim brings an urban planning and engineering background to the editorial team. (Each of them also contributed chapters to the book). Their training, research, fieldwork, and tradecraft have positioned them, collectively, to generate an outstanding and diverse group of papers that fit seamlessly into an integrated volume but also merit examining as singular contributions to what is one of the most impactful social and economic issue of the day, i.e., the impacts of COVID-19 on society – globally, regionally, and locally.

And just what are those impacts? And how are they manifest over space and time? That is what this volume of storytelling geographies attempts to explore. "Geographies" – because each has a scale of spatial analysis unique to the place or context of the pandemic's impact. "Storytelling" – because the COVID-19 pandemic has become a personal experience, either directly (primary impact) or indirectly (secondary impact) for millions of people worldwide. And many of these stories reveal trends and patterns that can be exceptionally explored though geospatial analyses. Analyses that are underpinned by geographic factors such as scale and location lay bare the power and influence of location, race, age, wealth, politics, media, and their collective consequences. Analyses that also expose and explore possible solutions that meet the challenges of addressing the pandemic's differential impact on communities, governments, industries, educational institutions, economic transactions, mobility, consumption patterns, and, not least of all, public health infrastructures.

What stands out more than anything in this collection of research topics, however, is the landscape of inequality on which it is painted: inequality that is regionally differentiated, dependent on scale, and thus ripe for geographic analysis. This volume of collected stories lays bare some of the truths of the pandemic that spatial analysis and maps effectively lay bare, the downstream consequences of which are yet to be told. The chapters herein not only address well-documented factors related to rich/poor, urban/rural, access to healthcare, race and ethnicity, educational, and digital divides but also delve into less explored differentials in transportation, housing, communication, access to open space, physical barriers, employment sector, legal norms, and the importance of neighborhoods. And, of course, maps and geospatial analysis provide a common language by which these stories are told. Indeed, maps generated by GIS tools and methods of spatial analysis often reveal patterns of inequity, particularly at different scales or via Earth observation processes, which might otherwise not be obvious by scientific inquiry alone, e.g., the impact the pandemic has had on small cities, its relationship to housing quality, and the relationships between mobility and deprived communities. In every case, further study is warranted that would utilize the power of geospatial technologies to develop place-specific tools, e.g., Francis Koti in his chapter calls for the development of an Africa-specific index for assessing socio-spatial vulnerabilities prompted by downstream impacts of the COVID-19 pandemic.

What this volume further underscores are the value of data – methods of their collection, tools for their analysis, veracity, completeness, quality, and integrity. Several of the chapters also highlight the fact that the quality and integrity of COVID-19 related data, and their interpretation and communication, also suffer from inequality and uncertainty. For an event as pervasive as a global pandemic, the challenges of collecting accurate data abound, making it difficult for policy makers to make timely decisions – to address care needs, allocate resources, devise communication strategies, and plan and implement mitigation strategies and actions. Data collection challenges are further compounded by privacy issues and ethical concerns, particularly as Albanese and Senesi point out, in an age of "surveillance capitalism."

In my role as the Geographer of the Department of State, I also serve as one of the co-chairs of the Worldwide Human Geography Data Working Group (WWHGD WG), which since 2011 has provided a platform for a discussion of human geography issues shaping our world, and a venue to better understand the importance of place in the ongoing development of culture and society. The goal of this community is to discover, share, and map human geography data to support decision making with the goal of improving human security. While reading the chapters of *Geospatial Stories of the Global COVID-19 Pandemic* I found myself contextualizing the themes of these stories with those of the WWHGD WG, which include demographics, ethnicity, education, health, economy, transportation, and communication, among others. It struck me how coincident these themes, underlying the range of human geography data essential both for understanding and sustaining communities, align with the geospatial "stories" in this collection, and how the pandemic touches so many distinct aspects of our collective human security that can be explored via geographic inquiry. Another one of my roles as the State Department Geographer is to maintain and disseminate the official U.S. government–recognized international boundaries for all countries of the world. It struck me as ironic that while in my job I demand the integrity of such borders in an unbending fashion, at the same time the world is mired in a pandemic that both shows little respect for borders yet is subject to policies that are strictly prescribed by governments defined by such borders and guide the "human geography" behavior of the citizens living within them.

The transdisciplinary nature of geography allows for an examination of the many facets of the coronavirus pandemic that are essential for understanding the COVID-19 story or, in this case, stories. Using case studies and examples of geospatial analyses, this volume examines several places around the world that have experienced the effects of the pandemic in different ways. As Melinda Laituri described in her opening chapter, the "consequential geography" of the pandemic exposes a landscape of inequality and vulnerable populations. The team of authors she and her co-editors put together for this volume were well positioned and well represented to explore the pandemic's inherent inequality from different regional, national, and academic backgrounds – but always with a geographic perspective grounded on differential scales of analysis. They themselves hail from every continent and represent a diverse and inclusive set of viewpoints, backgrounds, expertise, gender, race, and ethnicity – enriching their stories about the differential impact the COVID-19 pandemic had on diverse socioeconomic and demographic sectors of communities around the world.

Lee Schwartz
U.S. Department of State
Washington, DC, USA
28 February 2022

Disclaimer: The views in this report are solely of the authors and do not represent those of the U.S. government agencies or any of the organizations mentioned. Assumptions made within the analysis are not a reflection of the position of any U.S. government entity.

Preface

When we first considered this volume in early 2021, we assumed that by the time this book was published the pandemic would be behind us, and this would be a reflective volume. Instead, at the beginning of 2022, we find ourselves continuing to be in the midst of the pandemic globally. The COVID-19 pandemic is an on-going and persistent global experience. The keyword is *experience* – as we soon discovered – who you are and where you live have a profound impact on your experience with the pandemic. Using a geographic lens, we were interested in that juxtaposition of the commonality created by the pandemic, the people's pandemic experiences in different places, and myriad reflections of inequality exposed. The existing inequalities are revealed through the increasing differences of income inequality, the deepening digital divide highlighting access to technology – a necessary tool during the pandemic, and the availability of and access to basic healthcare services and vaccinations. This is the moment to examine these inequalities through sharing geospatial stories of the pandemic.

A geographic frame of reference that is both expansive and integrated is a useful approach to assessing the state of the pandemic around the world. Geography crosswalks physical, social, and economic themes to define and create a holistic understanding of place. Place is central to our collective experience because where and how we live is so disrupted – and we have all shared this experience in different ways. The spiraling impacts of the pandemic coupled with government policies and socio-economic disruption further reveal the landscape of vulnerability – who is vulnerable, who is becoming vulnerable, and what is the impact of vulnerability in terms of where people live and work? These are all questions that have a spatial dimension. As the pandemic and human behavior continue to evolve, these questions also have a temporal dimension.

The intersection of geography and technology is exhibited through the numerous online data dashboards that use maps to tell a story. Maps are accessible. Interactive maps are interesting. Highlighting geospatial platforms (interactive maps), data (demography), and applications (identifying hotspots of caseloads) demonstrate the value and importance of these tools to share stories and create visualizations. However, these tools also have their limitations. How do they circumscribe our own geography and sense of place? How can they be employed while also ensuring that personal privacy is protected?

We invited researchers, students, and practitioners from around the world to do two things: (1) address a critical issue related to the pandemic through a geospatial and geographic lens; and (2) share a personal story of the pandemic. The pandemic provides a moment of reflection where we are embedded in the growing realization that the world has changed. In fact, *change* has been the most fundamental shared experience we have all had – the changing conditions of our everyday lives, during the past two years, and into the future. Adaptations to the virus will be place specific and long term. We have much to learn from each other, and this volume contributes to the on-going dialogue.

Melinda Laituri — Fort Collins, CO, USA
Robert B. Richardson — East Lansing, MI, USA
Junghwan Kim — Blacksburg, VA, USA

Contents

List of Figures

List of Tables

About the Editors

Melinda Laituri is a professor emerita at Colorado State University (CSU), USA. She is also a research associate at Harvard University in the Center for Geographic Analysis. She was the founding director of the Geospatial Centroid at CSU and a co-director of the Center for Environmental Justice at CSU. As a Fulbright scholar, she visited Botswana, South Africa, and India. She was a Rachel Carson Fellow at the Rachel Carson Center for Environment and Society at Ludwig-Maximilians-Universitat, Munich. She was a Jefferson Science fellow with the U.S. Department of State.

Robert B. Richardson is a professor at Michigan State University, USA. He is an environmental economist with interests in the contribution of ecosystem services to socioeconomic well-being. He has conducted research on the value of ecosystem services in sub-Saharan Africa, Central America, and Southeast Asia, as well as throughout the USA. He was previously a member of the Board of Scientific Counselors with the U.S. Environmental Protection Agency. He was a Jefferson Science Fellow with the U.S. Department of State.

Junghwan Kim is an assistant professor at the Department of Geography at Virginia Tech. Before joining Virginia Tech, he was a postdoctoral fellow at the Center for Geographic Analysis (CGA) at the Institute of Quantitative Social Science (IQSS) at Harvard University. He received his Ph.D. in Geography from the University of Illinois at Urbana-Champaign in 2021. His research interests include human mobility, environmental health, GIScience methods, geoprivacy, and geospatial data ethics. His works have been published in several peer-reviewed journals, including the *Annals of the American Association of Geographers, Journal of Transport Geography, Cartography and Geographic Information Science, Environmental Research*, and *Transactions in GIS*. He has received many prestigious awards, including the Peter Gould Award from the American Association of Geographers (AAG) Health and Medical Geography Specialty Group, the John Odland Award from the AAG Spatial Analysis and Modeling Specialty Group, and the American Planning Association (APA) Transportation Planning Division Student Paper Award.

About the Authors

Valentina Albanese is a geographer and senior lecturer in the Department of Law, Economy, and Culture at the University of Insubria, Como (Italy), and a research fellow at Social Sciences Institute, University of Maynooth, Dublin (Ireland).

Fatima L. Benitez is a geographer and researcher in the Geography Institute and Galapagos Science Center at San Francisco de Quito University, Ecuador.

Brett Biles is a Murrawarri man and director of indigenous health education in the Medicine and Health Faculty at the University of New South Wales, Australia.

Nama Raj Budhathoki is the founder of Kathmandu Living Labs, Nepal, and regional director at Open Mapping Hub, Asia-Pacific.

Malcolm Campbell is Associate Professor of Geography in the School of Earth and Environment at the University of Canterbury, New Zealand, and a director of the GeoHealth Laboratory.

Enkhtungalag Chuluunbaatar is co-founder of Public Lab Mongolia in Ulaanbaatar, Mongolia, and oversees the organizations long-term planning, program development, and partnerships.

Laura Cline is a geographer and the director of the Participatory Mapping Partnerships (PMP) activities at the Office of the Geographer and Global Issues, U.S. Department of State, Washington, D.C., USA.

Lindsey Conrow is Lecturer in Geography in the School of Earth and Environment at the University of Canterbury, New Zealand.

Aída Guhlincozzi is a postdoctoral scholar in the Departments of Geography and Women's and Gender Studies and Cambio Center Fellow at the University of Missouri, USA.

Matthew Hobbs is a senior lecturer in the School of Health Sciences at the University of Canterbury, New Zealand, and a director of the GeoHealth Laboratory.

Xiao Huang is an assistant professor in the Department of Geosciences at the University of Arkansas and research affiliate in the Spatial Data Lab of Center in the Center for Geographic Analysis at Harvard University, USA.

Ranjit John is the founder and CEO of Hawkai Data, a technology company that enables enterprise digital transformations and the creation of new business and customer experiences.

Deepak Raj Joshi is former chief executive officer at Nepal Tourism Board, Nepal.

Asmod Karki is former chief operating officer at Kathmandu Living Labs, Nepal.

Harriet Kebirungi is Senior Lecturer at Kyambogo University, Uganda in the Department of Development Studies and Director of Gender Mainstreaming at the same University.

Simon Kingham is Professor of Geography in the School of Earth and Environment at the University of Canterbury, New Zealand, a chief science advisor at the Ministry of Transport, and a director of the GeoHealth Laboratory.

Jason H. Knouft is a professor in the Department of Biology at Saint Louis University and a Large River Ecologist at the National Great Rivers Research and Education Center, USA.

Francis T. Koti is a professor at Middle Tennessee State University in the Department of Global Studies and Human Geography, USA.

Byambatsetseg Lkhagvasuren is co-founder of Public Lab Mongolia, Ulaanbaatar, Mongolia, and spearheading the air quality initiatives such as air quality education and open data.

Meifang Li is a research scientist at Dartmouth College in the Department of Geography and has been granted the "Research Fellow" title by Jiangxi Normal University and Nanchang Disease Prevention and Control Center, China since 2020.

Xiao Li is an associate transportation researcher from the Mobility Analysis Program of the Texas A&M Transportation Institute.

Byron Lozada is an economist and researcher at Universidad del Pacífico in Guayaquil-Ecuador and former postdoc at Universidad San Francisco de Quito.

Lukas Marek is a postdoctoral research fellow in the GeoHealth Laboratory at the University of Canterbury, New Zealand.

Patricia Martinez is a geographer and researcher at the Geography Institute at San Francisco de Quito University, Ecuador.

Gaston Mbonglou is the CEO of U.S. and Africa Sourcing Growth Advisors (UASG Advisors), a company that focuses on process improvement and digital transformation. He is also director of the Africa Hub for Participatory Mapping Partnership (PMP).

Carlos F. Mena is Professor of Geography and Ecology in the School of Biological and Environmental Sciences in Universidad San Francisco de Quito, Ecuador. C.F. Mena is co-director of the Galapagos Science Center and director of the Institute of Geography at USFQ.

Maya V. Mishra is a senior undergraduate in the Department of Ecology and Evolutionary Biology at Princeton University, USA, researching how outer space can be leveraged for human health applications on Earth.

Hadijah Mwenyango is a lecturer in the Department of Social Work and Social Administration at Makerere University, Uganda, and coordinator in the East African Social Work Regional Resource Centre.

Xuru Peng is an undergraduate student majoring in geography at Dartmouth College, Hanover, New Hampshire, USA.

Ashley M. Pierce is an associate program director in the environmental engineering and sustainability cluster in the Division for Chemical, Bioengineering, Environmental, and Transport Systems in the Engineering Directorate at the U.S. National Science Foundation.

Nellie Pollard-Wharton is a Kooma woman, mother, social worker, and associate lecturer and PhD candidate in the School of Population Health at the University of New South Wales, Australia.

Marie Price is Professor of Geography and International Affairs at George Washington University, Washington D.C., USA, and the president of the American Geographical Society.

Sampath Rapuri is a high school senior at St. Mark's School of Texas and an affiliate at the Center for Geographic Analysis at Harvard University, USA.

Miri Raven is a senior scientia lecturer at the University of New South Wales, Australia, in the Social Policy Research Centre and Environment and Society Group.

Carolina Sampedro is a geographer and researcher in the Geography Institute at San Francisco de Quito University, Ecuador.

Lee Schwartz is a geographer in the U.S. Department of State and director of the Office of the Geographer, Washington D.C., USA, where he directs research, analysis, and fieldwork on global issues related primarily to complex emergencies and environmental sustainability.

Dara E. Seidl is Associate Professor of Geographic Information Systems at Colorado Mountain College in Leadville, Colorado, USA.

Giorgio S. Senesi is a geologist and researcher at National Council Research (CNR), Institute for Plasma Science and Technology (ISTP), Bari seat, Italy.

Xun Shi is a full professor at Dartmouth College in the Department of Geography.

Amanda R. Shores is an AAAS Science and Technology Policy Fellow at the U.S. National Science Foundation focused on managing transdisciplinary funding opportunities for global environmental change topics for the Belmont Forum.

Erdenetsogt Sumiyasuren is co-founder and the executive director of Public Lab Mongolia, Ulaanbaatar, Mongolia, with a background and experience in geographic information systems and remote sensing in public, private, and, currently, civil society sectors.

Nyamsuren Tsadmid is co-founder of Public Lab Mongolia, Ulaanbaatar, Mongolia, and manages the organization's finances and holds an MBA.

Aleksandar Valjarević is an associate professor in the Faculty of Geography at the University of Belgrade, Serbia.

Deshira Wallace is an assistant professor in the Department of Health Behavior in the Gillings School of Public Health at the University of North Carolina at Chapel Hill, USA.

Kevin Wang is a researcher in the field of applied and theoretical machine learning. He is currently finishing his undergraduate degree in computer science at the University of California, Berkeley, USA.

Siqin Wang is an associate lecturer in the School of Earth and Environment Sciences at the University of Queensland, Australia, and research affiliate in the Spatial Data Lab of the Center for Geographic Analysis at Harvard University, USA.

Jesse Wiki is a postdoctoral research fellow in the School of Population Health at the University of Auckland, New Zealand.

Aryati Yashadhana is a social scientist and research fellow in the Centre for Primary Health Care and Equity and the School of Population Health at the University of New South Wales, Australia.

María B. Zapata is a geographer in the Geography Institute at San Francisco de Quito University, Ecuador.

Abbreviations

AGA	American Geographical Society
C2M2	Cities' COVID Mitigation Mapping
CARES	Coronavirus Aid, Relief, and Economic Security Act
CDC	Centers for Disease Control and Prevention
EO	Earth Observations
GDP	Gross Domestic Product
GEDI	Global Ecosystem Dynamics Investigation
GIS	Geographic Information System
GPS	Global Positioning System
LiDAR	Light Detection and Ranging
MAUP	Modifiable Area Unit Problem
MODIS	Moderate Resolution Imaging Spectroradiometer
MTUP	Modifiable Temporal Unit Problem
NASA	National Aeronautics and Space Administration
NDVI	Normalized Difference Vegetation Index
PPE	Personal Protective Equipment
RNA	Ribonucleic Acid
SARS	Severe Acute Respiratory Syndrome
SDG	Sustainable Development Goals
UK	United Kingdom
UN	United Nations
US	United States of America
WFP	World Food Programme
WHO	World Health Organization

Chapter 1
How COVID Changed Our Daily Geographies

Marie Price

Personal Story
I study human migration because of the many ways human mobility shapes and changes places. On March 1, 2020, I flew to Prague, Czechia, to begin a stint as a visiting scholar at Charles University working with geographers who study migration. As I left the United States, COVID was spreading, but the numbers were low. Taking precautions, I brought masks and hand sanitizers with me. I assumed my research would continue. In less than 2 weeks, the enchanting city of Prague, with its streetcars, cobbled squares, and steeples, began to shut down. By the second week of March, the university had closed to in-person instruction. On the night of March 12th, President Trump's news conference reported that international travel to the United States would end at midnight Friday March 13th. It was time to book a return flight.

The eeriest moment of my return trip was walking through the virtually empty corridors of Frankfurt Airport. Two weeks earlier, the place buzzed with suited business travelers, families, and students. Now I felt alone. As I walked down a long corridor, I spied a lone traveler seated quietly on a bench. I walked closer, hoping to exchange a friendly word, only to realize that this 'stranger' was not a person at all, but a life-size statue of Albert Einstein. It was at that moment I realized the world of mobility that I had taken for granted had halted. I flew home in a half-empty plane and did not board an aircraft again for 6 months when I visited my ailing mother before she passed away.
Marie Price

Much of the scientific effort to alter the trajectory of COVID-19 and produce vaccines and medical treatments has focused on the biomedical sciences. This book highlights the obvious and subtle ways that geography and geographical sciences informed our collective understanding of and response to COVID-19. Key concepts such as scale, space, diffusion, environmental context, and place were and are

M. Price (✉)
George Washington University, Washington, DC, USA

American Geographical Society, New York, NY, USA
e-mail: mprice@gwu.edu

M. Laituri et al. (eds.), *The Geographies of COVID-19*, Global Perspectives on Health Geography, https://doi.org/10.1007/978-3-031-11775-6_1

critical in guiding behaviors to slow COVID-19's transmission and understand the uneven impacts of the virus over time. Geospatial tools such as maps, remotely sensed imagery, dashboards, and location tracking technology all played a role in visualizing the virus's impact and developing human responses to it and will be detailed in this volume.

There is geography in the practices of social distancing, the privileging of outdoor settings, the stunning decline in mobility, and the attributes of place that help to explain various pandemic reactions and outcomes. This essay argues that an explicit set of geographic concepts and tools shaped our everyday geographies during the pandemic. Geospatial tools, especially maps, were invaluable in helping us understand the shifting extent and impact of the pandemic but as Melinda Laituri makes clear in Chap. 2, there is both power and limits to our geospatial tools and the data they rely upon. Authors in this volume apply a framework of first order impacts (immediate responses to the virus itself) and second order ones which derive from human responses to the virus. But as geographer Barney Warf suggests, there were also preconditions that, in the case of the United States, led to higher infection rates and deaths (Warf, 2021).

Now 2 years since the pandemic began, we better understand first order impacts. Yet there is a growing list of second order impacts that influence diverse race/class/age groups and the places they reside/work as popular headlines report declining school enrollments and struggling students (Day et al., 2021; June, 2021) or the Great Resignation of workers in 2021 (Thompson, 2021; Richardson, 2022). The long-term social, economic, and political consequences of COVID-19 will be the subject of scholarly investigation for decades; this volume sheds light on what we have learned so far, especially from a geo-spatial perspective. While the scope of this book is global, many of the examples in this chapter come from the US experience.

Daily Geographies Altered

The pandemic provided a dramatic example of how quickly human mobility and daily patterns can shift, remaking our daily geographies with varying intensity. In the early stages of the pandemic, some states and localities imposed extreme lockdown measures aimed at keeping people in their homes or reducing the distance they could travel from their homes. Others shifted to remote work and education strategies, dramatically altering the daily activity paths of household members. In short order, millions of home spaces became office spaces for workers and classrooms for children thrust into online education. In the United States, an essential worker class emerged (from grocery clerks to police officers and hospital staff) who went to actual places of work rather than work remotely from the relative safety of their home. Even so, millions lost jobs. Federal stimulus monies were issued to soften the economic blow of the pandemic in the United States. For those who did not have homes, cities turned unoccupied hotel rooms into homeless shelters. Even innocuous tasks such as opening mail or shopping for groceries were stress inducing activities before COVID transmission was fully understood. For most parts of the world, mixed levels of fear infused and altered daily activities in some ways.

Before masking became a widely accepted—albeit contested—method to slow transmission, social distancing emerged as a time-honored low-tech approach to reduce the risk of infection. People isolated alone or in small groups, often referred to as COVID pods. Even walking down streets turned into a COVID dance of dodging and street crossing to avoid close and/or direct contact with others. By the summer of 2020, gathering outdoors emerged as the safest form of socializing, so people and companies expanded their outdoor gathering spaces, resulting in some cities turning sidewalks and even entire blocks into socially distanced and outdoor spaces to dine. Outdoor activities such as camping, hiking, canoeing, boating, biking, and walking experienced upticks in popularity. As social interaction declined, people turned to animal companionship; in the United States, dog and cat adoptions surged during COVID-19 as one in five households acquired a new pet during the pandemic (ASPCA, 2021).

The pivot to digital and virtual interactions also happened suddenly as a result of the pandemic. In particular, daily activities shifted to the home and neighborhood scales, especially in the early stages of the pandemic. Digital meetings and collaborations surged, and there was growing dependence upon platforms such as Zoom and later the phenomenon of Zoom fatigue. A reliance on virtual interactions introduced many second order impacts: from the nature of work, the shift to remote education, to the use of different systems of transport and the uncertainty of supply chains. Work, education, and transport continue to be impacted by COVID-19 although the intensity of these impacts has oscillated over time and space. With people not going to work or school or business meetings, planes, trains, buses, and subways were markedly underutilized. The private car, always popular in the United States, became the preferred safe mode of travel. With the widespread use of smart phones as sensors, telemetry analysis showed a stunning decline in daily mobility whether tracking public transportation usage or movement in general, especially in the spring of 2020 (Mapbox, 2020). As we enter 2022, human circulation has returned, in some places, to near pre-pandemic levels, but work settings continue to be affected.

One area where human circulation is still disrupted is international travel and migration. A report written by geographer Alan Gamlen (2020) for the International Organization of Migration notes that by March 2020, 91% of the world's population lived in countries where there were border restrictions imposed on people arriving from other countries who were not citizens or residents of that country. While many of these restrictions have been lifted, human movement across international borders has declined, especially for some of the most vulnerable people fleeing conflict or oppression. For example, during the first year of the pandemic, the resettlement of refugees virtually stopped. Gamlen concludes with a provocative second order impact question: might the pandemic mark the end of the current age of migration with its reliance on international migrant labor?

Led and Misled by the Map

Maps mattered during the COVID-19 pandemic and became the most public geographic tool deployed to inform our understanding of the changing scope of the pandemic. Freely available dashboards such as the ones created by Johns Hopkins University (JHU, 2022) and the New York Times (NYT, 2022) reported daily infections, hospitalizations, deaths, and vaccination rates at various scales from institutions (hot spots such as nursing homes, prisons, or universities) to county, state, and country totals around the globe. To deal with spatial and temporal differences, dashboards presented 2-week and 4-week averages, rates of infection (that allowed comparisons with smaller and larger demographic units), as well as overall counts during particular time frames. Being able to drill down to smaller units of geography, say a county instead of an entire state, provided a more precise understanding of the real-time pandemic threat. Remarkably this information was freely available to anyone with internet access. There is a long-established history of geographers mapping disease (Paul, 1985), yet the speed with which maps were created and updated during the pandemic is a stunning example of the geospatial digital revolution.

Visualizations of data in near real-time were readily available during this pandemic. Yet the choices made in data selection, areal units, and time period studied can greatly influence how the data are presented and interpreted. Geographers are keenly aware of how the areal unit of analysis can create bias or uncertainty. The modifiable area unit problem (MAUP), and its corollary, the modifiable temporal unit problem (MTUP), are recognized by geographers using geospatial data (Helbich et al., 2021). For example, data by county in the United States is often viewed as a refinement over state data. Yet counties vary enormously by size and density, resulting in patterns on the map that could visually overstate the problem in one area (a large but lightly populated western county) and downplay it in another (a small densely populated county in the east). Likewise, the time period mapped could yield radically different maps as infection spikes rose and fell unevenly. Since 80% of people live in cities in the United States, there is a tendency to think of COVID-19 having a greater impact on cities. The earliest images of New York City overrun with COVID patients in the spring of 2020 come to mind (Rothfeld et al., 2020). Yet a geographical analysis by Zhang and Schwartz (2020) based on US data from the first months of the pandemic concluded that the incidence of mortality in some small cities and nonmetropolitan counties was similar to that in New York City. When we see maps of the absolute mortality numbers, COVID-19 appears to be an urban problem, but when mortality rates are mapped, the pattern is more mixed.

For global COVID-19 data, country boundaries were the unit of convenience, but this tended to emphasize the most populous and larger territorial states, especially when total numbers of cases or deaths were considered. The tendency to privilege state boundaries over all others, especially in international affairs, has been termed the ‘territorial trap’ by political geographer John Agnew (1994). Seeing national borders as barriers of defense is a standard political response. By April 2020 most countries had closed their borders to non-citizens (Connor, 2020). Models suggest that in the early stages of the pandemic, strict border closures slowed the spread of transmission, but once the virus had entered into a territorial state, border closures made little difference (Mallapaty, 2021). In federal systems of governance, responses to COVID-19 devolved to smaller units such as states or provinces, which may be more appropriate to manage public health responses. Yet state-driven responses to the pandemic prevail amid growing tensions among other units of government that want more localized control (Agnew, 2021).

Preconditions That Intensified the Negative Impacts of COVID-19

In addition to the first-order and second-order impacts of COVID-19, an appreciation of various preconditions is helpful. For the United States, geographer Barney Warf makes the case that certain aspects of America society allowed for rapid inroads of the virus. These include distrust of the state, individualism, anti-intellectualism, high degrees of religiosity, lack of universal healthcare, high levels of obesity, the world's highest prison incarceration rate, homelessness, and systemic racism and racial inequality (Warf, 2021, p. 497). Warf argues that US fetishism towards a neoliberal ethos that emphasizes market solutions with only minimal state interventions made the United States uniquely vulnerable to the virus. As of late January 2022, the United States is the country with the most recorded deaths, nearing 900,000, and it has one of the highest death rates (1 in 376 people) among developed countries (New York Times, 2022). The concept of preconditions that made localities (at various scales) more or less vulnerable to COVID-19 deserves further consideration.

Another major second order impact is the unprecedented level of government spending to combat COVID-19. Between the CARES Act ($2.2 trillion) signed by the Trump Administration in March 2020 and the COVID Relief bill signed by the Biden Administration ($1.9 trillion) a year later at least $4 trillion dollars were directed to address health, testing, and socioeconomic needs in the United States. Given the severity of the threat posed by the pandemic, countries that had the resources, spent vast amounts of capital to reduce the negative impacts of COVID-19. Understanding the effectiveness of these economic investment and public reactions to these efforts will be critical going forward.

Questioning Location Tracking

At the American Geographical Society, where I serve as president, we felt the need to respond to the pandemic by offering information and insight. There were some obvious opportunities to communicate with our members and the general public. We retooled *DailyGeo,* our weekday update about geography in the news, to include a different map of COVID-19 every day for nearly 2 years (AGS, 2020b). This allowed our viewers to learn about different ways the pandemic was spreading and being visualized. We also developed several podcasts that focused on the pandemic and the responses by geographers using geospatial technology.

The major initiative we undertook was the Location Tech Task Force funded by a grant from the Henry Luce Foundation. In 2020 many believed that location tracking, which relies upon smart phones, could be easily deployed (Mushahwar et al., 2020). South Korea demonstrated considerable success using this technology for early detection of exposure to the virus (Kim & Kwan, 2021). The objective of the Location Tech Task Force was to bring together specialists to discuss not just the feasibility of the technology but the larger concerns about geo-privacy, surveillance, vulnerable publics, management of the data and accountability (AGS, 2020a). The final report highlights the various governance and integrity issues that undercut trust in the technology. It also pointed out individual and group vulnerability to geo-tracking and the concerns that the technology could be easily expanded into other forms of location tracking once the threat of COVID-19 had subsided. There were also many ethical issues surrounded the security and management of the data collected. In the end, location tracking was not widely adopted in the United States or elsewhere. The technology worked but its use posed many thorny issues that have to be resolved before gaining public trust in such a system.

As the location tracking issue demonstrates, such geospatial tools were actively developed to address the pandemic. And there may be a role for them in the future if concerns about geo-privacy, trust, and governance are addressed. But many other technologies, especially making dashboards and maps widely accessible to the public, were fundamental in responding to the pandemic. Using geographical analysis will be equally valuable as we seek to understand the complex and long-term second order impacts of this twenty-first century pandemic.

References

Agnew, J. (1994). The territorial trap: The geographical assumptions of international relations theory. *Review of International Political Economy, 1*(1), 53–80. https://doi.org/10.1080/09692299408434268

Agnew, J. (2021). Anti-federalist federalism: American "populism" and the spatial contradictions of us government in the time of COVID-19. *Geographical Review, 111*(4), 510–527. https://doi.org/10.1080/00167428.2021.1884982

American Geographical Society (AGS). (2020a). *The location tech task force final report*. https://ethicalgeo.org/wpcontent/uploads/2020/12/LocationTechTaskForce_FinalReport_12-20.pdf

American Geographical Society (AGS) (2020b). *DailyGeo*. https://ubique.americangeo.org/daily-geo/

ASPCA. (2021). *New ASPCA survey shows overwhelming majority of dogs and cats acquired during the pandemic are still in their homes*. https://Www.Aspca.Org/about-Us/Press-Releases/New-Aspca-Survey-Shows-Overwhelming-Majority-Dogs-and-Cats-Acquired-During

Connor, P. (2020). *More than nine-in-ten people worldwide live in countries with travel restrictions covid Covid-19*. https://www.pewresearch.org/fact-tank/2020/04/01/more-than-nine-in-ten-people-worldwide-live-in-countries-with-travel-restrictions-amid-covid-19/

Day, T., Chang, I.-C. C., Chung, C. K. L., Doolittle, W. E., Housel, J., & McDaniel, P. N. (2021). The immediate impact of COVID-19 on postsecondary teaching and learning. *The Professional Geographer, 73*(1), 1–13. https://doi.org/10.1080/00330124.2020.1823864

Gamlen, A. (2020). *Migration and mobility after the 2020 pandemic: The end of an age?* https://publications.iom.int/books/covid-19-and-transformation-migration-and-mobility-globally-migration-and-mobility-after-2020

Helbich, M., Mute Browning, M. H. E., & Kwan, M.-P. (2021). Time to address the spatiotemporal uncertainties in COVID-19 research: Concerns and challenges. *Science of the Total Environment, 764*, 142866. https://doi.org/10.1016/j.scitotenv.2020.142866

Johns Hopkins University (JHU). (2022). *Coronavirus resource center*. https://coronavirus.jhu.edu/map.html

June, A. W. (2021). Undergraduate enrollment continues its slide, dipping 3.2 percent from last year. *The Chronicle of Higher Education, 68*(6). https://www.chronicle.com/article/undergraduate-enrollment-continues-its-slide-dipping-3-2-percent-from-last-year

Kim, J., & Kwan, M.-P. (2021). An examination of people's privacy concerns, perceptions of social benefits, and acceptance of COVID-19 mitigation measures that harness location information: A comparative study of the US and South Korea. *ISPRS International Journal of Geo-Information, 10*(1), 25. https://doi.org/10.3390/ijgi10010025

Mallapaty, S. (2021). What the data say about border closures and COVID spread. *Nature, 589*(7841), 185–185. https://doi.org/10.1038/d41586-020-03605-6

Mapbox. (2020, March 24). *Where and when local travel decreased from COVID-19 around the world*. https://blog.mapbox.com/movement-changes-around-the-world-from-covid-19-cc79db7e04c7

Mushahwar, A., Felz, D., & Knight, J. (2020). *Location and Mobile Data in the Fight against COVID-19: What to Expect in the U.S. and Around the Globe*. 10. https://www.alstonprivacy.com/location-and-mobile-data-in-fight-against-covid-19/

New York Times (NYT). (2022). *Coronavirus world map: Tracking the global outbreak*. https://www.nytimes.com/interactive/2021/world/covid-cases.html

Paul, B. K. (1985). Approaches to medical geography: An historical perspective. *Social Science & Medicine, 20*(4), 399–404. https://doi.org/10.1016/0277-9536(85)90015-2

Richardson, R. B. (2022). Defining first- and second-order impacts through maps. In M. Laituri, R. B. Richardson, & J. Kim (Eds.), *The Geographies of COVID-19: Geospatial Stories of a Global Pandemic*. Springer Nature.

Rothfeld, M., Sengupta, S., Goldstein, J., & Rosenthal, B. M. (2020). 13 deaths in a day: An 'apocalyptic' Coronavirus surge at an N.Y.C. Hospital. *New York Times*. https://www.nytimes.com/2020/03/25/nyregion/nyc-coronavirus-hospitals.html

Thompson, D. (2021). Three myths of the great resignations. *The Atlantic*. https://www.theatlantic.com/ideas/archive/2021/12/great-resignation-myths-quitting-jobs/620927/

Warf, B. (2021). The coronavirus pandemic and American neoliberalism. *Geographical Review, 111*(4), 496–509. https://doi.org/10.1080/00167428.2021.1884981

Zhang, C. H., & Schwartz, G. G. (2020). Spatial disparities in coronavirus incidence and mortality in the United States: An ecological analysis as of may 2020. *The Journal of Rural Health, 36*(3), 433–445. https://doi.org/10.1111/jrh.12476

Part I
Geographies of a Pandemic: A Place-Based Approach (Overview)

Junghwan Kim

Part I provides an overview of the role of geography in the COVID-19 pandemic. It consists of five chapters, including geospatial tools and technologies in pandemic research, maps to understand the pandemic's impacts, quantitative geography methods in COVID-19 research, privacy and ethics issues in COVID-19 policies, and the Cities' COVID Mitigation Mapping (C2M2) program.

In Chap. 2, Dr. Laituri provides a historical and geographical overview of the use of geospatial tools and technologies to address pandemics in general and COVID-19 specifically and longer-term adaptation to extreme events. Referring to the term "consequential geography" coined by Edward Soja (2010), Dr. Laituri illustrates that the consequential geography of the pandemic is manifested in myriad places around the world with uneven and inequitable impacts on economies, environments, societies, and cultures. Dr. Laituri further illustrates that a key theme of the pandemic is the exposure of inequality around the world exacerbated by not only the virus but compounded by the government and social responses. Dr. Laituri argues that the consequential geography of the pandemic exposes a landscape of inequality and vulnerable populations.

In Chap. 3, Dr. Richardson defines the first- and the second-order impacts of the pandemic through mapping. Throughout the pandemic, maps have helped us to understand the spread of the virus and visualize rapidly changing information about its impacts. Dr. Richardson presents several examples and cases from published research to identify the limitations and challenges associated with geospatial applications, such as time, space, and scale. Dr. Richardson also identifies and describes how maps have been used to measure the impacts of the pandemic.

In Chap. 4, Dr. Kim, Mr. Wang, and Mr. Rapuri review the role of quantitative geographical approaches in COVID-19 research. By reviewing 331 papers that adopted quantitative geospatial approaches, they identified three research themes:

J. Kim
Virginia Tech, Blacksburg, VA, USA

(1) investigating geographical disparities in COVID-19 cases and deaths as well as the accessibility to COVID-19-relevant facilities such as testing and vaccination sites, (2) examining various factors that affect COVID-19 cases and deaths and building a model to predict those in the future, and (3) other topics that are relatively less studied than the aforementioned two themes. They also observe fewer papers on second-order impacts than first-order impacts.

In Chap. 5, Dr. Seidl examines the values and ethics involved in the debates over whether and/or how to apply mobile location data to prevention efforts through contact tracing, quarantine enforcement, symptom checking, and flow modelling. Mobile location data became a center point of debate in the midst of the COVID-19 pandemic. Unlike pandemics of the past, the advancement of cellular phone technology and its adoption by a large proportion of the world population caused many governments, health authorities, and companies to explore how phone-based geospatial data could be applied to prevent disease transmission. Dr. Seidl argues that the precedents set during the COVID-19 pandemic for the use of personal geospatial data and the infrastructure established to support its collection will follow us into the next world challenge.

In Chap. 6, Ms. Cline and Dr. Laituri introduce the Cities' COVID Mitigation Mapping (C2M2) program. It is a program led by the US Department of State from summer 2020 through fall 2021 illustrating the reach of participatory mapping partnerships applied to second-order impacts of the COVID-19 pandemic, and the geospatial model used offers lessons applicable for recovery from the pandemic as well as a wide range of issues that will continue to impact developing cities, to include resilience and sustainability.

Reference

Soja, E. (2010). *Seeking spatial justice*. University of Minnesota Press.

Chapter 2
Geography of the Pandemic

Melinda Laituri

Personal Story
I used to travel almost every month to far and distant places – and I loved it. I loved going to a new country, experiencing the culture, food, music, but, most importantly, the people. My last trip was in February 2020. I was in Kyiv, Ukraine, watching the unfolding of the first stages of the global shutdown and wondering if I was going to get back to the United States. The only trip I have been on in the past 2 years is to Cleveland, Ohio, to visit family in July 2021 when we thought the pandemic was starting to ebb. Instead, it has continued to flow, and I am reluctant to travel. I made an investment in adopting a cautious approach to the pandemic and wonder when I will decide to compromise. However, I have been exploring the world at my fingertips through virtual travel to exotic locations. I have visited places I never thought to explore before – Chernobyl, Mars, and Iceland's oldest shipwreck. It isn't quite the same but does fill a bit of the void. *Melinda Laituri*

Introduction

Geography is central to the COVID-19 story, the underlying science of pandemic exposure and spread, and the societal responses to control and manage the virus and its effects. The current pandemic is a spatial story that crosses boundaries, scales, and cultures and can be told through maps – the language of geography. Descriptions of the pandemic are rife with references to geographies of location, wayfinding, and direction – navigating health responses, uncharted territory of a rapidly expanding pandemic, waves of transmission, invasion of habitats by humans, the shifting

The original version of this chapter was revised. The correction to this chapter is available at https://doi.org/10.1007/978-3-031-11775-6_24

M. Laituri (✉)
Colorado State University, Fort Collins, CO, USA
e-mail: melinda.laituri@colostate.edu

M. Laituri et al. (eds.), *The Geographies of COVID-19*, Global Perspectives on Health Geography, https://doi.org/10.1007/978-3-031-11775-6_2

epicentre of the virus, shelter in place, remote learning, mapping the virus genome, global research roadmap, paths of transmission, and downstream or cascading effects of the pandemic. This volume of case studies focuses on the geographies of COVID-19 around the world. These geographies are located in both time and space, revealing impacts that are both immediate and long-term. The story of the pandemic is dynamic, in constant flux, and flush with ephemeral observations. The COVID-19 pandemic will be prolonged due to the onset of variants as well as setting the stage for similar future events. This chapter provides a synopsis of how geography and geospatial approaches are used to understand this event and the emerging "new normal."

Maps are created through mapping – the process of using the tools of geography that include geospatial data, remote sensing, mobile devices, and interactive visualizations yielding dynamic results. These products derived from a geographic perspective – grounded in a location and using a geospatial approach for analysis – are central to understanding the pandemic. What do maps – or more importantly – geography tell us about the pandemic? There are four general themes to consider. First, geography and questions of *where* are central to this story. *Where* is it present? *Where* is it spreading? *Where* is it most pronounced? *Where* am I in relation to it? Second, the geographic concept of scale is important in how the story is depicted – a global, generalized picture tells a significantly different story than the story in my hometown and is dependent upon the availability of and access to data at the appropriate resolution and scale. Third, place matters – where we live defines access to resources and essential services, revealing racial, social, and economic disparities. And finally, what has emerged about the pandemic is a complex story of uncertainty and vulnerability where the virus defies generalities, manifesting itself differently in different places with long term impacts that we will have to contend with (see The Geography of COVID-19, Parts 1–4 blog post, Laituri et al., 2020).

This chapter provides a historical and geographical overview of the use of geospatial tools and technologies – geographic information science (GIS) – to address pandemics and disasters in general, COVID-19 specifically, and longer-term adaptation to extreme events. Situating such events in place and time emphasizes the importance of using these tools and how they shape the stories we tell. The geography of the pandemic is twofold: the virtual geography of the pandemic and the consequential geography of the pandemic in places across the globe. The virtual pandemic is told in maps, data, and dashboards, creating living, dynamic maps of a global catastrophe. The ecosystem of the virtual pandemic includes big data, dashboards, dynamic maps, algorithms, coding, satellite imagery, and social media. Data dashboards – built upon a suite of geospatial and technical tools – use base maps of political geography (countries, states, and provinces) to display and visualize data about the pandemic (statistics on case numbers, deaths, and transmission) to showcase distributional impacts through concepts of human and medical geography. These dashboards are simplifications of the complex landscape of the pandemic and highlight the need for regional, subnational, city, and sub-city data or specifically, local data to tell the local story (Bakari et al., 2021).

Consequential Geographies

In *Seeking Spatial Justice*, Edward Soja coined the term "consequential geography" in reference to understanding the "social production of space" comprised of the social, political, and economic organization of places (2010, p 31). The consequential geography of the pandemic is manifested in myriad places around the world with uneven and inequitable impacts on economies, environments, societies, and cultures. This consequential geography reveals the relationship between places through the connectivity exhibited by the infrastructure (i.e., physical infrastructure of wired networks and cables of the Internet; transportation networks that support supply chains; medical facilities that provide access to vaccines) of the functional geography of who and how people are connected (Khanna, 2016). Connectivity is reflected through financial flows, integrated economies, supply chains, and mobile populations that transcend boundaries. Similarly, the consequential geography of the pandemic defines how pathogens are spread due to the intersection of social relations, economic flows, environmental factors, and human behaviours.

The geographic legacy of past pandemics is evident in multiple ways. In 2015, the World Health Organization issued new guidelines for naming conventions of human infectious diseases to minimize negative effects on nations, economies, and people (WHO, 2015). Prior to this best practice numerous pathogens were named for places; examples include Rocky Mountain spotted fever, West Nile virus, Middle East respiratory syndrome, and the Spanish flu (Christakis, 2020, p. 174). This geographic legacy is also reflected in the many monuments, churches, hospitals, and cemeteries built in response to past epidemics. In 1423, Venice established one of the first quarantine sites or lazarettos to manage bubonic plague; a stone pillar in Vienna commemorates the Black Death of 1679. A cemetery, Disco Hill, was established in Liberia to allow burial teams to dispose of the dead due to the Ebola virus in 2014; and a memorial grove in San Francisco commemorates those lost to AIDS since the 1980s (Ben-Ami, 2021).

Geography, Maps, and Technology

Maps and charts are essential to understanding human relationships to the environment and disease. Two examples demonstrate how geographic data were used to create maps and charts of disease and transmission. An early example of understanding the consequential geography of a pandemic was John Snow and his collaboration with Henry Whitehead to map the cholera epidemic in London in 1854. John Snow created a map of cholera cases and water pumps that when coupled with local knowledge supplied by Henry Whitehead supported the theory of cholera as a water-borne disease, leading to reforms in public infrastructure and waste management (Johnson, 2006; Tufte, 1997). Once scientific opinion accepted the waterborne theory of cholera, Snow's map became an important demonstration of the integration of science and local knowledge, linking an artifact of the built environment to a pattern of disease and disaster. Minard's 1896 statistical graph of Napoleon's march to Moscow (1812) and return is the representation of two dimensions of six different types of data (the number of Napoleon's troops, distance, temperature, latitude and longitude, direction of travel, and location relative to specific dates) (Tufte, 2002). This rendering of time and space offered an innovative way to integrate multiple types of geographic data. Reading Snowden's narrative account (2019) of Napoleon's misadventures illuminated the role of disease – dysentery and typhus – in the diminutive numbers that returned to Paris, which is not part of the data of the original graph, revealing a story of disease and death in time and place.

The full trajectory of geography is beyond the scope of this chapter; however, there are some pivotal technical developments that significantly contribute to using a geospatial approach in understanding the consequential geography of the pandemic. In the 1980s, geography underwent a second quantitative revolution[1] with the development of geographic information systems (GIS). Location-based data, georeferenced information, and computer modelling combined to develop applications for place-based decision-making. Increasing availability of satellite imagery and the use of global positioning systems (GPS) to assess land use/land cover change and sensors to track environmental conditions furthered the use of GIS in university curriculums, business applications, and government uses. Geographical approaches embedded in the tools of GIS demonstrated the fundamental nature of integrated analysis of multiple and disparate types of data to yield new information and new insights (Tomlinson, 2003). Characterized as the spatial turn (Torre, 2008), the 1990s was a period of further development of geospatial techniques, increasing accessibility of computer-based software tools, and development of spatial concepts (i.e., place, landscapes, relative space, or spatiality), fuelling transdisciplinary approaches across the natural and social sciences as well as the humanities (NRC, 2006) and strengthening partnerships between universities, government, and industry (Craglia & Shanley, 2015).

Geospatial approaches enabled transdisciplinary research to examine disasters, environmental change, and the geography of vulnerable populations (Laituri, 2021). A key contribution in the late 1980s was the United Church of Christ's Commission for Racial Justice report "Toxic Waste and Race in the United States," which revealed the disproportionate environmental impacts on people of color and low-income communities of hazardous waste sites. Maps overlaid demographic data (US Bureau of Census) and Environmental Protection Agency's (EPA) locations of hazardous facilities to reveal "communities with the greatest number of commercial hazardous waste facilities had the highest composition of racial and ethnic residents" (Chavez & Lee, 1987, pg. xiii). These maps were part of ground-breaking efforts to depict inequitable landscapes and launched the environmental justice movement in the United States (Bullard, 1993). In 2015, the Global Atlas of

[1] The first quantitative revolution occurred in the 1960s, ushering in an emphasis on science and mathematics using statistical and other quantitative methods such as multivariate analysis, spatial and simulation modeling (Preston, 1972).

Environmental Justice (https://ejatlas.org/) provided descriptions of locations and types of environmental justice case studies around the world. These descriptions include categories of waste management, climate justice, biodiversity, and water management. Early in the COVID-19 pandemic, it was evident that the virus impacts people of color disproportionately, reaffirming that fundamental tenet of environmental justice (Winskill et al., 2020). Where poor people live and the preconditions to getting sick are critical questions that are part of current efforts to assess the pandemic and understand the underlying geography.

Peter Gould's The Slow Plague (1993) described the AIDS pandemic using maps and geographic concepts – diffusion and spatial modelling – to trace the transmission in Africa, Thailand, and the United States. He applied a geographic perspective to examine the cultural underpinning of the AIDS pandemic through the interplay of sex, poverty, transportation (i.e., air travellers carrying AIDS to distant locations) and place. Gould documented the rapid spread, pathways and trajectories across different spatial scales, applying analysis of socio-economic data and AIDS cases to trace those most impacted – poor and ethnic minorities concentrated in urban centres.

In 2011, Emory University developed the first interactive online map of HIV – AIDSvu – that maintains current data on AIDS in the United States at the state, county, and city levels. These data include demographics, transmission patterns, and data comparisons of socio-economic characteristics. The interactive maps are representations of the diversity and complexity spanning the global reach of the AIDS pandemic, laying the groundwork for future geographic examinations of infectious disease.

The US Disaster Mitigation Act of 2000 supported efforts to better plan for disasters through hazard mapping, community participation, and risk. Susan Cutter et al. (2003) and her team created the social vulnerability index (SoVI) to environmental hazards using county-level socio-economic and demographic data. This approach facilitated the development of such tools as the Social Vulnerability Index (2006–2014) for multi-hazard vulnerability assessments using GIS to spatially combine vulnerability of populations, physical geographic characteristics, socio-economic data, and social fabric across multiple hazards. (Tate et al., 2010). Cutter's work was foundational for the development of the Centre for Disease Controls' (CDC) social vulnerability index (SVI) (Flanagan et al., 2011) for disaster management. The Agency for Toxic Substances and Disease Registry hosts an interactive map of the SVI in the United States with prepared maps at the census tract level of four thematic areas (socioeconomic status, household composition, housing type, and ethnicity) that are then combined to calculate the SVI of a community's capacity to respond to disasters. These web-based platforms exemplify the nature of interactive maps using datasets from multiple sources, representing the geography of vulnerability.

In the early 2000s, multiple disasters drove further advancements in using the suite of integrated spatial technologies and extending the reach of web-based platforms. The 2004 Indonesian tsunami saw the establishment of methods for increasing access to and sharing of high-resolution satellite imagery across humanitarian agencies to assess damages and provide information to first responders. In 2005, Hurricane Katrina demonstrated the use of mashups to create message boards and web portals, to locate shelters and missing persons (Laituri & Kodrich, 2008). The Haitian earthquake of 2010 identified how remote mappers could use high resolution satellite imagery and the platform – OpenStreetMap – to map urban areas, road networks, buildings, and other infrastructure to share with on-site first-responders and local mappers to assess and ground truth the rapidly generated new data (Radford, 2020). The Humanitarian OpenStreetMap team and Red Cross Missing Maps offer venues in response to disasters for remote and local community mappers to provide baseline data for under-mapped locations. These efforts have expanded who makes the map. Citizen science projects solicit participation from the public in scientific investigations. For example, the Cornell Lab of Ornithology and the National Audubon Society sponsor the Great Backyard Bird Count annually – a four-day event to identify, count, and share birding observations. Crowdsourcing refers to users contributing to specific mapping tasks. One example is OpenStreetMap, a platform to map the world – particularly under-mapped areas – created by volunteers termed a "community of mappers" (Craglia & Shanley, 2015, pg. 4).

Virtual Geography: Interactive Maps and Data Dashboards

These types of interactive, web-based platforms have set the stage for the numerous data dashboards developed during the pandemic, creating an expansive virtual geography of the pandemic. Ivanković et al. (2021) examine 158 dashboards from 53 countries developed by governmental agencies using both maps and graphs. Data dashboards use the tools of geospatial analysis to access big data, such as social media feeds and satellite imagery, and statistics. Interactive maps are embedded in interfaces for querying data across time and space. Creating these data dashboards demands a number of unique skills: coding, statistical understanding of data (COVID-19 case numbers, demographic data), methods to extract data from multiple databases, access to real-time data, and visualization (choropleth maps, graphic design). Multiple software companies have created off-the-shelf templates to rapidly create data dashboards – Esri, Tableau, and Microsoft Power BI (Pantino, 2021). This intersection of government data, private software companies, the Internet, and public need demonstrates how geospatial approaches create linkages across sectors.

The virtual geography of the pandemic exposes several limitations of geospatial tools, platforms, and dashboards in describing local and regional geographies. Data are the backbone of stories about a virus that transcends political boundaries as it ebbs and flows around the world. However, data are limited. While data at the appropriate scale enables the telling of local stories that represent people and places, most data dashboards depict information at the country, state, and major city scale. Scale and resolution matter particularly when data that are too coarse can hide and obscure not only variables, but people represented by those variables (O'Neal,

2016). Our stories are only as good as the data we have. The data are constrained by what is collected (or not), how numbers are aggregated, the level of precision of data collection instruments, and algorithms. Maps and associated models are simplifications – "opinions embedded in mathematics" (O'Neal, 2016). The analyses undertaken and shared on data dashboards need to include data sources, disclaimers, and limitations in the results presented as these representations of data are increasingly ubiquitous.

Use of off-the-shelf data dashboards has enabled multiple applications and enhanced access to pandemic information. However, these dashboards have built in defaults which may not reflect regional conditions and local needs requiring coding skills to modify and customize the interface. Coders have been in increased demand during the pandemic – but coders with multiple skills to analyze statistical data and understand the nuances of geospatial statistics are also needed and in limited supply. Open Watch Data describes the "data value chain" – availability, openness, dissemination, use and uptake – tracking best practices and guidelines from multiple organizations and users. Rudow, et al. (2021) provide an overview of lessons learned that include: the collection of timely data for rapid dissemination; adequate demographic data with an emphasis on gender, disabled people, and indigenous communities; attention to privacy; and improvement in data interoperability, standards, and data management.

Geography of the Digital Divide

The virtual pandemic-scape also exposes the digital divide – a key driver of inequality in the digital age (CNUCED, 2020). This digital divide is characterized by a lack of access to basic technology such as computers and cell phones, inadequate Internet infrastructure of wires and networks, and a reliance on spatial modelling dependent on social media. The digital divide is still a problem where a lack of Internet services and infrastructure limits access to schools, places of work, and other services (Lai & Widmar, 2021). Least developed countries lag behind in terms of digital readiness in terms of Internet accessibility, broadband quality, and mobile data costs. This digital divide inhibits and limits how geospatial tools, applications, and solutions can be used to address data-driven decision-making in many areas of the Global South. Additionally, the reliance of many models and algorithms on social media (a source of big data) emphasizes those who are connected. An examination of networks illuminated by twitter feeds, such as the COVID tracking project, not only reveals connectivity provided by mobile technologies (i.e., cell phones and global positioning systems) but also offers an avenue to identify gaps and exposing the emphasis on the built environment – often where the infrastructure exists vs. those areas where there may be limited infrastructure. How do the algorithms about these data impact those not adequately represented? Understanding and identifying under-represented populations in big data is essential for improved planning and data driven decision-making in a "post-virus" world.

An established avenue of geographical research has been environmental monitoring (land use/land cover change) and modelling (climate change predictions) using Earth observation satellite imagery and geospatial analysis (Phiri & Morgenroth, 2017). Since 1972, Landsat has provided satellite imagery of the earth, enabling analysis of land use and land cover change. Improvements in imaging capability and enhanced sensors have expanded the use of remotely sensed data to analyse environmental conditions over time (Loveland & Dwyer, 2012). There are two examples of how these approaches have contributed to understanding this pandemic: the impacts of land use induced zoonotic "spillover" in opening avenues for disease transmission from wildlife to humans (Plowright et al., 2021); and the environmental impacts of the "anthropause" due to the global lockdown and reduced human activity (Rutz et al., 2020).

Land use/land cover change analysis in the form of urban growth, hydroelectric dam construction, mining, and deforestation increase human/wildlife interaction (Grossman, 2021). The turn of the century saw a rapid expansion of the human footprint with intensifying development and urbanization in areas of rich biodiversity and sensitive habitats (tropical forests) (Venter et al., 2016). Illegal bush meat markets and trafficking in wildlife across international borders yield lucrative economic flows of cash and viral flows of disease with new avenues of movement (Swift et al., 2007). The use of satellite imagery, GPS tracker collars, and drones has been used to track wildlife movement patterns that may indicate suspicious activity and enforcement response due to wildlife trafficking (Wheeler, 2014). While wild animals are threatened by the bushmeat trade and wildlife trafficking, so too are humans who handle, prepare, and cook certain kinds of animals. Bats and pangolins have been cited as potential hosts for the type of coronavirus, a zoonotic disease that is also found in international bush markets (Quammen, 2020). In addition to mobile and satellite technology, phylogeny graphs – a form of genetic mapping, transmission diagrams of viral diffusion, and maps of contact tracing and disease spread contribute to illustrating the complex, geographic nature of epidemics. Other analytical tools include identifying hotspots of outbreaks of contagion through using integrated data about the numbers of disease cases, population density, settlement patterns, and transportation networks. Such an approach was used to create the Ebola Mapping Tools to create potential risk maps of future Ebola outbreaks based on environmental variables, such as bat habitat and Ebola outbreaks (Fortunati, 2016).

Geographies of Transformation

The pandemic ushered in a year of lockdowns, quarantines, and isolation – the Great Pause or anthropause (Rutz et al., 2020). An outcome of the Anthropause was the application of multiple geospatial applications to track changing patterns of the natural world in response to the pandemic. The consequential geography of a "human pause" had dramatic impacts on the natural world. Rutz et al. (2020) examine the impact on reduced human mobility and wildlife activity through creating a bio-logging application using social media, sensor networks, and citizen scientists. The World Forum for Acoustic Ecology captures COVID-19 Soundscapes from around the world, integrating sound and space. Satellite imagery of night lights was used to assess the impacts on energy consumption, transportation, and social interactions, linking spatio-temporal changes, population density, and anthropogenic light emissions (Rubinyi et al., 2020). The Great Pause gave us a moment to consider the unintended consequences of slowing down the world where the pandemic constrained our physical space while expanding our virtual space.

The pandemic is a transformative event on the world stage. The consequential geography of the pandemic is central to examining the intertwined dynamics of biological and social phenomenon to identify impacts across different spatial scales, from the molecular to the global. Epidemics set the stage where social conditions – social and economic inequalities over time and place – amplify biological factors and exacerbate conditions of oppression. Initial waves of the Black Death (1347–1351) resulted in large scale population loss where famines, high levels of poverty, and low wages led to precarious conditions of health. Wealthier people could travel to the countryside; poorer people sheltered in place with death. Successive waves of the Black Death resulted in adaptive practices such as quarantines, isolation, and social distancing. However, the resulting decline in population meant that labor was in demand and wages increased as Europe rebounded from the pandemic of the plague, changing social and economic patterns (Wright, 2020). Pandemics lead to discovery and changing attitudes and practices. In the early 1800s, the coincidence of germ theory (infectious diseases caused by microbes) and the sanitary reform movement (clean water systems, sanitation, and safe housing regulation) reduced the transmission of deadly pathogens in European and US cities (Shah, 2020). New institutional structures were formed to manage sanitation practices such as boards of health and hospitals organized into specific wards (Shah, 2020). We will experience multiple changes due to the COVID-19 pandemic which, as of this writing, is still apparent around the world.

The consequential geography of the pandemic exposes a landscape of inequality and vulnerable populations. The pandemic illuminates differential access to critical infrastructure for remote learning, availability of health care, and access to basic services. This crisis magnifies fundamental inequalities that require robust data to track the virus in at-risk populations as well as identify innovative solutions for both economic and community health at a local scale. Crime (Ceccato et al., 2021), domestic violence (Brink et al., 2021), and corruption (Gallego et al., 2020) are all facets of the pandemic that are mapped, exposing the geographies of risk. The identification of vulnerable populations (who are they?), the landscape of inequity (where are they?), and what do they need (where are basic services?) are examples of how examining the contributions of geography and geospatial technologies can support place-based strategies for solutions.

We will need innovative approaches to address the long-term impacts of the pandemic. In May 2021, UNESCO declared environmental education for sustainable development as essential to the core curriculum to transform society and ensure human and planetary health and well-being. An integrated approach to education built on environmental stewardship, civic engagement, and collaborative skills centred on place-based learning. Geography is central to this approach and the skillset associated with geospatial tools is inherently collaborative, enabling a robust workforce. Additionally, the pandemic also reveals the need for conversations across disciplines for comprehensive, innovative solutions. The COVID-19 virus has catalyzed an international research effort tracked and coordinated by WHO to conduct research that is cross-cutting and inclusive of both natural and social sciences. WHO has also identified 15 international laboratories that coordinate with national labs around the world to increase connectivity within the science community. Essential to this effort is the virtual geography that has emerged from the pandemic of sharing data, demonstrating interoperability, and using geospatial tools for place-based and data-driven decision-making. Our responsibility as geographers and geospatial students, practitioners, and scientists is to ensure grounded, ethical, and sound scientific approaches in addressing the profound problems we face.

References

Bakari, O., Zando, M., & Blaser, M. (2021). Opinion: COVID-19 lesson – local data skills matter. *DevEx World 2020: Global Views*. https://www.devex.com/news/sponsored/opinion-covid-19-lesson-local-data-skills-matter-98910

Ben-Ami, Y. (2021). How plagues shape the landscape. *New York Times*. https://www.nytimes.com/2021/02/23/travel/epidemics-landmarks.html

Brink, J., Cullen, P., Beek, K., & Peters, S. (2021). Intimate partner violence during the COVID-19 pandemic in Western and Southern European countries. *The European Journal of Public Health., 31*(5), 1058–1063. https://doi.org/10.1093/eurpub/ckab093

Bullard, R. (Ed.). (1993). *Confronting environmental racism: Voices from the grassroots* (1st ed.). South End Press.

Ceccato, V., Kahn, T., & Herrmann, C. (2021). Pandemic restrictions and spatiotemporal crime patterns in New York, Sao Paulo, and Stockholm. *Journal of Contemporary Criminal Justice*. https://doi.org/10.1177/10439862211038471

Chavez, B., & Lee, C. (1987). *Toxic Wastes and race in the United State: A national report on the racial and socio-economic characteristics of communities with hazardous waste sites*. Commission for Racial Justice, United Church of Christ. Public Data Access, Inc. http://uccfiles.com/pdf/ToxicWastes&Race.pdf

Christakis, N. (2020). *Apollo's Arrow: The profound and enduring impact of the coronavirus on the way we live*. Little, Brown Spark.

Conference des nations unies sur le commerce et le developpement (CNUCED). (2020. April 6). *Coronovirus reveals need to bridge the digital divide*. https://unctad.org/fr/node/2368

Craglia, M., & Shanley, L. (2015). Data democracy – Increase supply of geospatial information and expanded participatory processes in the production of data. *International Journal of Digital Earth*. https://doi.org/10.1080/17538947.2015.1008214

Cutter, S., Boruff, B., & Shirley, W. (2003). Social vulnerability to environmental hazards. *Social Science Quarterly, 84*(2), 242–261. https://doi.org/10.1111/1540-6237.8402002

Flanagan, B., Gregory, E., Hallisey, E., Heitgerd, J., & Lewis, B. (2011). A social vulnerability index for disaster management. *Journal of Homeland Security and Emergency Management, 8*(1). https://doi.org/10.2202/1547-7355.1792

Fortunati, R. (2016). Mapping Ebola to prepare for future outbreaks. *IHME (Institute for Health Metrics and Evaluation): Measuring What Matters*. https://www.healthdata.org/acting-data/mapping-ebola-prepare-future-outbreaks

Gallego, J., Prem, M., & Vargas, J. (2020). *Corruption in the times of pandemic*. https://www.iq.harvard.edu/files/harvard-iqss/files/juan-vargas_corruptioncovid_v2.pdf

Gould, P. (1993). *The Slow Plague: A geography of the AIDS pandemic*. Wiley-Blackwell.

Grossman, D. (2021). The sentinels. *Science, 6541*. https://doi.org/10.1126/science.372.6541.450

Ivanković, D., Barbazza, E., Bos, V., Fernandes, Ó. B., Gilmore, K. J., Jansen, T., et al. (2021). Features constituting actionable COVID-19 dashboards: Descriptive assessment and expert appraisal of 158 public web-based COVID-19 dashboards. *Journal of Medical Internet Research, 23*(2), e25682. https://doi.org/10.2196/25682

Johnson, S. (2006). *The Ghost Map: The story of London's most terrifying epidemic – and how it changed science, cities and the modern world*. Riverhead Books.

Khanna, P. (2016). *Connectography: Mapping the future of global civilization*. Random House.

Lai, J., & Widmar, N. O. (2021). Revisiting the digital divide in the COVID-19 era. *Applied Economic Perspectives and Policy, 43*(1), 458–464. https://doi.org/10.1002/aepp.13104

Laituri, M. (2021). Mapping spatial justice for marginal societies. In J. P. Wilson (Ed.), *The geographic information science & technology body of knowledge*. https://doi.org/10.22224/gistbok/2021.1.6

Laituri, M., & Kodrich, K. (2008). Online disaster response community: People as sensors of high magnitude disasters using internet GIS. *Sensors, 8*(5). https://doi.org/10.3390/s8053037

Laituri, M., Linn, S., Carver, D., Brazenwood, A., Chamberlain, L., Gudmestad, S., Norris, C. (2020). *The geography of COVID-19, Parts 1 – 4*. https://gis.colostate.edu/blog/

Loveland, T., & Dwyer, J. (2012). Landsat: Building a strong future. *Remote Sensing of Environment, 122*. https://doi.org/10.1016/j.rse.2011.09.022

National Research Council (NRC). (2006). *Learning to think spatially* (p. 10.17226/11019). The National Academies Press.

O'Neal, C. (2016). *Weapons of math destruction: How big data increases inequality and threatens democracy*. Crown.

Pantino, M. (2021). The rise of the pandemic dashboard. *Bloomberg CityLab*. https://www.bloomberg.com/news/features/2021-09-25/why-every-government-needs-a-covid-dashboard

Phiri, D., & Morgenroth, J. (2017). Developments in landsat land cover classification methods: A review. *Remote Sensing, 9*. https://doi.org/10.3390/rs9090967

Plowright, R. K., Reaser, J. K., Locke, H., Woodley, S. J., Patz, J. A., Becker, D. J., et al. (2021). Land use-induced spillover: A call to action to safeguard environmental, animal, and human health. *The Lancet Planetary Health, 5*(4), e237–e245. https://doi.org/10.1016/S2542-5196(21)00031-0

Preston, J. (1972). *All possible worlds: A history of geographical ideas*. Odyssey Press.

Quammen, D. (2020). Did pangolin trafficking cause the coronavirus pandemic? *The New Yorker*. https://www.newyorker.com/magazine/2020/08/31/did-pangolins-start-the-coronavirus-pandemic

Radford, T. (2020) Reflecting on HOT and the humanitarian mapping community. Closing remarks form 2020 Humantarian OpenStreetMap Summit. https://www.hotosm.org/updates/reflecting-on-hot-and-the-humanitarian-mapping-community-hot-executive-director-tyler-radfords-closing-remarks-from-the-2020-humanitarian-openstreetmap-summit/

Rubinyi, S., Goldblatt, R., & Park, H. (2020). Nighttime lights are revolutionizing the way we understand COVID-19 and our world. *World Bank Blogs*. https://blogs.worldbank.org/sustainablecities/nighttime-lights-are-revolutionizing-way-we-understand-covid-19-and-our-world

Rudow, C., Baldi, E., & the Open Data Watch Team. (2021). A year of data insights in the time of COVID-19. *Open Data Watch blog*. https://opendatawatch.com/blog/a-year-of-data-insights-in-the-time-of-covid-19/

Rutz, C., Loretto, M., Bates, A., Davidson, S., Duarte, C., Jetz, W., Johnson, M., Kato, A., Kays, R., Mueller, T., Primack, R., Ropert-Coudert, Y., Tucker, M., Wikelski, M., & Cagnacci, F. (2020).

COVID-19 lockdown allows researchers to quantify the effects of human activity on wildlife. *Nature Ecology & Evolution, 4*. https://doi.org/10.1038/s41559-020-1237-z

Shah, S. (2020). It's time to tell a new story about the coronavirus—Our lives depend on it. *The Nation*. https://www.thenation.com/article/society/pandemic-definition-covid/

Snowden, F. (2019). *Epidemics and society: From the Black Death to the present*. Yale Press.

Soja, E. (2010). *Seeking spatial justice*. University of Minnesota Press.

Swift, L., Hunter, P. R., Lees, A. C., & Bell, D. J. (2007). Wildlife trade and the emergence of infectious diseases. *EcoHealth, 4*(1), 25. https://doi.org/10.1007/s10393-006-0076-y

Tate, E., Cutter, S., & Berry, M. (2010). Integrated multihazard mapping. *Environment and Planning B: Urban analytics and City Science*. https://doi.org/10.1068/b35157

Tomlinson, R. (2003). *Thinking about GIS: Geographic information system planning for managers*. Esri Press.

Torre, A. (2008). A "spatial turn" in history? Landscapes, visions, resources. *Annales. Histoire, Science Sociales, 63*(5). https://www.cairn-int.info/journal-annales-2008-5-page-1127.htm?contenu=article

Tufte, E. (1997). *Visual explanations: Images and quantities, evidence and narrative*. Graphics Press.

Tufte, E. (2002). *The visual display of quantitative information*. Graphics Press.

Venter, O., Sanderson, E. W., Magrach, A., Allan, J. R., Beher, J., Jones, K. R., et al. (2016). Sixteen years of change in the global terrestrial human footprint and implications for biodiversity conservation. *Nature Communications, 7*(1), 1–11. https://doi.org/10.1038/ncomms12558

Wheeler, J. (2014). Declaration of illegal wildlife trade: Opportunities for satellites. *Via Satellite*. https://www.satellitetoday.com/uncategorized/2014/03/26/declaration-on-illegal-wildlife-trade-opportunities-for-satellite/

Winskill, P., Whittake, C., Walker, P., Watson, O., Laydon, D., … & Ghani, A. (2020). Report 22: Equity in responses to the COVID-19 Pandemic: An assessment of the direct and indirect impacts on disadvantaged and vulnerable populations in low- and lower middle-income countries. *Imperial College London*. https://doi.org/10.25561/78965

World Health Organization (WHO). (2015). *WHO issues best practices for naming new human infectious diseases*. https://www.who.int/news/item/08-05-2015-who-issues-best-practices-for-naming-new-human-infectious-diseases

Wright, L. (2020). Crossroads: A scholar of the plague things that pandemics wreak havoc – and open minds. *The New Yorker*. https://www.newyorker.com/magazine/2020/07/20/how-pandemics-wreak-havoc-and-open-minds

Chapter 3
Defining First- and Second-Order Impacts Through Maps

Robert B. Richardson

Personal Story

My personal experience during the COVID-19 pandemic is delineated into two distinct phases. The first phase was a lockdown, but with the companionship of a pet. I had just relocated to Washington, DC, USA for a one-year fellowship, and I was accompanied by my 14-year-old dog and quarantine buddy, Toohey. Typically, I would have been at work at my university campus for around 8 hours a day, but in this new era, I worked from home every day with Toohey at my side. She and I went for long walks in our new city, visiting parks and trails, and meeting the neighbors, and we grew more attached to each other in this new lockdown lifestyle. However, a year into the pandemic, Toohey's physical health and quality of life diminished quickly, and it became apparent that I had to say goodbye to her. At 15 years old and in frail condition, it was time to let her go. Thus began the second phase of the pandemic, which for me has been marked by mixed emotions, including grief, gratitude, and loneliness, but also relief in knowing that rapidly developed vaccines will help us end this global crisis. In this second phase, I miss Toohey a great deal, and I still feel her absence. The one-year fellowship in Washington, DC, has ended, and I now look forward to a new chapter of life, without Toohey, even as the story of the pandemic continues to unfold.

Introduction

The impacts of the COVID-19 pandemic have been significant and wide-ranging, and they have touched nearly every dimension of the human experience globally, even as the effects have been felt unevenly. It is important to emphasize that the impacts of the pandemic are also still ongoing, and our understanding of its effects on our lives—across time and space—is still emerging as the novel coronavirus continues to spread globally. This chapter introduces a characterization of the

---

R. B. Richardson (✉)
Michigan State University, East Lansing, MI, USA
e-mail: rbr@msu.edu

M. Laituri et al. (eds.), *The Geographies of COVID-19*, Global Perspectives on Health Geography, https://doi.org/10.1007/978-3-031-11775-6_3

impacts of the COVID-19 pandemic and defines both the direct and indirect effects of the virus and of efforts to contain it and mitigate its negative consequences. In addition, the chapter describes how these impacts can be better understood through maps.

The virus initially emerged with the first known case in Wuhan, China, in December 2019, and it quickly became a global concern. On January 30, 2020, the World Health Organization (WHO) declared the coronavirus a Public Health Emergency of International Concern (WHO, 2020a). Soon after, on March 11, 2020, the WHO Director-General announced that the COVID-19 disease could be characterized as a pandemic because of the alarming levels of spread, severity, and inaction. Since that time, numerous institutions have provided near real-time data dashboards with information about daily new cases and deaths (e.g., CDC, 2020; JHU, 2020; WHO, 2020b), and each of these resources uses maps at different scales to depict the spread of known infections, and the locations where severe cases have led to deaths. These institutions provide a range of practical online and mobile mapping dashboards and applications for tracking the COVID-19 pandemic and associated events as they unfold around the world over time (Boulos & Geraghty, 2020). These data and their associated maps communicate information about the direct effects of the virus and cases of the COVID-19 disease—that is, the first-order impacts of the pandemic. This information has been critical to efforts to contain the outbreak of the virus and mitigate its effects.

As the virus spread globally, governments, communities, businesses, and individuals responded with a variety of actions, policies, and behaviors aimed at containing the spread of the virus. Local responses included the rapid deployment of testing capacity and contact tracing. Other measures ranged from travel restrictions and border closures to the suspension of in-person education at all levels, the closure of office activities, face covering requirements, and social distancing guidelines. These responses to the COVID-19 pandemic, in turn, sparked a wave of second-order impacts that will likely continue to be felt globally for years, including a global economic crisis. Second-order impacts have specific ramifications and tangible consequences that derive from human responses to the pandemic, rather than from the virus itself (Laituri et al., 2021). For example, the pandemic has exacerbated pre-existing inequities in areas such as income inequality, poverty, and food insecurity, effectively reversing previous gains and setting back hard-earned progress toward sustainable development. Some of these second-order impacts can be easily measured, analyzed, and mapped, such as economic impacts, migration, shifts in demographic trends, and other effects for which data are available. Other second-order impacts may be more difficult to measure, such as the effects of the disruption in traditional learning in schools, the psychological effects of increased social isolation, and the effects of misinformation about the virus on trust in science and institutions. Nevertheless, measuring and mapping the second-order impacts of the COVID-19 pandemic will be crucial in mitigating negative effects and informing investments in recovery and resilience.

Many research questions related to the impacts of the COVID-19 pandemic are geographic in nature, as emphasized in Chap. 1 of this volume. The first- and second-order impacts of the pandemic are seasonal, spatial, and scalar across multiple sectors of society, and geospatial approaches to examining these impacts can be useful to policy makers and inform their efforts to mitigate their negative effects (Laituri et al., 2021). Not surprisingly, the virus and the COVID-19 disease have inspired the rapid publication of research related to the pandemic across many scientific disciplines, and many researchers have used geospatial approaches for examining, tracking, and projecting the many impacts of the pandemic. Much of this published research acknowledges the limitations and challenges associated with geospatial applications, such as time, space, and scale, as well as other challenges associated with data collection during a pandemic, and the need to minimize contact in order to contain the spread of the disease.

Early in the pandemic, the scientific and medical communities quickly recognized that a global response to contain the spread of the virus would certainly involve some sort of immunization program. Developing, testing, approving, and deploying effective vaccines to increase immunity to the virus became a global priority. Vaccine development is typically a lengthy and arduous process that has required 10–15 years to accomplish; the fastest a vaccine has ever been developed, the mumps vaccine in 1967, took 4 years (Roos, 2021). By July 2020, two biopharmaceutical companies—Pfizer and Moderna—emerged as leaders in the race to develop a COVID-19 vaccine using rapidly-deployed clinical trials. The Pfizer vaccine—developed in collaboration with the German-based BioNTech—was approved for WHO Emergency Use Listing in December 2020, and this approval was soon followed by the listing for the vaccines developed by AstraZeneca, Johnson & Johnson, Moderna, and others in early 2021 (WHO, 2021). These actions sparked a global movement to vaccinate the public, and maps depicting vaccination rates by country were soon part of many COVID-19 data dashboards. Vaccinations continue to be tracked and mapped, although vaccination rates vary widely by country, depending on vaccine availability, deployment capacity, and public willingness to receive the injection.

Mapping First-Order Impacts

Maps have been central to the story of the COVID-19 pandemic and its first-order impacts. The proliferation of interactive maps related to the first-order impacts of the pandemic has been an important contribution towards better understanding the evolving threat of the virus. Since the onset of the pandemic, information related to daily new cases, hospitalizations, and deaths from COVID-19 has been collected and published on the websites of numerous public health institutions (e.g., CDC, 2020; JHU, 2020; WHO, 2020b), and each of them portrays this information spatially using maps at different scales—from county to nation to world maps (Boulos & Geraghty, 2020). These choropleth maps illustrate the prevalence of the disease using shaded colors of countries by rates of transmission. Daily rates of new infections are reported as average daily cases per 100,000 people, which allows for comparison across jurisdictions with different population sizes. Average daily cases are provided as the average number of new cases a day for the past 7 days, as a seven-day rolling average is calculated to smooth out fluctuations in daily case count reporting (CDC, 2021; Mayo Clinic, 2021). These maps and data dashboards highlight the importance of sharing and visualizing public data in the interests of public health (Laituri et al., 2021). Tracking these maps of daily new cases and deaths from COVID-19 became a daily routine for many people, and the maps and data were provided on news media sites around the world. In addition, these maps were used by policymakers to identify hot spots and trends and respond with actions aimed at protecting public health.

The COVID-19 pandemic has enabled and accelerated scientific research across many disciplines, and numerous publications using geospatial approaches have examined the spread of the virus, mapped infection hot spots, and analyzed death rates. Incorporating geographic information systems (GIS) into COVID-19 pandemic surveillance, modeling, and response enhances understanding and control of the disease. Applications of GIS include (i) surveillance and data sharing, (ii) infectious disease forecasting, (iii) digital contact tracing, (iv) integrating geographic data in COVID-19 modeling, (v) investigating geographic social vulnerabilities and health disparities, and (vi) communicating the status of the outbreak or status of facilities for return-to-normal operations (Smith & Mennis, 2020). The locations and availability of personal protective equipment, ventilators, hospital beds, and other items can be optimized with the use of GIS and other geospatial approaches to examining and managing the virus. There are numerous challenges in applications of GIS, including the protection of individual privacy and civil liberties and the integration of knowledge from geography with the fields of medicine, public health, and public policy (Kim et al., 2021). However, geospatial technologies may provide a balanced solution for enabling effective use of location-based data while protecting citizens' privacy. For example, developing innovative spatio-temporal computing and aggregation algorithms could enable the efficient extraction of population flows from less sensitive social media data (Yang et al., 2020).

In a scoping review of the geospatial techniques and associated findings in relation to the characteristics of the pandemic, Fatima et al. (2021) provided a synthesis of published, peer-reviewed journal articles based on the spatial analysis of COVID-19. The authors found that researchers used a wide range of spatial and statistical software to apply spatial analysis for the purpose of disease mapping, exposure mapping, and epidemiological modeling. Factors limiting the use of these spatial techniques included the limited availability and bias of COVID-19 data—as well as the scarcity of fine-scaled demographic, environmental, and socioeconomic data—which restrained most of the researchers from exploring causal relationships of potential influencing factors of COVID-19. Most of the studies found centered on Asia and the Americas, which highlights the need for more comparable spatial studies using geographically fine-scaled data in other areas of the world.

Geospatial technologies have been critical tools for tracking the spread of the virus, examining transmission rates, and mapping death counts worldwide (Ahasan et al., 2020). These technologies have aided local and national decision makers in their efforts to slow the spread of the virus and contain the COVID-19 disease. Multiple data dashboards and maps have been developed to track global and country-level data and depict geographic visualizations of cases, hospitalizations, and deaths that allow for the dissemination of vital information in the interests of public health.

Mapping Second-Order Impacts

Early in the pandemic, responses to the novel coronavirus were implemented swiftly and ranged from executive actions such as business closures, travel restrictions, and lockdowns, to individual actions such as changes in consumption patterns, the use of face coverings, and maintaining physical distance from others. These responses triggered wide-ranging second-order impacts of the pandemic, such as increased unemployment, migration, and food insecurity, among others (Laituri et al., 2021). The COVID-19 pandemic has interrupted the functioning of social and economic systems worldwide in ways that are evolving and are not yet fully understood. Disruptions to education, income losses, mental health effects, and socio-political divisions related to the virus and the efficacy of vaccines are just a few examples of second-order impacts of the pandemic. There is evidence that COVID-19 has had significantly negative effects on psychological well-being, in part because of loneliness, social isolation, and the added anxiety and stress related to coping with the changes associated with the pandemic (Groarke et al., 2020; Okruszek et al., 2020). Furthermore, research has shown that the COVID-19 pandemic has exacerbated gender, racial, and economic inequality (Alon et al., 2020; Abedi et al., 2021).

The World Bank maintains the COVID-19 Household Monitoring Dashboard, one of the most comprehensive sources of second-order impacts of the COVID-19 pandemic across developing regions (World Bank, 2021). The Dashboard provides data and maps depicting the socioeconomic impacts of COVID-19 on households and individuals across 72 countries, based on 142 harmonized indicators across 16 topics ranging from education and income to food security, health, and housing, among others. The indicators are based on high-frequency phone surveys, and the Dashboard allows for comparisons among countries and over time. Figure 3.1 provides an example of a map of these countries depicting the share of the working population that stopped working since the COVID-19 outbreak. The Dashboard reveals a complex spatio-temporal story about the pandemic, including information about the actions taken to cope with the impacts of the pandemic, such as reducing consumption or selling assets to pay for basic living expenses.

The coronavirus outbreak led to severe consequences in significant sectors of the global economy (Sharma et al., 2021). The economic impacts of the COVID-19 pandemic can be measured and mapped to portray spatial and temporal information about trends in unemployment, wages, spending, and prices for basic goods and services, among others (Chetty et al., 2020). A group of scholars, in collaboration with Opportunity Insights, developed an online economic tracker that monitors the

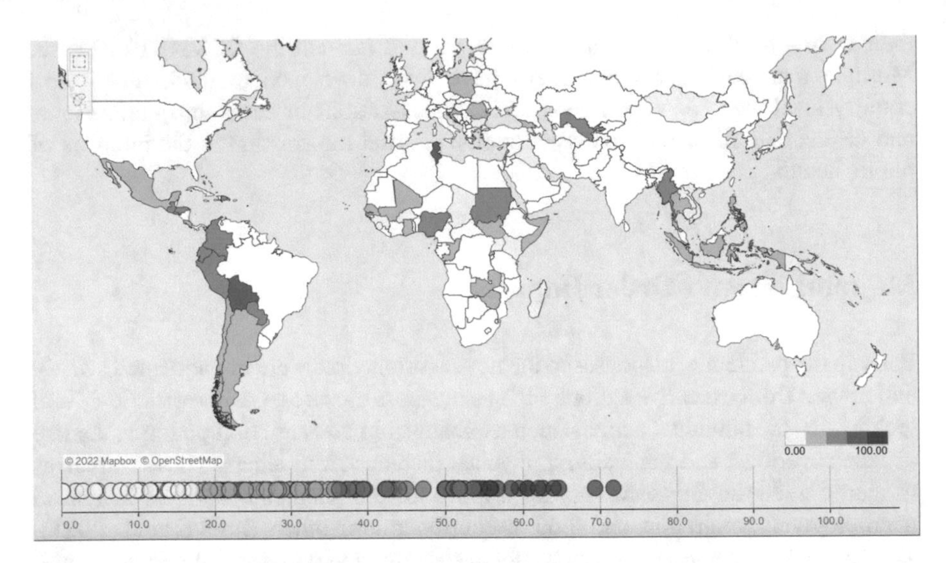

Fig. 3.1 Map of 72 countries depicting the share of respondents to a high-frequency phone survey that stopped working since the COVID-19 outbreak. (Source: World Bank, 2021 © 2022 Mapbox, © OpenStreetMap)

economic impacts of COVID-19 on people, businesses, and communities across the United States in real time (Opportunity Insights, 2021). The Economic Tracker provides data and visualizations that depict trends in economic indicators since the beginning of the pandemic, including consumer spending, small business revenue, and employment, among others. The Economic Tracker is interactive, and indicators can be mapped to allow for visual comparison between states or counties.

The economic impacts of the COVID-19 pandemic have been most acutely and widely faced in areas where the hospitality and tourism sectors are important to regional economies (Laituri et al., 2021). Indicators of these impacts include trends in air travel, hotel bookings, restaurant revenues, and visitation to attractions, and this information can be collected as data and mapped to understand trends in the travel and tourism industries, including their impacts from the pandemic and rates of recovery. Research on the impacts of the COVID-19 pandemic on the tourism sector has expanded rapidly, and the literature on this topic has been summarized in an integrative review that identifies important gaps and highlights a future research agenda (Zopiatis et al., 2021). Geospatial approaches have been applied to map and assess the economic impacts of the COVID-19 pandemic on tourism that has expanded rapidly to include studies on tourism in China (Qiu et al., 2020), India (Chandel et al., 2021; Singh et al., 2021), Nepal (Sah et al., 2020), and South Africa (Rogerson & Rogerson, 2020), among others.

Food insecurity has risen sharply during the pandemic, particularly in urban areas. The United Nations World Food Programme (WFP) estimates that 768 million people faced chronic hunger in 2020, up nearly 20% from 2019 (WFP, 2021). Urban food security has deteriorated during the COVID-19 pandemic in part because of disruptions to national and globalized food supply chains that drove up food prices, highlighting concerns about the vulnerability of food systems (Laborde et al., 2020). WFP has developed the HungerMap^{LIVE} dashboard to track and predict key aspects of food insecurity every day, including key indicators such as the number of people with insufficient food consumption and those employing crisis-level or above coping strategies. HungerMap provides maps and other visualizations depicting various dimensions of food security at global, national, and subnational levels. Indicators include the prevalence of insufficient food consumption and acute malnutrition, along with spatial markers representing some of the drivers of food insecurity, such as conflict, hazards, drought, and economic shocks.

In addition, the NASA Harvest COVID-19 Dashboard provides worldwide data and maps depicting regional trends in virus cases by country, as well as measures of food security, crop conditions, trade, and market prices for food and other commodities (NASA, 2021). The Dashboard is part of the NASA Harvest Portal, which aims to enable and advance adoption of satellite Earth observations to benefit food security, agriculture, and human and environmental resilience worldwide.

One of the most significant social impacts of the COVID-19 pandemic is the disruption to education at all levels. The suspension of in-person learning in schools prompted changes in the delivery of education that ranged from digital learning platforms and online classes to the provision of learning resources via radio and television (Tadesse & Muluye, 2020). School closures have been substantial and uneven across countries. The United Nations maintains data dashboards on the global monitoring of school closures caused by the COVID-19 pandemic, including visualizations depicting regional and country-level data. Early in the pandemic, on April 1, 2020, schools and higher education institutions were closed in 185 countries, affecting more than 1.5 billion learners, or 89.4% of total enrolled students (UNESCO, 2021). Their data show that poorer countries reported the longest average duration of closures, widening the learning gap between countries. Availability of remote learning modalities has varied widely by socioeconomic levels. Research has suggested that the global lockdown of educational institutions during the pandemic will cause long-term interruptions in students' learning and development, and that learning losses due to COVID-19 school closures could continue to accumulate even after students return (Kaffenberger, 2021).

The outbreak of COVID-19 has affected global mobility in complex and unprecedented ways in the form of various travel restrictions, suspensions of air travel, and border closures imposed by governments and authorities since the onset of the pandemic. To better understand this, the UN International Organization for Migration (IOM) has developed a global mobility database to map these impacts on human mobility, across global, regional. and country levels (IOM, 2022). Figure 3.2 provides a map of countries depicting the operational status at various points of entry, including airports, land borders, and maritime borders.

Since human movement is an important driver of the transmission of infectious diseases (Kraemer et al., 2019), geospatial data sources, such as public transportation use, geotagged social media data, and mobile phone data, can be used to detect and map such movement (Yang et al., 2020), although using such data to track human mobility prompts privacy concerns (de Montjoye et al., 2018; Kim & Kwan, 2021). The COVID-19 pandemic has significant implications for migrant workers,

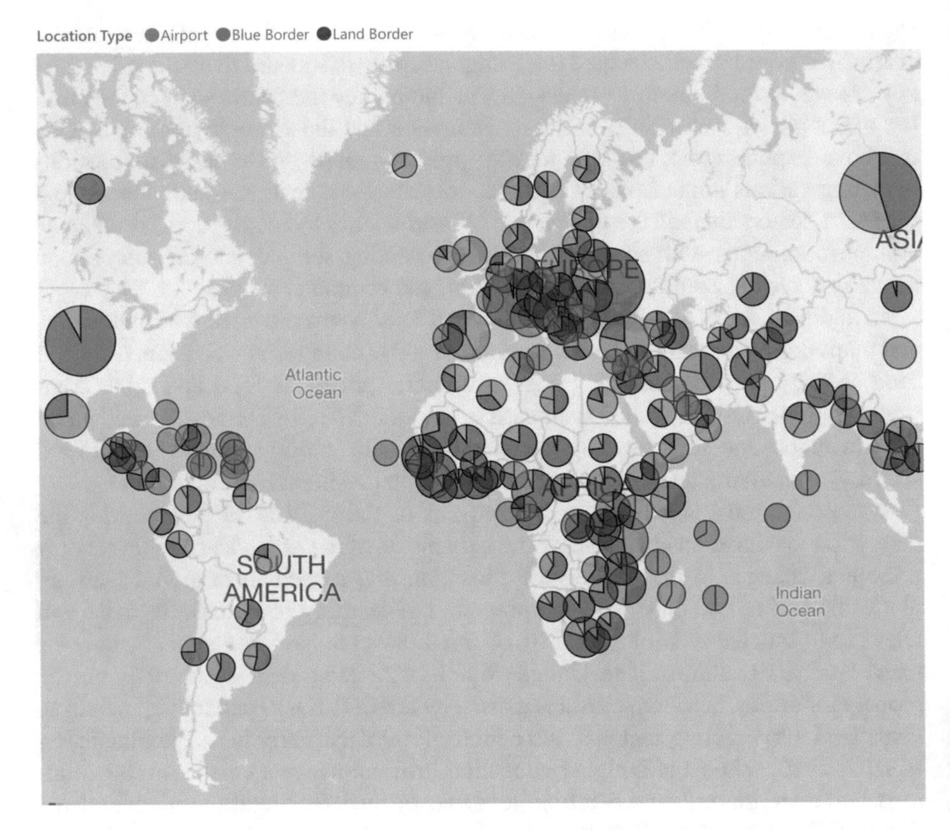

Fig. 3.2 Map of countries depicting the operational status of points of entry. (Source: IOM, 2022 © 2022 TomTom, © 2022 Microsoft Corporation, © OpenStreetMap)

livelihoods, and remittance flows in numerous regions throughout the world (Laituri et al., 2021). Reverse migration—from urban to rural areas—has been a major second-order impact of COVID-19 primarily because of job losses, and one study described the effect as the largest mass migration in India in nearly 75 years (Mukhra et al., 2020). Figure 3.3 provides a map of India depicting the flows of migrants across states in response to the outbreak of COVID-19.

The environmental impacts of the COVID-19 pandemic have largely reflected the changes in human behavior globally. Numerous case studies have documented improvements in air quality, particularly in cities where lockdown measures were implemented in response to the spread of the virus. As such, these areas have been marked by a decrease in industrial activity and automobile travel, and a decline in greenhouse gas emissions in the short-term. Geospatial applications have been used to assess the effects of pandemic-induced lockdown measures on air quality in regional, country, and city scales in the South and Southeast Asian regions using satellite-based data (Roy et al., 2021). Other studies have found similar improvements in air quality from lockdown measures in Brazil (Nakada & Urban, 2020), China, (Xu et al., 2020), Ecuador (Zalakeviciute et al., 2020; Zambrano-Monserrate

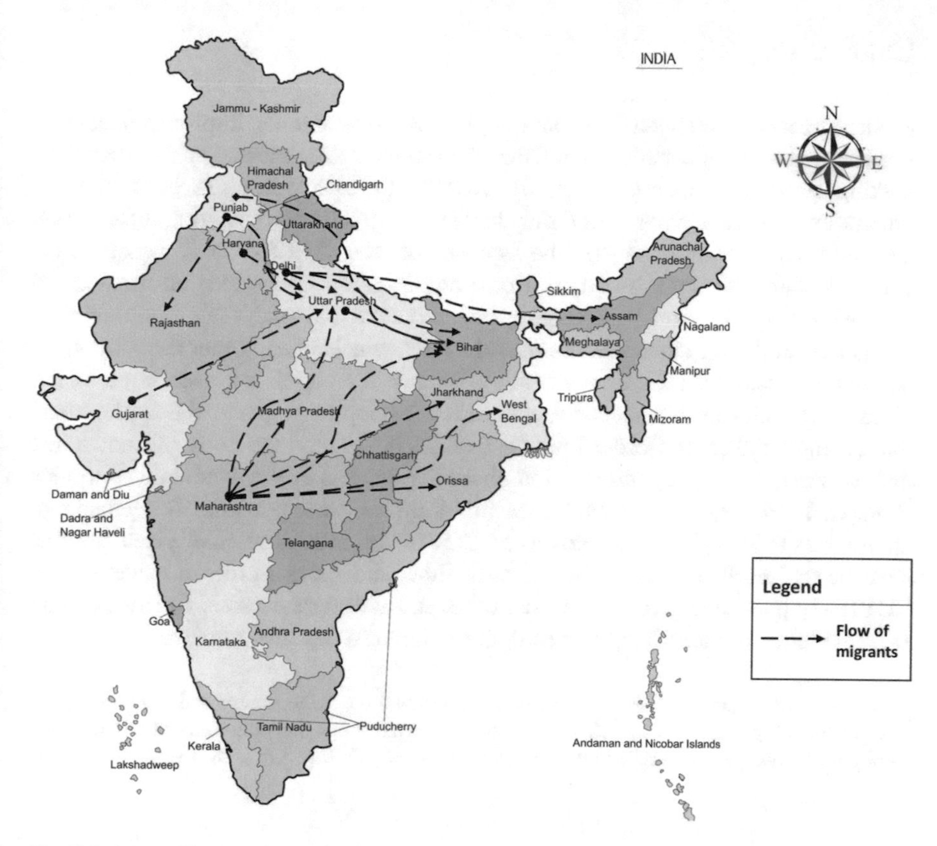

Fig. 3.3 Map of India depicting the migration trends between states. (Source: Mukhra et al., 2020)

& Ruano, 2020), Egypt (Mostafa et al., 2021), India (Singh & Chauhan, 2020), and the United Kingdom (Ropkins & Tate, 2021).

Research has also shown that the rapid decline in tourism activity in many regions has been associated with temporarily improved environmental conditions, such as reduced noise pollution, litter, and coastal water pollution (Cecchi, 2021; Ormaza- González et al., 2021). However, some studies have provided evidence that changes in household consumption during the pandemic have generated greater volumes of municipal waste, including packaging from shipments, household waste, personal protective equipment, and medical waste (Mostafa et al., 2021), which has created additional burden on waste management systems and recycling centers (Kulkarni & Anantharama, 2020).

Other research has focused on the flows of information during the pandemic, including the political polarization and social divisions that have been exacerbated during the COVID-19 outbreak (Jiang et al., 2020). Some studies have used geospatial applications to map the worldwide spread of misinformation and conspiracy theories about coronavirus (Stephens, 2020). Such misinformation has contributed to a rejection of COVID-19 containment measures, vaccine hesitancy, and eroding trust in science (Morgan et al., 2021).

Conclusion

Modern geospatial technologies that integrate web-based tools, improved data sharing, and real-time information provide critical support to public and private decision-making (Boulos & Geraghty, 2020). COVID-19 dashboards are based on the integration of these elements, and they have been vital tools for sharing information and advancing awareness of the spread of the SARS-CoV-2 coronavirus. Communication through map-based dashboards offers accessible information to people around the world eager to protect themselves and their communities. Furthermore, these dashboards are useful to policymakers and other decision makers who aim to lessen the downstream effects of the ongoing transmission of the disease. This chapter introduced merely a few examples of geospatial applications that examine the second-order impacts of COVID-19, and the body of knowledge that uses maps to convey information about the impacts of the pandemic continues to expand. However, these examples highlight the vast potential for geospatial approaches to investigate the impacts of COVID-19. Maps and other visualizations have been central to understanding both first- and second-order impacts of the COVID-19 pandemic across space and time, and at different scales, and they can be valuable tools for identifying and mitigating the most negative impacts.

Disclaimer The views expressed in this report are solely those of the authors and do not represent those of the US government agencies or any of the organizations mentioned. Assumptions made within the analysis are not a reflection of the position of any U.S. government entity.

References

Abedi, V., Olulana, O., Avula, V., Chaudhary, D., Khan, A., Shahjouei, S., Li, J., & Zand, R. (2021). Racial, economic, and health inequality and COVID-19 infection in the United States. *Journal of Racial and Ethnic Health Disparities, 8*(3), 732–742. https://doi.org/10.1007/s40615-020-00833-4

Ahasan, R., Alam, M. S., Chakraborty, T., & Hossain, M. M. (2020). Applications of GIS and geospatial analyses in COVID-19 research: A systematic review. *F1000Research, 9*, 1379. 10.12688/f1000research.27544.1.

Alon, T., Doepke, M., Olmstead-Rumsey, J., & Tertilt, M. (2020). *The impact of COVID-19 on gender equality*. National Bureau of Economic Research, Working Paper 26947.. https://doi.org/10.3386/w26947

Boulos, M. N., & Geraghty, E. M. (2020). Geographical tracking and mapping of coronavirus disease COVID-19/Severe Acute Respiratory Syndrome Coronavirus 2 (Sars-cov-2) epidemic and associated events around the world: How 21st century GIS technologies are supporting the global fight against outbreaks and epidemics. *International Journal of Health Geographics, 19*(1), 8. https://doi.org/10.1186/s12942-020-00202-8

CDC (Centers for Disease Control and Prevention). (2021). *COVID data tracker*. Centers for Disease Control and Prevention. Available at: https://covid.cdc.gov/covid-data-tracker/

Cecchi, T. (2021). Analysis of volatiles organic compounds in Venice lagoon water reveals COVID 19 lockdown impact on microplastics and mass tourism related pollutants. *Science of the Total Environment, 783*, 146951. https://doi.org/10.1016/j.scitotenv.2021.146951

Chandel, R. S., Kanga, S., & Singh, S. K. (2021). Impact of COVID-19 on tourism sector: A case study of Rajasthan, India. *Aims Geosciences, 7*(2), 224–243. https://doi.org/10.3934/geosci.2021014

Chetty, R., Friedman, J. N., Hendren, N., Stepner, M., & The Opportunity Insights Team. (2020). *How did COVID-19 and stabilization policies affect spending and employment? A new real-time economic tracker based on private sector data* (pp. 1–109). National Bureau of Economic Research, Working Paper 27431.. https://doi.org/10.3386/w27431

De Montjoye, Y. A., Gambs, S., Blondel, V., Canright, G., De Cordes, N., Deletaille, S., Engø-Monsen, K., Garcia-Herranz, M., Kendall, J., Kerry, C., Krings, G., Letouzé, E., Luengo-Oroz, M., Oliver, N., Rocher, L., Rutherford, A., Smoreda, Z., Steele, J., Wetter, E., et al. (2018). On the privacy-conscientious use of mobile phone data. *Scientific Data, 5*(1), 1–6. https://doi.org/10.1038/sdata.2018.286

Fatima, M., O'Keefe, K. J., Wei, W., Arshad, S., & Gruebner, O. (2021). Geospatial analysis of COVID-19: A scoping review. *International Journal of Environmental Research and Public Health, 18*, 2336. https://doi.org/10.3390/ijerph18052336

Groarke, J. M., Berry, E., Graham-Wisener, L., McKenna-Plumley, P. E., McGlinchey, E., & Armour, C. (2020). Loneliness in the UK during the COVID-19 pandemic: Cross-sectional results from the COVID-19 Psychological Wellbeing Study. *PLoS One, 15*(9), e0239698. https://doi.org/10.1371/journal.pone.0239698

IOM (International Organization for Migration). (2022). COVID-19 mobility impacts. Available at: https://migration.iom.int/

JHU (Johns Hopkins University). (2020). *Coronavirus Resource Center*. Johns Hopkins University & Medicine. Available at: https://coronavirus.jhu.edu/

Jiang, J., Chen, E., Yan, S., Lerman, K., & Ferrara, E. (2020). Political polarization drives online conversations about COVID-19 in the United States. *Human Behavior and Emerging Technologies, 2*(3), 200–211. https://doi.org/10.1002/hbe2.202

Kaffenberger, M. (2021). Modelling the long-run learning impact of the Covid-19 learning shock: Actions to (more than) mitigate loss. *International Journal of Educational Development, 81*, 102326. https://doi.org/10.1016/j.ijedudev.2020.102326

Kim, J., & Kwan, M. P. (2021). An examination of people's privacy concerns, perceptions of social benefits, and acceptance of COVID-19 mitigation measures that harness location information: A comparative study of the US and South Korea. *ISPRS International Journal of Geo-Information, 10*(1), 25. https://doi.org/10.3390/ijgi10010025

Kim, J., Kwan, M. P., Levenstein, M. C., & Richardson, D. B. (2021). How do people perceive the disclosure risk of maps? Examining the perceived disclosure risk of maps and its implications for geoprivacy protection. *Cartography and Geographic Information Science, 48*(1), 2–20. https://doi.org/10.1080/15230406.2020.1794976

Kraemer, M. U. G., Golding, N., Bisanzio, D., Bhatt, S., Pigott, D. M., Ray, S. E., Brady, O. J., Brownstein, J. S., Faria, N. R., Cummings, D. A. T., Pybus, O. G., Smith, D. L., Tatem, A. J., & Reiner, R. C. (2019). Utilizing general human movement models to predict the spread of emerging infectious diseases in resource poor settings. *Scientific Reports, 9*(1), 1–11. https://doi.org/10.1038/s41598-019-41192-3

Kulkarni, B. N., & Anantharama, V. (2020). Repercussions of COVID-19 pandemic on municipal solid waste management: Challenges and opportunities. *Science of the Total Environment, 743*, 140693. https://doi.org/10.1016/j.scitotenv.2020.140693

Laborde, D., Martin, W., Swinnen, J., & Vos, R. (2020). COVID-19 risks to global food security. *Science, 369*(6503), 500–502. https://doi.org/10.1126/science.abc4765

Laituri, M., Richardson, R. B., Kim, J., Cline, L. V., Viscuso, S., & Schwartz, L. (2021). Examining second-order impacts of COVID-19 in urban areas. *Annals of GIS*, 1–12. https://doi.org/10.1080/19475683.2021.1954087

Mayo Clinic. (2021). *COVID-19 patient and visitor guidelines*. Mayo Foundation for Medical Education and Research. Available at: https://www.mayoclinic.org/coronavirus-covid-19

Morgan, V., Auskova, A., & Janoskova, K. (2021). Pervasive misinformation, COVID-19 vaccine hesitancy, and lack of trust in science. *Review of Contemporary Philosophy, 20*, 128–138. https://doi.org/10.22381/RCP2020218

Mostafa, M. K., Gamal, G., & Wafiq, A. (2021). The impact of COVID 19 on air pollution levels and other environmental indicators-A case study of Egypt. *Journal of Environmental Management, 277*, 111496. https://doi.org/10.1016/j.jenvman.2020.111496

Mukhra, R., Krishan, K., & Kanchan, T. (2020). COVID-19 sets off mass migration in India. *Archives of Medical Research, 51*(7), 736–738. https://doi.org/10.1016/j.arcmed.2020.06.003

Nakada, L. Y. K., & Urban, R. C. (2020). COVID-19 pandemic: Impacts on the air quality during the partial lockdown in São Paulo State, Brazil. *Science of the Total Environment, 730*, 139087. https://doi.org/10.1016/j.scitotenv.2020.139087

NASA (National Aeronautics and Space Administration). (2021). NASA harvest COVID-19 dashboard, NASA Harvest Portal. Available at: https://harvestportal.org/dashboard/

Okruszek, Ł., Aniszewska-Stańczuk, A., Piejka, A., Wiśniewska, M., & Żurek, K. (2020). Safe but lonely? Loneliness, anxiety, and depression symptoms and COVID-19. *Frontiers in Psychology, 11*, 3222. https://doi.org/10.3389/fpsyg.2020.579181

Opportunity Insights. (2021). *The opportunity insights economic tracker: Supporting the recovery from COVID-19*. Opportunity Insights. Available at: https://opportunityinsights.org/

Ormaza- González, F. I., Castro-Rodas, D., & Statham, P. J. (2021). COVID-19 impacts on beaches and coastal water pollution at selected sites in Ecuador, and management proposals post-pandemic. *Frontiers in Marine Science, 8*, 710. https://doi.org/10.3389/fmars.2021.669374

Qiu, R. T., Park, J., Li, S., & Song, H. (2020). Social costs of tourism during the COVID-19 pandemic. *Annals of Tourism Research, 84*, 102994. https://doi.org/10.1016/j.annals.2020.102994

Rogerson, C. M., & Rogerson, J. M. (2020). COVID-19 and tourism spaces of vulnerability in South Africa. *African Journal of Hospitality, Tourism and Leisure, 9*(4), 382–401. https://doi.org/10.46222/ajhtl.19770720-26

Roos, D. (2021). How a new vaccine was developed in record time in the 1960s. *History.com*. Available at: https://www.history.com/news/mumps-vaccine-world-war-ii

Ropkins, K., & Tate, J. E. (2021). Early observations on the impact of the COVID-19 lockdown on air quality trends across the UK. *Science of the Total Environment, 754*, 142374. https://doi.org/10.1016/j.scitotenv.2020.142374

Sah, R., Sigdel, S., Ozaki, A., Kotera, Y., Bhandari, D., Regmi, P., Rabaan, A. A., Mehta, R., Adhikari, M., Roy, N., Dhama, K., Tanimoto, T., Rodríguez-Morales, A. J., & Dhakal, R. (2020). Impact of COVID-19 on tourism in Nepal. *Journal of Travel Medicine, 27*(6), taaa105. https://doi.org/10.1093/jtm/taaa105

Sharma, N., Yadav, S., Mangla, M., Mohanty, A., Satpathy, S., Mohanty, S. N., & Choudhury, T. (2021). Geospatial multivariate analysis of COVID-19: A global perspective. *GeoJournal*, 1–15. https://doi.org/10.1007/s10708-021-10520-4

Singh, R. P., & Chauhan, A. (2020). Impact of lockdown on air quality in India during COVID-19 pandemic. *Air Quality, Atmosphere & Health, 13*(8), 921–928. https://doi.org/10.1007/s11869-020-00863-1

Singh, A. L., Jamal, S., & Ahmad, W. S. (2021). Impact assessment of lockdown amid covid-19 pandemic on tourism industry of Kashmir Valley, India. *Research in Globalization, 3*, 100053. https://doi.org/10.1016/j.resglo.2021.100053

Smith, C. D., & Mennis, J. (2020). Incorporating geographic information science and technology in response to the COVID-19 pandemic. *Preventing Chronic Disease, 17*, E58. https://doi.org/10.5888/pcd17.200246

Stephens, M. (2020). A geospatial infodemic: Mapping Twitter conspiracy theories of COVID-19. *Dialogues in Human Geography, 10*(2), 276–281. https://doi.org/10.1177/2043820620935683

Tadesse, S., & Muluye, W. (2020). The impact of COVID-19 pandemic on education system in developing countries: A review. *Open Journal of Social Sciences, 8*(10), 159–170. https://doi.org/10.4236/jss.2020.810011

UNESCO (United Nations Educational, Scientific and Cultural Organization). (2021). *COVID-19 education response*. UNESCO. Available at: http://covid19.uis.unesco.org/global-monitoring-school-closures-covid19/

World Bank. (2021). *COVID-19 household monitoring dashboard*. The World Bank Group. Available at: https://www.worldbank.org/en/data/interactive/2020/11/11/covid-19-high-frequency-monitoring-dashboard

WFP (World Food Programme). (2021). HungerMap^{LIVE}. Available at: https://hungermap.wfp.org/

WHO (World Health Organization). (2020a). *Coronavirus disease (COVID-19) pandemic: Interactive timeline*. World Health Organization. Available at: https://www.who.int/emergencies/diseases/novel-coronavirus-2019/interactive-timeline/#!

WHO (World Health Organization). (2020b). *WHO coronavirus (COVID-19) dashboard*. World Health Organization. Available at: https://covid19.who.int/

WHO (World Health Organization). (2021). *Coronavirus disease (COVID-19): Vaccines*. World Health Organization. Available at: https://www.who.int/news-room/questions-and-answers/item/coronavirus-disease-(covid-19)-vaccines

Xu, K., Cui, K., Young, L. H., Hsieh, Y. K., Wang, Y. F., Zhang, J., & Wan, S. (2020). Impact of the COVID-19 event on air quality in central China. *Aerosol and Air Quality Research, 20*(5), 915–929. https://doi.org/10.4209/aaqr.2020.04.0150

Yang, C., Sha, D., Liu, Q., Li, Y., Lan, H., Guan, W. W., Hu, T., Li, Z., Zhang, Z., Thompson, J. H., Wang, Z., Wong, D., Ruan, S., Yu, M., Richardson, D., Zhang, L., Hou, R., Zhou, Y., Zhong, C., et al. (2020). Taking the pulse of COVID-19: A spatiotemporal perspective. *International Journal of Digital Earth, 13*(10), 1186–1211. https://doi.org/10.1080/17538947.2020.1809723

Zalakeviciute, R., Vasquez, R., Bayas, D., Buenano, A., Mejia, D., Zegarra, R., Diaz, A., & Lamb, B. (2020). Drastic improvements in air quality in Ecuador during the COVID-19 outbreak. *Aerosol and Air Quality Research, 20*(8), 1783–1792. https://doi.org/10.4209/aaqr.2020.05.0254

Zambrano-Monserrate, M. A., & Ruano, M. A. (2020). Has air quality improved in Ecuador during the COVID-19 pandemic? A parametric analysis. *Air Quality, Atmosphere & Health, 13*(8), 929–938. https://doi.org/10.1007/s11869-020-00866-y

Zopiatis, A., Pericleous, K., & Theofanous, Y. (2021). COVID-19 and hospitality and tourism research: An integrative review. *Journal of Hospitality and Tourism Management, 48*, 275–279. https://doi.org/10.1016/j.jhtm.2021.07.002

Chapter 4
Quantitative Geographical Approaches in COVID-19 Research: A Review on First- and Second-Order Impacts

Junghwan Kim, Kevin Wang, and Sampath Rapuri

Personal Stories
Since the beginning of the COVID-19 pandemic (January 2020), I have not been able to visit my home country (South Korea) because of travel restrictions. I hope to visit my home country soon when this pandemic is over. Also, I would like to take this opportunity to express my sincere gratitude to my late grandfather, Mr. Yeong-guk Heu, for his endless support and encouragement to me. *Junghwan Kim.*

From remote learning to being unable to visit family members outside the country, I have missed the normalcy of being together with friends and family. When restrictions are lifted, I hope to resume my normal routine and catch up on the missed time with everyone else at school and at home. I would also like to thank my parents for their unending support and guidance during these trying times. *Sampath Rapuri.*

Introduction

Quantitative geographical approaches, such as spatial and spatiotemporal modeling and geo-visualizations, have played a pivotal role in COVID-19 research (Boulos & Geraghty, 2020; McLafferty et al., 2021). For example, COVID-19 data and map dashboards have served as an effective means of providing important COVID-19 pandemic information to the public as well as to policymakers. Moreover,

J. Kim (✉)
Virginia Tech, Blacksburg, VA, USA
e-mail: Junghwankim@vt.edu

K. Wang
University of California, Berkeley, CA, USA

S. Rapuri
Johns Hopkins University, Baltimore, MD, USA

M. Laituri et al. (eds.), *The Geographies of COVID-19*, Global Perspectives on Health Geography, https://doi.org/10.1007/978-3-031-11775-6_4

researchers have developed and utilized various spatial and spatiotemporal models to enhance our understanding of virus transmission and changes in the behaviors of individuals (e.g., policy compliance). These quantitative geographical approaches are particularly important for public health policymakers to better formulate evidence-based COVID-19 mitigation measures, such as mobility restrictions.

It has become important to provide an overview of the role of quantitative geographical approaches in COVID-19 research. It allows researchers to understand the critical role of quantitative geographical approaches in COVID-19 research comprehensively. Moreover, it enables researchers to identify critical knowledge gaps that may pertain to developing effective pandemic mitigation methods. In this light, some previous studies provided an overview of the role of quantitative geographical approaches in COVID-19 research. For example, Franch-Pardo et al. (2021) discovered that a growing number of studies have been utilizing quantitative geographical methods to investigate virus transmission during the COVID-19 pandemic. Fatima et al. (2021) provided an in-depth review of 38 papers and identified clustering, hotspot analysis, space-time scan statistics, and regression modeling methods as popular approaches.

Although these previous studies provide a valuable overview of the role of quantitative geospatial approaches in COVID-19 research, we argue that there is a significant research gap. These previous studies largely focused on first-order impacts of the COVID-19 pandemic while overlooking second-order impacts. First-order impacts indicate immediate health-related impacts of the COVID-19 pandemic, such as COVID-19 transmission and increases in confirmed cases and deaths (Laituri et al., 2021). Second-order impacts indicate secondary impacts of the COVID-19 pandemic on various sectors of our daily life, including social, economic, and environmental aspects (Laituri et al., 2021). Since the COVID-19 pandemic has substantially impacted almost every aspect of our lives and society, it is critical to review how quantitative geospatial methods have been used to investigate second-order impacts in addition to the first-order impacts that previous studies have reviewed. In other words, despite a few exceptions (Laituri et al., 2021), many previous studies (e.g., Franch-Pardo et al., 2020, 2021; Fatima et al., 2021) did not pay much attention to reviewing studies on second-order impacts.

This chapter fills this significant research gap by reviewing studies that adopted quantitative geographic approaches for COVID-19 research on both first-order and second-order impacts. Specifically, we ask two research questions: RQ1. What is the role of quantitative geospatial approaches in COVID-19 research focusing on first-order impacts? RQ2. What is the role of quantitative geospatial approaches in COVID-19 research focusing on second-order impacts? To answer these questions, we calculated metrics, such as the number of peer-reviewed academic research papers in terms of the journal, the study area, and the research topic. We also utilized a text analysis approach to create word clouds illustrating the frequency of words used in the titles of the papers.

Data and Methods

We created a database of peer-reviewed academic research papers that adopted quantitative geographical approaches for COVID-19 research. We focused on 45 journals in the fields of geography, geographic information science (GIScience), and urban and regional planning (Table 4.1). We selected general geography journals (e.g., *Applied Geography*) rather than journals that focus on specific topics (e.g., *Journal of Transport Geography*) because we aimed at investigating diverse research topics that quantitative geographers in general widely study. However, as an exception, we purposefully included journals directly related to health and medical geography (e.g., *Health and Place*) because not doing so would exclude important papers that adopted quantitative geographical approaches for COVID-19 research.

From the 45 journals, we found 331 COVID-19 research papers on quantitative geographical approaches and created the list. For each journal, we used "COVID" as a search keyword to find any COVID-19 research papers. We conducted the search process in late October 2021. We did not include commentary, opinions, or review papers. We focused on empirical papers that adopted quantitative geographical approaches. For each paper, we recorded the journal name, paper title, study area, whether the study was about COVID-19 first-order or second-order impacts, research topic, and digital object identifier (DOI). If the study involved multiple countries, we recorded "multiple countries" as the study area. Regarding the research topic, we had four categories: (1) human mobility, (2) health and well-being, (3) environment, and (4) others. We selected these categories after an initial screening of the papers.

The average number of papers among the journals is 7.4 (papers per journal). *Sustainable Cities and Society* published the most papers (46 papers, 13.9%), followed by *ISPRS International Journal of Geo-Information* (44 papers, 13.3%,

Table 4.1 A list of academic peer-reviewed journals in geography, GIScience, and urban and regional planning that were included in our database

Annals of GIS; Annals of the American Association of Geographers; Applied Geography; Cities; Frontiers in Built Environment; Geocarto International; Geographical Review; GeoJournal; Geo-spatial Information Science; Health and Place; International Journal of Digital Earth; International Journal of Geographical Information Science; International Journal of Health Geographics; International Journal of Urban Sciences; ISPRS International Journal of Geo-Information; Journal of Geographical Sciences; Journal of Regional Science; Journal of Urban Health; Journal of Urban Management; Landscape and Urban Planning; Papers in Applied Geography; Population, Space and Place; Professional Geographers; Regional Studies, Regional Science; Spatial and Spatio-temporal Epidemiology; SSM Population Health; Sustainable Cities and Society; Transactions in GIS

This list illustrates selected journals that have at least three papers. The list is in alphabetical order by title open-access option only), *Health and Place* (27, 8.2%), *Cities* (25, 7.6%), and *SSM Population Health* (25, 7.6%, open-access option only). These five journals account for approximately 50% of the papers in our database. Moreover, the average number of papers per country is 6.2 (papers per country). The U.S. published the most papers (79 papers, 27.6%), followed by China (54 papers, 18.9%). These two countries account for approximately 47% of the papers in our database. Given the large body of scholars in these two countries, this result aligns with our expectations. However, we have fewer papers focusing on India (16 papers, 5.6%) than we expected, which is an interesting observation considering the large population of scholars in India. Please note that we excluded 46 papers that included multiple study areas or did not have a specific study area.

Next, we analyzed the database by obtaining descriptive statistics of metrics, including the number of papers, regarding the journal, the study area, and the research topic. Moreover, we created word frequency tables of the papers' titles by adopting a text analysis method using an ad-hoc Python program (e.g., Bird et al., 2009; Oesper et al., 2011). The word frequency table presents how frequently each word is used in the text. We also created word clouds that visualize the word frequency tables. Word clouds are effective visualization tools to illustrate the relative importance of words. Words that are larger are more important (i.e., more frequently appearing in the text) than other words. We did not include words that are trivial or only appear once in the text.

Results

Papers on Quantitative Geographical Approaches: First-Order Impacts

This subsection reviews 186 papers that focused on COVID-19 first-order impacts. There are 186 papers (56% of the 331 papers) on COVID-19 first-order impacts. *ISPRS International Journal of Geo-Information* published the most papers (32, 17.2%), followed by *Sustainable Cities and Society* (22, 11.8%), *Health and Place* (16, 8.6%), *Spatial and Spatio-temporal Epidemiology* (14, 7.5%), and *SSM Population Health* (12, 6.5%). These five journals account for approximately 50% of the 186 papers on first-order impacts. Regarding the study area, the United States and China account for approximately half of the 186 papers, which is not substantially different from our previous findings obtained from reviewing all 331 papers.

Note that *Spatial and Spatio-temporal Epidemiology* is included in the most-published journals. This journal accounts for 7.5% of papers on first-order impacts but only 5% of total papers (including ones on second-order impacts). This observation implies that spatial and spatiotemporal methods play a significant role in modeling the COVID-19 pandemic compared to other quantitative geographical methods.

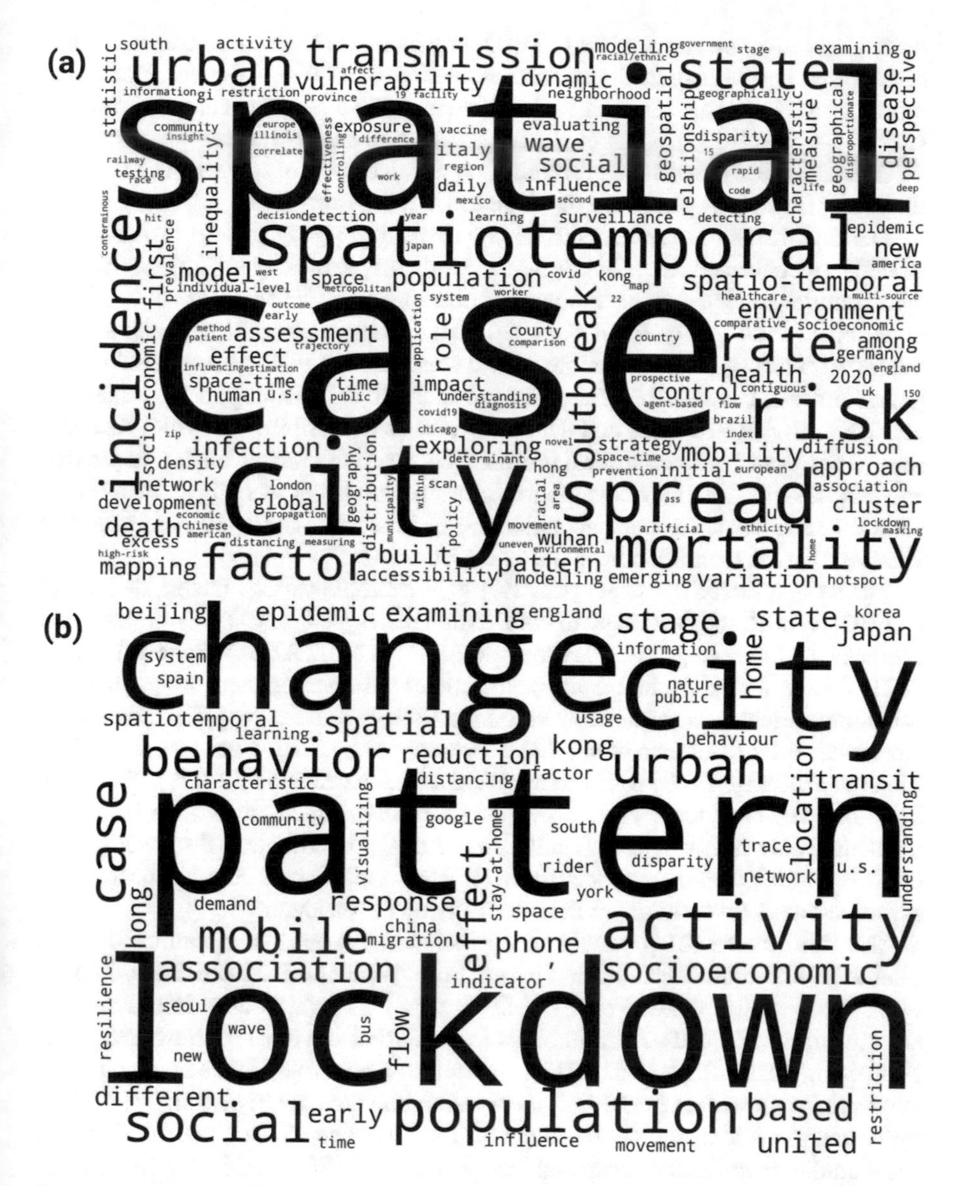

Fig. 4.1 Word clouds illustrate the relative frequency of words used in the titles of (**a**) 186 papers on COVID-19 first-order impacts and (**b**) 59 papers on COVID-19 second-order impacts focusing on human mobility

Next, we created word clouds that visualize the relative frequency of words used in the titles of 186 papers on first-order impacts (Fig. 4.1a). The figure illustrates that the following words are frequently used in the titles: *cases, spatial, city, spatio-temporal, spread, risk, mortality,* and *urban*. The result indicates that cities and urban areas are important in studies on first-order impacts (e.g., Ghosh et al., 2020; Liu, 2020; Mishra et al., 2020), corroborating our general expectations that cities have been major epicenters of the COVID-19 pandemic. The word cloud result also reveals that spatial and spatiotemporal modeling of COVID-19 cases, risks, and mortality are important research topics for quantitative geographers (e.g., Appiah-Otoo & Kursah, 2021; Kianfar et al., 2021; Zhang & Li, 2021).

Overall, we identified three overarching research themes of studies on COVID-19 first-order impacts by observing the database and the word cloud.

The first theme (68 papers, 36.6%) is to investigate geographical disparities in COVID-19 confirmed cases and deaths as well as accessibility to COVID-19 relevant facilities, such as testing and vaccination sites (e.g., Kamis et al., 2021; Kang et al., 2020; Park, 2021; Wu et al., 2020). This theme also includes many studies aiming to detect spatial and spatiotemporal clusters of COVID-19 cases and deaths (e.g., Cordes & Castro, 2020; Desjardins et al., 2020). Examining socio-spatial disparities in COVID-19 cases and deaths as well as accessibility to health facilities (e.g., testing and vaccination sites) is crucial because the research outcomes provide essential insights into understanding existing socio-spatial inequity in health outcomes. It also enhances our understanding of how the COVID-19 pandemic has widened those disparities in people's health outcomes.

The second theme (97 papers, 52.2%) is to examine various factors that affect COVID-19 cases and deaths and to build a model that predicts COVID-19 cases and deaths in the future (e.g., Jaya & Folmer, 2021; Solis et al., 2021; Valente & Laurini, 2021). These factors include built environment variables, sociodemographic and economic variables, environmental variables, policy variables (e.g., lockdown policies), and so on (e.g., Guo et al., 2021; Hamidi et al., 2020; Huang et al., 2021a; Meng et al., 2021; Tribby & Hartmann, 2021; Viezzer & Biondi, 2021). This is an important research topic as it provides an essential theoretical foundation for explaining how viruses spread. It will also eventually contribute to the development of COVID-19 prediction models that can play a significant role in formulating evidence-based public health policies to mitigate the pandemic.

The third theme (21 papers, 11.3%) includes other research themes that are relatively less studied than the first two themes. We identified several sub-themes. First, some studies focused on COVID-19 data quality/assurance issues and the effectiveness of COVID-19 dashboards (e.g., Carballada & Balsa-Barreiro, 2021; Fang et al., 2021; Kolak et al., 2021). Given the critical role of data in quantitative geographic approaches in COVID-19 research, the data quality/assurance topic is important to avoid potential *"Garbage In, Garbage Out (GIGO)"* issues. Second, some studies investigated geospatial data privacy and ethics in COVID-19 research and policies, such as digital contact tracing and COVID-19 dashboards (e.g., Houfaf-Khoufaf et al., 2021; Huang et al., 2021b; Kim & Kwan, 2021). Detailed geospatial data (e.g., a person's mobile phone location trajectory) is critical in quantitative geographical studies on COVID-19. However, it also involves risks, such as geoprivacy breaches, legal issues, and ethical concerns. Third, other studies have developed and proposed new algorithms and methods (e.g., indoor contact tracing) to control the pandemic (e.g., Ojagh et al., 2021; Usui et al., 2021). In sum, we observed that less attention had been paid to these three sub-themes.

Thus, we recommend that future research pay more attention to these areas to fill the significant gaps.

Papers on Quantitative Geographical Approaches: Second-Order Impacts

This subsection reviews 145 papers (44% of the 331 papers) related to COVID-19 second-order impacts. We observed that there are more papers on first-order impacts (56%) than on second-order impacts (44%). One possible explanation for this is that data needed to investigate first-order impacts can be accessed relatively more easily than those needed to investigate second-order impacts. For example, key variables for investigating first-order impacts include new confirmed COVID-19 cases and deaths, sociodemographic variables, and so on. Researchers can easily access these datasets through open-source data repositories, such as those provided by the Centers for Disease Control and Prevention. However, data capturing various second-order impacts (e.g., immediate income loss, change in mental health, and so on) are not widely and promptly made available to researchers.

Regarding the journals we covered, *Sustainable Cities and Society* published the most papers (24, 16.6%), followed by *Cities* (14, 9.7%), *SSM Population Health* (13, 9.0%), *ISPRS International Journal of Geo-Information* (12, 8.3%), and *Health and Place* (11, 8.6%). These five journals account for approximately 50% of the papers on COVID-19 second-order impacts. Regarding the study area, the United States and China account for approximately 42% of the papers on COVID-19 second-order impacts, which is not substantially different from our previous findings obtained from reviewing all papers.

We observed four research topics in studies on second-order impacts. The topic of human mobility has the most papers (59, 40.7%), followed by health and well-being (37, 25.5%), environment (26, 17.9%), and others (23, 15.9%). This finding is in line with our general expectation that "*how do people travel and move around different places?*" is one of the fundamental research questions of quantitative geographers.

Perhaps the result – having the highest number of papers on human mobility – reflects good accessibility of the most up-to-date human mobility data. Specifically, we observed that many studies utilized human mobility data obtained from cell phone signal-based location data, which can be collected and processed much faster than traditional survey-based data. Moreover, private data providers – such as Google Community Mobility Reports (e.g., McKenzie & Adams, 2020), Apple Mobility Trends Reports (e.g., Trasberg & Cheshire, 2021), SafeGraph (e.g., Huang et al., 2021c), StreetLight (e.g., Kar et al., 2021), and other smart card and location data providers (e.g., Bhin & Son, 2021; Eom et al., 2021; Long & Ren, 2022) – have widely distributed pre-processed and aggregated mobility data. The availability of mobility data that immediately captures human mobility trends has provided significant research opportunities to quantitative geography researchers. This might eventually result in the publishing of many studies on COVID-19 second-order impacts focusing on human mobility.

In this chapter, we focused only on the topic of human mobility (i.e., the topic with the most appearances) in creating a word cloud. Figure 4.1b is a word cloud illustrating the frequency of words that appear in the titles of these 59 papers on human mobility. The figure illustrates that "*pattern, lockdown, change, city, activity, population, behavior, urban, case,* and *social*" frequently appear in the titles. Similar to what we observed in the word cloud of studies on COVID-19 first-order impacts (e.g., Fig. 4.1a), the result reveals that cities and urban areas are critical geographic areas of interest. The word cloud also suggests that many quantitative geographers are interested in investigating changes in human mobility caused by COVID-19 lockdown policies.

Although these studies provide significant insight into how human mobility patterns have been impacted by the COVID-19 pandemic, there are several caveats, including algorithmic uncertainty of mobility data based on cell phone signal-based location data (Kwan, 2016; Mooney & Pejaver, 2018). First, cell phone users might not capture the distribution of the entire population as certain sociodemographic segments, such as low-income and senior people, might not be included. Second, cell phone signal-based location data might not capture short-distance trips or trips undertaken in certain areas where the signal quality is poor, such as rural areas or high-density building areas. Lastly, it might not provide meaningful information on people's transportation mode choice behaviors, which is an important topic in human mobility studies. Therefore, we recommend that future studies fill these significant gaps by actively utilizing traditional data, such as household activity-travel surveys, which can provide a more nuanced and comprehensive picture of our understanding of changes in human mobility during the COVID-19 pandemic.

Conclusion

This chapter reviewed the role of quantitative geographical approaches in COVID-19 research. We particularly focused on two types of COVID-19 research: (1) studies on first-order impacts that investigate immediate health-related impacts of the COVID-19 pandemic (e.g., virus transmission, confirmed cases, and deaths) and (2) studies on second-order impacts that examine COVID-19's secondary impacts, including social, economic, and environmental aspects (Laituri et al., 2021). We focused on 331 papers that adopted quantitative geospatial approaches. We selected these papers from 45 journals in geography, GIScience, and urban and regional planning.

Our result revealed that *Sustainable Cities and Society* published the most papers adopting quantitative geographical approaches for COVID-19 research, followed by *ISPRS International Journal of Geo-Information*, *Health and Place*, *Cities*, and *SSM Population Health*. These five journals account for approximately 50% of the

331 papers on COVID-19 research. Regarding the study area, the United States published the most papers, followed by China. These two countries account for approximately 50% of the 331 papers.

Regarding studies on COVID-19 first-order impacts, we identified three research themes. The first theme is investigating geographical disparities in COVID-19 cases and deaths and accessibility to COVID-19 relevant facilities, such as testing and vaccination sites. The second theme is examining various factors that affect COVID-19 cases and deaths and building a model to predict those in the future. The third theme includes other topics (e.g., geospatial data privacy and ethics, data quality and assurance, and so on) that are relatively less studied than the first two themes. Therefore, we recommend that future research pay more attention to the third theme, which is crucial to diversifying and broadening the role of quantitative geographical approaches in COVID-19 research.

We observed fewer papers on second-order impacts than on first-order impacts. The topic of human mobility was investigated the most, followed by health and well-being and the environment. We discussed that the recently growing availability of human mobility datasets (that are largely based on cell phone signal-based location data in most cases) might contribute to the number of COVID-19second-order impact papers on the topic of human mobility.

However, there are several limitations to our study. First, we did not consider papers from other journals that are not classified as geography research but do contain papers adopting quantitative geographical approaches for COVID-19 research (e.g., *Environmental Research*, *Science of the Total Environment*, and so on). Future research can benefit from including more papers from these journals to provide a more comprehensive review of the role of quantitative geographical approaches in COVID-19 research. Second, regarding text analysis methods, we only created word clouds. However, many other innovative text analysis methods (e.g., TF-IDF analysis) can provide additional interesting and significant insights. We recommend future studies adopt various text analysis methods to analyze papers' titles and abstracts to enhance our understanding of the role of quantitative geographical approaches in COVID-19 research.

Despite these limitations, this chapter significantly contributes to the literature as it provides a relatively comprehensive review of the role of quantitative geospatial approaches in COVID-19 research and highlights important future research directions.

References

Appiah-Otoo, I., & Kursah, B. (2021). Modelling spatial variations of novel coronavirus disease (COVID-19): Evidence from a global perspective. *GeoJournal*, 1–15. https://doi.org/10.1007/s10708-021-10427-0

Bhin, M., & Son, S. (2021). Reduction and reallocation of bus use under COVID-19: An analysis of bus card data of Gyeonggi Province, South Korea. *International Journal of Urban Sciences*, 1–21. https://doi.org/10.1080/12265934.2021.1936137

Bird, S., Klein, E., & Loper, E. (2009). *Natural language processing with Python: Analyzing text with the natural language toolkit*. O'Reilly Media, Inc.

Boulos, K., & Geraghty, E. M. (2020). Geographical tracking and mapping of coronavirus disease COVID-19/severe acute respiratory syndrome coronavirus 2 (SARS-CoV-2) epidemic and associated events around the world: How 21st century GIS technologies are supporting the global fight against outbreaks and epidemics. *International Journal of Health Geographics, 19*(8). https://doi.org/10.1186/s12942-020-00202-8

Carballada, A., & Balsa-Barreiro, J. (2021). Geospatial analysis and mapping strategies for fine-grained and detailed COVID-19 data with GIS. *ISPRS International Journal of Geo-Information, 10*(9), 602. https://doi.org/10.3390/ijgi10090602

Cordes, J., & Castro, M. C. (2020). Spatial analysis of COVID-19 clusters and contextual factors in New York City. *Spatial and Spatio-Temporal Epidemiology, 34*, 100355. https://doi.org/10.1016/j.sste.2020.100355

Desjardins, M., Hohl, A., & Delmelle, E. (2020). Rapid surveillance of COVID-19 in the United States using a prospective space-time scan statistic: Detecting and evaluating emerging clusters. *Applied Geography, 118*, 102202. https://doi.org/10.1016/j.apgeog.2020.102202

Eom, S., Jang, M., & Ji, N. (2021). Human mobility change pattern and influencing factors during COVID-19, from the outbreak to the deceleration stage: A study of Seoul Metropolitan City. *The Professional Geographer*, 1–15. https://doi.org/10.1080/00330124.2021.1949729

Fang, H., Xin, S., et al. (2021). Evaluating the effectiveness and efficiency of risk communication for maps depicting the hazard of COVID-19. *Transactions in GIS*. https://doi.org/10.1111/tgis.12814

Fatima, M., O'Keefe, K., et al. (2021). Geospatial analysis of COVID-19: A scoping review. *International Journal of Environmental Research and Public Health, 18*(5), 2336. https://doi.org/10.3390/ijerph18052336

Franch-Pardo, I., Napoletano, B., et al. (2020). Spatial analysis and GIS in the study of COVID-19. A review. *Science of the Total Environment, 739*, 140033. https://doi.org/10.1016/j.scitotenv.2020.140033

Franch-Pardo, I., Desjardins, M. R., et al. (2021). A review of GIS methodologies to analyze the dynamics of COVID-19 in the second half of 2020. *Transactions in GIS*. https://doi.org/10.1111/tgis.12792

Ghosh, A., Nundy, S., et al. (2020). Study of COVID-19 pandemic in London (UK) from urban context. *Cities, 106*, 102928. https://doi.org/10.1016/j.cities.2020.102928

Guo, Y., Yu, H., et al. (2021). Exploring the impacts of travel-implied policy factors on COVID-19 spread within communities based on multi-source data interpretations. *Health & Place, 69*, 102538. https://doi.org/10.1016/j.healthplace.2021.102538

Hamidi, S., Sabouri, S., & Ewing, R. (2020). Does density aggravate the COVID-19 pandemic? Early findings and lessons for planners. *Journal of the American Planning Association, 86*(4), 495–509. https://doi.org/10.1080/01944363.2020.1777891

Houfaf-Khoufaf, W., Touya, G., & Le Guilcher, A. (2021). Geographically masking addresses to study COVID-19 clusters. *Cartography and Geographic Information Science*, 1–15. https://doi.org/10.1080/15230406.2021.1977709

Huang, J., Kwan, M.-P., & Kan, Z. (2021a). The superspreading places of COVID-19 and the associated built-environment and socio-demographic features: A study using a spatial network framework and individual-level activity data. *Health & Place, 72*, 102694. https://doi.org/10.1016/j.healthplace.2021.102694

Huang, J., Kwan, M.-P., & Kim, J. (2021b). How culture and sociopolitical tensions might influence people's acceptance of COVID-19 control measures that use individual-level georeferenced data. *ISPRS International Journal of Geo-Information, 10*(7), 490. https://doi.org/10.3390/ijgi10070490

Huang, X., Lu, J., et al. (2021c). Staying at home is a privilege: Evidence from fine-grained mobile phone location data in the United States during the COVID-19 pandemic. *Annals of the American Association of Geographers*, 1–20. https://doi.org/10.1080/24694452.2021.1904819

Jaya, I., & Folmer, H. (2021). Bayesian spatiotemporal forecasting and mapping of COVID-19 risk with application to West Java Province, Indonesia. *Journal of Regional Science*. https://doi.org/10.1111/jors.12533

Kamis, C., Stolte, A., et al. (2021). Overcrowding and COVID-19 mortality across US counties: Are disparities growing over time? *SSM-Population Health*, 100845. https://doi.org/10.1016/j.ssmph.2021.100845

Kang, J., Michels, A., et al. (2020). Rapidly measuring spatial accessibility of COVID-19 healthcare resources: A case study of Illinois, USA. *International Journal of Health Geographics, 19*(1), 1–17. https://doi.org/10.1186/s12942-020-00229-x

Kar, A., Le, H., & Miller, H. (2021). What is essential travel? Socioeconomic differences in travel demand in Columbus, Ohio, during the COVID-19 lockdown. *Annals of the American Association of Geographers*, 1–24. https://doi.org/10.1080/24694452.2021.1956876

Kianfar, N., Mesgari, M., et al. (2021). Spatio-temporal modeling of COVID-19 prevalence and mortality using artificial neural network algorithms. *Spatial and Spatio-Temporal Epidemiology*, 100471. https://doi.org/10.1016/j.sste.2021.100471

Kim, J., & Kwan, M.-P. (2021). An examination of people's privacy concerns, perceptions of social benefits, and acceptance of COVID-19 mitigation measures that harness location information: A comparative study of the US and South Korea. *ISPRS International Journal of Geo-Information, 10*(1), 25. https://doi.org/10.3390/ijgi10010025

Kolak, M., Li, X., et al. (2021). The US COVID Atlas: A dynamic cyberinfrastructure surveillance system for interactive exploration of the pandemic. *Transactions in GIS, 25*(4), 1741–1765. https://doi.org/10.1111/tgis.12786

Kwan, M.-P. (2016). Algorithmic geographies: Big data, algorithmic uncertainty, and the production of geographic knowledge. *Annals of the American Association of Geographers, 106*(2), 274–282. https://doi.org/10.1080/00045608.2015.1117937

Laituri, M., Richardson, R., et al. (2021). Examining second-order impacts of COVID-19 in urban areas. *Annals of GIS*, 1–12. https://doi.org/10.1080/19475683.2021.1954087

Liu, L. (2020). Emerging study on the transmission of the Novel Coronavirus (COVID-19) from urban perspective: Evidence from China. *Cities, 103*, 102759. https://doi.org/10.1016/j.cities.2020.102759

Long, J., & Ren, C. (2022). Associations between mobility and socio-economic indicators vary across the timeline of the Covid-19 pandemic. *Computers, Environment and Urban Systems, 91*, 101710. https://doi.org/10.1016/j.compenvurbsys.2021.101710

McKenzie, G., & Adams, B. (2020). A country comparison of place-based activity response to COVID-19 policies. *Applied Geography, 125*, 102363. https://doi.org/10.1016/j.apgeog.2020.102363

McLafferty, S., Guhlincozzi, A., & Winata, F. (2021). Counting COVID: Quantitative geographical approaches to COVID-19. In G. Andrews, V. Crooks, J. Pearce, & J. Messina (Eds.), *COVID-19 and similar futures*. Springer. https://doi.org/10.1007/978-3-030-70179-6_54

Meng, Y., Wong, M., et al. (2021). Yearly and daily relationship assessment between air pollution and early-stage COVID-19 incidence: Evidence from 231 countries and regions. *ISPRS International Journal of Geo-Information, 10*(6), 401. https://doi.org/10.3390/ijgi10060401

Mishra, S., Gayen, A., & Haque, S. (2020). COVID-19 and urban vulnerability in India. *Habitat International, 103*, 102230. https://doi.org/10.1016/j.habitatint.2020.102230

Mooney, S. J., & Pejaver, V. (2018). Big data in public health: Terminology, machine learning, and privacy. *Annual Review of Public Health, 39*, 95–112. https://doi.org/10.1146/annurev-publhealth-040617-014208

Oesper, L., Merico, D., et al. (2011). WordCloud: A cytoscape plugin to create a visual semantic summary of networks. *Source Code for Biology and Medicine, 6*(1), 1–4. https://doi.org/10.1186/1751-0473-6-7

Ojagh, S., Saeedi, S., & Liang, S. (2021). A person-to-person and person-to-place COVID-19 contact tracing system based on OGC IndoorGML. *ISPRS International Journal of Geo-Information, 10*(1), 2. https://doi.org/10.3390/ijgi10010002

Park, J. (2021). Who is hardest hit by a pandemic? Racial disparities in COVID-19 hardship in the US. *International Journal of Urban Sciences, 25*(2), 149–177. https://doi.org/10.1080/12265934.2021.1877566

Solis, P., Dasarathy, G., et al. (2021). Understanding the spatial patchwork of predictive modeling of first wave pandemic decisions by US governors. *Geographical Review, 111*(4), 592–615. https://doi.org/10.1080/00167428.2021.1947139

Trasberg, T., & Cheshire, J. (2021). Spatial and social disparities in the decline of activities during the COVID-19 lockdown in greater London. *Urban Studies*. https://doi.org/10.1177/00420980211040409

Tribby, C., & Hartmann, C. (2021). COVID-19 cases and the built environment: Initial evidence from New York City. *The Professional Geographer*, 1–12. https://doi.org/10.1080/00330124.2021.1895851

Usui, H., Asami, Y., & Yamada, I. (2021). A normative model to estimate the number of persons not social distancing in a 3D complex built space. *International Journal of Geographical Information Science*, 1–21. https://doi.org/10.1080/13658816.2021.1938076

Valente, F., & Laurini, M. (2021). Robust trend estimation for COVID-19 in Brazil. *Spatial and Spatio-Temporal Epidemiology, 39*, 100455. https://doi.org/10.1016/j.sste.2021.100455

Viezzer, J., & Biondi, D. (2021). The influence of urban, socio-economic, and eco-environmental aspects on COVID-19 cases, deaths and mortality: A multi-city case in the Atlantic Forest, Brazil. *Sustainable Cities and Society, 69*, 102859. https://doi.org/10.1016/j.scs.2021.102859

Wu, Y., Yan, X., et al. (2020). Association of time to diagnosis with socioeconomic position and geographical accessibility to healthcare among symptomatic COVID-19 patients: A retrospective study in Hong Kong. *Health & Place, 66*, 102465. https://doi.org/10.1016/j.healthplace.2020.102465

Zhang, T., & Li, J. (2021). Understanding and predicting the spatio-temporal spread of COVID-19 via integrating diffusive graph embedding and compartmental models. *Transactions in GIS*. https://doi.org/10.1111/tgis.12803

Chapter 5
COVID-19's Impact on Geospatial Data: Ethics and Values

Dara E. Seidl

Personal Story

In many ways, the COVID-19 pandemic made it feel like time had stopped. I spent the first few weeks hunkered down with my partner, somewhat worried and alert, but also enjoying many cozy days of delicious bread-making. Like most, we thought it would be a brief period to get through, rather than a years-long experience. As time went on, we found ourselves fortunate to not be touched directly by illness or family sorrow, but there were other losses: the enduring calculation of risk for any outing, not seeing loved ones for years at a time, and the brief horror of watching a TV character walk into a public place without a mask. Most surprising, though, were the ways in which life continued. Four months into the pandemic, I accepted a new position which brought our family from the Central Coast of California to Leadville, Colorado, a small city situated at over 10,000 feet in elevation. It has been strange to move to a new place, live there for over a year, but not yet have a real sense of the town. We are fortunate to have this new adventure waiting for us when the pandemic risk eases, along with one more pandemic addition: our new collie puppy. *Dara Seidl.*

Introduction

When COVID-19 hit the world stage, it necessitated a swift public health response. The pandemic response invoked numerous debates about how geographical data should be collected and used. Many of the tactics governments and health agencies employed to track the spread of the virus were geographical in nature, including the collection and sharing of geospatial data about individuals. Widespread phone-based location tracking particularly incited debate for running counter to prevailing ethics and values regarding privacy and freedom of movement. This chapter examines the ethics and values positioned as at stake in the COVID-19 response.

D. E. Seidl (✉)
Colorado Mountain College, Leadville, CO, USA
e-mail: deseidl@coloradomtn.edu

M. Laituri et al. (eds.), *The Geographies of COVID-19*, Global Perspectives on Health Geography, https://doi.org/10.1007/978-3-031-11775-6_5

While sometimes used interchangeably as concepts, *values* have a distinct meaning from *ethics*. Rather than prescribing specific ways of behaving, values are ideals to which individuals or communities aspire (Barsky, 2019). While all humans have values, we may order them differently, viewing some as more important than others. This value ordering reflects our priorities as individuals. Examples of commonly held values are safety, social justice, happiness, and peace. Values inherently related to the COVID-19 pandemic include health, safety, privacy, equity, and freedom of movement. By contrast to these categories of ideals underpinning values, ethics are guidelines for types of behaviors considered appropriate or inappropriate. In this way, ethics essentially provides rules for acceptable behavior. For example, the Geographic Information Systems (GIS) Code of Ethics (Huxhold & Craig, 2003) provides guidelines for mapping professionals to follow in their work, including striving to avoid harm to individuals by protecting individual privacy. The GIS Certification Institute shares a similar ethical code,[1] under which obligations to society include allowing people to know whether they are in a database and to view and correct information about themselves. These ethical guidelines for behavior make clear an adherence to favored values, including honesty, respect, and privacy.

Often it seems as though values are placed in competition with each other during a crisis. For example, the passage of the United States Patriot Act in 2001 following the September 11th attacks allowed the government to collect sweeping phone, email, internet, and banking data on American residents. This law was passed in the name of national security, as its stated purpose was to identify and catch terrorists, but the practice and law have been decried as running counter to the values of privacy and civil liberties (Mell, 2002). In the current pandemic crisis, privacy and civil liberties are again positioned by popular media as at odds with a common good: public health. With the rollout of tracking technology to aid in identifying and stopping the spread of COVID-19, the predominant debate was framed as a binary choice between saving more lives with digital contact tracing or forgoing the use of location tracking, thereby respecting privacy and civil liberties (Gasser et al., 2020).

However, a number of scholars have debunked this binary framing as too simplistic, or even as a false tradeoff, as the benefits of digital tools against COVID-19 remain unproven. Gasser et al. (2020) advocate for the examination of digital public health technologies with a focus on public health outcomes and the ethics guiding these outcomes, rather than a solutionist approach focusing on the benefits technology can bring to public health management. Kitchin (2020) suggests that if geospatial health surveillance cannot deliver on its promised or intended public health outcomes, then the trade between public health and civil liberties is a false one. Goldenfein et al. (2020) trace the juxtaposition of privacy and public health as the latest in a familiar line of "privacy vs. X" debates leveraged by government and corporate entities to increase surveillance. The argument made by these entities is that data surveillance is key to solving each crisis, X, justifying the development of

[1] GIS Certification Institute. Rules of Conduct for Certified GIS Professionals (GISPs). Available at https://www.gisci.org/Ethics/RulesofConduct.aspx (Last accessed 08 January 2022).

extensive surveillance infrastructures which remain in place beyond the end of each crisis. Sadowski (2020) specifically calls this privacy rights-public health debate an authoritarian trade-off, as the companies providing data surveillance position themselves as a universal solution, undergoing little scrutiny in each emergency situation. The next sections of this chapter outline the technologies introduced in response to COVID-19 and the tensions between them and established geographic ethics, privacy, and civil liberties.

Smartphone Ownership

Public health researchers have identified four categories of digital tools deployed against COVID-19, all of which include the collection or sharing of device location data. These categories are contact tracing, symptom checkers, quarantine compliance, and flow modelling (Gasser et al., 2020). These categories will be discussed in detail in the proceeding section. An important backdrop for the enthusiasm to deploy these individual-tracking tools is the worldwide growth in cell phone ownership, and in particular smartphone ownership. Compared to prior pandemics, the rise of smartphones has equipped a large proportion of the world population with a personal tracking device. A Pew Research Center study found that in the United States, by the time the COVID-19 pandemic began, 81% of American adults were reported to own a smartphone, and 96% were reported to own a cell phone (Anderson, 2019).

Smartphone ownership also has an inverse relationship with the age group in the United States. Americans aged 65 and older are less likely to own a smartphone, at just 53%, compared to 96% of Americans between the ages of 18 and 29 (Anderson, 2019). Lower rates of smartphone adoption are found among residents of rural areas, adults with lower incomes, and those whose educational attainment was a high school diploma or less. Similarly, smartphone adoption rates a year before the pandemic varied widely between countries. According to another Pew Research Center study (Taylor & Silver, 2019), practically 100% of adults in South Korea have a cell phone, with 95% of adults having a smartphone, the highest adoption rate of any country. Besides the United States with smartphone adoption at 81% of adults, Spain, Australia, Sweden, the Netherlands, and Israel all have adult smartphone rates of 80% or higher. Taylor and Silver (2019) found a stark difference in cell phone adoption between countries deemed advanced economies and emerging economies, as determined by World Bank income group and gross domestic product (GDP) rates. The median smartphone adoption rate for emerging economy countries is 45%, compared to 76% in advanced economies. Some of the countries in the emerging economy list are India (smartphone adoption at 24%), Nigeria (39%), Mexico (52%), Brazil (60%), and South Africa (60%). While smartphone adoption worldwide was higher than ever before, the variable rates across countries and demographic groups called into question whether any smartphone-based health surveillance measures could be effectively or equitably harnessed to mitigate the spread of COVID-19.

Existing Location Data Infrastructure

Despite uneven adoption across populations, the precision with which cell phones and smartphones could be applied to detect travel patterns seemed to make them an irresistible target in the quest to stop the spread of the virus, particularly pre-vaccine and before the factors behind transmission were fully understood. A large body of research harnesses location data from mobile devices to study and predict human behavior (Rout et al., 2021), including to assess travel patterns (Vich et al., 2017), tourist locations (Shoval & Ahas, 2016), and even spatial behaviors related to mental health (Chow et al., 2017; Gong et al., 2019), including trajectories of depression (Canzian & Musolesi, 2015). In addition to a healthy research body on individual location sensing with participants willing and consenting to be tracked, there was also widespread pre-pandemic data collection infrastructure tracking individual locations without their knowledge. The location intelligence or location data broker industry is an estimated $12 billion market (Keegan & Ng, 2021), whose main players include at least 47 companies, including Near, Mobilewalla, PlaceIQ, Cuebiq, Place.ai, SafeGraph, X-Mode (Outlogic), and Unacast. These location data brokers collect, aggregate, and sell individual location data, often scouring it through phone apps requesting permission for user locations, such as with map, weather, coupon, video, and gaming applications, with many companies not disclosing their sources of location data. In one example of these behind-the-scenes location data flows, data broker X-Mode, now called Outlogic, was found selling location data from Muslim prayer apps Muslim Pro and Qibla Compass to military contractors (Cox, 2021). In a more targeted example, a Catholic news organization purchased "anonymized" data from a mobile location broker that outed a priest who was found to be visiting gay bars through location data collected in the Grindr dating app (Boorstein et al., 2021). These examples signal the widespread existence of location data flows where mobile users might not consent if they fully understood the potential for location to be used against them.

When COVID-19 started spreading, many location data brokers saw an opportunity to advertise their close-to-real-time location repositories as instrumental to the pandemic response. The company SafeGraph advertised its social distancing metrics with maps showing percentages of phones by block group traveling away from home, with locations collected from undisclosed mobile applications. Unacast similarly created a social distancing scorecard, assigning letter grades to counties and states based on phone Global Positioning System (GPS) data and detections of changed travel behavior (Morrison, 2020). In a notorious example, a partnership between X-Mode Social and Tectonix GEO released a geographic visualization of 5000 devices at a beach in Fort Lauderdale, Florida, during spring break, then traveling across much of the United States, flaunting social distancing recommendations in March 2020 (Holmes, 2020). These data companies capitalized on the popularity garnered through free COVID-related geographic datasets to forge partnerships with researchers for future projects. Hu et al. (2021) use SafeGraph location data to detect changes in visits to bars and liquor stores. Althoff et al. (2021) similarly partner with SafeGraph for data to assess the geography of remote work. Such studies reflect the normalization of incorporating location data from mobile users who are likely unaware they are being tracked for such purposes. Mobile users are routinely exposed to dark patterns, in which application interfaces are designed to confuse users into accepting settings that are inconsistent with their preferences, such as accepting higher levels of tracking (Luguri & Strahilevitz, 2021).

The rollout of phone-based contact-tracing apps in service of pandemic mitigation is an example of technological solutionism, in which technology is presented as a solution before a problem is fully investigated and defined, or before the potential pitfalls and unintended consequences of the technology are studied (Morozov, 2013). An example of a topic treated in this way is the use of smartphone brands and usage to predict human altruism and compassion (Anand et al., 2021). As Taylor (2021) writes, the hasty deployment of contact-tracing apps by higher-income countries demonstrated a belief that technology could be used to solve a problem as complex as the pandemic. Taylor notes that before the summer of 2020 in the Global North, the failure to stop the spread of the virus was framed by policy-makers as an issue of technology acceptance and individuals unwilling to install tracking applications. Kitchin (2020) argues that the widespread solutionism of the app-based pandemic response is in part due to years of lobbying by technology companies and states' desire to stimulate high-tech economic sectors. Under technological solutionism, the rush to deploy tracking technology as a primary means of stopping virus spread is not pared back when the technology's shortcomings become apparent. The solutionism also reinforces a dichotomy between privacy and public health, as if location data were the missing link in the ability to effectively contact trace.

Location-Based Technologies for the COVID-19 Pandemic

With this background of widespread (but variable) smartphone adoption, existing structures of tangled personal location data collection, and high interest in technology-based solutions, the stage was set for rollouts of location-based tools to combat COVID-19. A first category of these tools is *proximity and contact tracing*, which track interactions between individuals based on their estimated spatial proximity (Gasser et al., 2020). Proximity tracing may leverage GPS, Wi-Fi, cell tower location, or Bluetooth with mobile phones (Kedron & Trgovac, 2021), or it may also take advantage of facial recognition in video feeds to identify individuals who are not maintaining two meters of distance. For example, Singapore's TraceTogether app applies Bluetooth to detect proximity between phones and alerts users found to be sufficiently close to a COVID-19-positive individual (Gasser et al., 2020). India deployed the Aarogya Setu application, which harnesses Global Navigation Satellite

System (GNSS) and Bluetooth to notify users if they have come into contact with someone who has tested positive (Gupta et al., 2020). The Aarogya Setu app also collects personal details such as name, age, sex, and medical conditions. Israel moved to access a trove of cell phone data collected by its security agency, Shin Bet, for the purposes of contact tracing (Halbfinger et al., 2020). Kitchin (2020) writes that neither Bluetooth nor GNSS has the spatial precision required to effectively establish contact tracing, which would involve consistently and accurately determining proximity of less than 2 meters. GNSS, including GPS, is accurate to between 5 and 20 m (Lee et al., 2016). Combining Bluetooth and Wi-Fi positioning can locate devices to under 1 m of accuracy but cannot determine whether a wall or window is separating individuals, which is important for accurate proximity detection.

In the United States, the use of targeted contact tracing apps for COVID-19 is coordinated by state. One reason public health officials cite for the deployment of digital contact tracing apps is the insufficiency of traditional contact tracing methods to keep up with cases (Simmons-Duffin, 2021). This slowness was in large part due to a lack of workers, with a secondary reason being a lack of trust by Americans; many on the contact list refused to pick up the phone. A study of traditional contact tracing in the United States found that two-thirds of individuals with COVID-19 were either not reached for contact tracing or named no known contacts, thereby failing to stop transmission (Lash et al., 2021). As of March 2021, more than 28 million Americans, or approximately 8.5%, had downloaded smartphone mobile apps for contact tracing using Bluetooth technology (Van Ness, 2021). In total, 24 states and Washington, D.C. have developed apps or exposure notification systems for use by residents, most relying on the Apple-Google Bluetooth exposure notification.

A second location data tool deployed against COVID-19 is the *symptom checker*. Users report their symptoms online and possibly obtain a diagnosis of COVID-19 (Gasser et al., 2020). The user inputs are stored, including location, for use in disease modelling and the allocation of resources. Examples include the CoronaMadrid symptom checking application from Spain and the COVID-19 Symptom Tracker serving the U.K. and the United States. Symptom checkers collect sensitive health data as well as IP (Internet Protocol), cell-tower-derived, or GPS locations. Aside from specific phone applications where users self-report their symptoms, thermal cameras and biometric wearables have also been deployed to detect COVID-19 symptoms and geographic patterns (Kitchin, 2020). The Kinsa HealthWeather map[2] based on temperature reads from smart thermometers is an example of a sensed symptom checker, where location and human temperature data are aggregated to show geographical trends.

A third geographical COVID-19 response is *quarantine enforcement*. This involves the monitoring in real time of individuals to ensure compliance with

[2] Kinsa HealthWeather COVID Map. Available at https://healthweather.us/ (Last accessed 08 January 2022).

quarantine restrictions. Mandatory quarantine enforcement apps were deployed in parts of China, Poland, and Taiwan. In China, in order to access particular buildings or transportation systems, visitors were required to scan their phones to obtain a color-based response as to whether they could enter or had to leave and quarantine (Goh, 2020). Poland required residents with confirmed diagnoses to send out a geo-tagged selfie within 20 min of receiving a text message from a health authority to ensure that they were quarantining at home (Nielsen, 2020). In Taiwan, a compulsory GPS tracking application was deployed to detect when infected individuals were leaving isolation and to issue fines of $33,000 (Timberg & Harwell, 2020). Tan et al. (2020) developed the SignatureHome system, deployed in Hong Kong as StayHomeSafe, which leverages a wristband-smartphone pairing to notify authorities when individuals assigned to quarantine have exited the geofenced home region. These enforcements are said to conflict with the value of freedom of movement and the personal autonomy that comes with it (Spadaro, 2020).

The fourth category of geographical tools deployed against COVID-19 is *flow modelling*, also known as mobility reports. The social distancing scorecards released by data brokers Unacast and SafeGraph are examples of these types of reports, which aggregate data on the movements of users. Public health researchers use flow modelling to show the general effectiveness of social distancing or quarantine policies across larger geographers (Gasser et al., 2020). Another widely considered flow model is Google's COVID-19 Mobility Reports, for which user locations are automatically detected from mobile users who have activated a location history setting. Because individual contributors to flow modelling do not need to download separate applications or specifically opt in, as they may for proximity-based contact tracing, there may be less awareness of how their locations are being shared for COVID research. Nevertheless, all of these tools rely on the ability to detect location from individual devices, collecting personal location at a very granular level and sharing it with both government and private entities.

Outside of applications that systematically collect location from mobile devices, maps of disease spread with traditionally collected geographic statistics became widely available. One of the most well-regarded platforms for viewing geospatial data related to the pandemic was the Johns Hopkins University COVID-19 Dashboard (Dong et al., 2020). The dashboard includes total cases, deaths, and administered vaccine doses when vaccines became available. The tool continues to be updated at the time of writing and is currently at over five million deaths and eight billion vaccine doses administered. Frequently cited by government and public health officials, the Johns Hopkins researchers confirm the case numbers with regional and local health departments across the world before publishing the statistics. Many smaller regions and local governments developed their own public health dashboards to track cases more closely in their immediate areas and share information with their constituents.

Even COVID tracking maps and dashboards became subject to debate as the spread of COVID became politicized. Rebekah Jones, a geographic information systems manager for the Florida Department of Health at the start of the COVID-19 pandemic, had her employment there terminated for insubordination in 2020. Jones claims that the firing was based on her refusal to manipulate the state's COVID-19 data in a way that would support the state's reopening and lifting of pandemic restrictions (Mazzei, 2020). Forbes Magazine named Jones the Technology Person of the Year (2020) for the COVID-19 tracking dashboard she established after the health department termination and subsequent police raid at her home (Popkin, 2020). The case attracted the attention of many geographers and news outlets, and Jones was invited as a keynote speaker at the 2021 American Association of Geographers Meeting with a talk entitled "A Geographer's Perspective on High-Stakes Public Science" (Loomar, 2021). Jones' personal and professional background came under scrutiny in the publicized controversy, which gained public interest in the narrative that a scientist was resisting pressure to skew statistics in favor of a government's stance on a public health issue. The story was initially framed as a matter of public accountability and holding fast to objectivity in the face of political pressure, a value espoused by many scientists.

Ethical Considerations for Digital COVID-19 Tools

While deployed in support of the values of public health and safety, the geographical response to the COVID-19 pandemic challenged professional codes of ethics for public health and GIS practitioners and ran counter to the established values of privacy, civil liberties, and freedom of movement. The GIS Code of Ethics calls for the privacy and confidentiality of data subjects, such that individuals are protected from secondary uses of their information and are able to inspect and correct their records. In the data broker industry, where collection of location data is not transparent, data subjects are seldom given the opportunity to understand or even view their personal data flows (Kedron & Trgovac, 2021). Kedron and Trgovac (2021) identify five core principles to measure the ethical standing of new digital contact tracing technologies: efficacy, privacy, equity, transparency, and accountability. Similarly, public health ethics seek to protect the values of privacy, autonomy, and justice (Gasser et al., 2020). With uneven adoption rates of smartphone technology, equity and justice are at stake as surveillance is applied unevenly to different groups. Another ethical principle challenged with COVID location tracking technology is data minimization, which keeps the data collected to only the data necessary to complete a task and minimizes the use of such data to just the original intended purpose (OECD, 1980). However, now that location data have been collected for the pandemic crisis and the digital infrastructure has been developed, scope creep is emerging. For example, Singapore used tracking data from its TraceTogether app to investigate a murder after the minister promised it would only be used for COVID mitigation purposes (Tarabay, 2021).

Even where location tracking technology has been made mandatory in the pandemic response, there are still occasions for individuals to exert personal choice and opt in or out. Throughout the United States, proximity-tracing applications are optional, and even participants who have opted in can turn off the device's locating receivers for satellite-based GPS and Bluetooth, thereby engaging in personal location masking (Seidl et al., 2020). Exploring privacy concerns and social acceptance of COVID-19 protocols in the United States and South Korea, Kim and Kwan (2021) find lower privacy concerns among South Koreans along with higher levels of perceived social benefits for COVID location-tracking protocols. The researchers hypothesize that this is because South Koreans, with a collectivist, rather than individualist orientation, are already used to surveillance systems and previously witnessed the success of the measures taken to control COVID-19. Despite offering a personal choice, optional tracking apps put the onus on the individual to decide whether to opt in or out, rather than receive uniform treatment and uniform data protection from the collecting agency. This arguably leads to less effective virus mitigation, but also to less equitable data representation across a population.

Discussion

In the deployment of applications meant to mitigate the spread of COVID-19, some government entities have voiced an adherence to the value of privacy, while others have not. South Korea initially published detailed location histories for individuals testing positive, which led others to uncover their identities and publicly shame them (Singer & Choe, 2020). This led health officials to stop the practice of publishing location histories in the name of human rights and privacy. In addition, officials were concerned that over-disclosure of such location histories would cause symptomatic individuals to avoid testing (Zastrow, 2020). Stigma over a positive COVID-19 diagnosis appears to be especially acute in India, leading to fear and a loss of social capital (Bhanot et al., 2021). The potential for stigma and discrimination is a driver for protecting personal privacy in the geospatial response to COVID-19. In some regions, privacy is not considered a highly-ranked value, and so the required collection of location data does not raise many ethical flags. For example, a COVID-19 close contact detector deployed in China in 2020 was well-received by the Chinese public because of high appreciation for the power of data for good (Kamel Boulos & Geraghty, 2020). Public opinion of location-based COVID-19 technologies likewise differs between countries with different value priorities. With similar conclusions to Kim and Kwan (2021), discussed previously, Altmann et al. (2020) measured the acceptability of a theoretical contact-tracing application in five countries: France, Germany, Italy, the United Kingdom, and the United States. The researchers found high support for the app overall, but less support in Germany and the United States, where there were higher concerns for cybersecurity and privacy, and less trust in government.

The deployment of location-dependent technologies in pandemic response will have implications for the norms surrounding location data flows moving forward. Future research in this area should focus on the extent of scope creep for the location-tracking infrastructure. Are the location data collected through government mobile apps being re-directed for unrelated activities? What unintended consequences have arisen from the expedited rollout of these technologies? How has public opinion changed over the course of the pandemic on location-tracking technologies? In what ways has pandemic-related location tracking had a chilling effect on individual behavior and personal autonomy? Researchers in this area should also examine any inequities in treatment arising from uneven adoption of these technologies.

Conclusion

The rollout of the geospatial response to COVID-19 was rapid and commonly justified as a necessary solution in stopping the spread of the disease. The technological solutionism of the first few months of the pandemic, captured in the expedited rollout of contract tracing apps, symptom checkers, quarantine apps, and mobility reports, positioned privacy protection as the enemy of public health with pervasive location data as the key to ending the crisis. This wedging between two values—privacy and public health—is a similar conceptualization to the post-9/11 debates positioning privacy versus national security in the eagerness to find terrorists through the vast surveillance networks of American telecommunications. In reality, the relationship between these values is more complex, and there are many opportunities to build privacy by design into technologies that serve the ends of public health and national security.

This chapter reviewed the backdrop and precedent for the rollout of geospatial applications in response to COVID-19: namely, the widespread adoption of mobile devices and existing digital infrastructure for pervasive location data collection. The geospatial response, reliant on personal mobile devices, was divided into four categories: contact tracing, symptom checkers, quarantine enforcement, and flow modelling. A strong eagerness to apply digital technology to solve the pandemic is reminiscent of technological solutionism, where challenges in procedure and efficacy are overlooked in the enthusiasm to solve the problem. Global backlash towards mandatory location tracking and refusal to opt-in where options were given are evidence of adherence to the values of privacy, autonomy, and individualism. The COVID-19 pandemic shined a light on existing location data broker infrastructure and the extent to which, without government intervention, individual trajectories are captured on a regular basis through a plethora of apps and sold to myriad third parties. With little opportunity for data subjects to dissent or opt out of specific uses, these practices present an ongoing challenge to geographers' professional codes of ethics. As the COVID-19 crisis continues, geographers will need to assess how location data practices fit in with their stated ethics and values.

Acknowledgement I would like to thank Dr. Robert Richardson and an anonymous reviewer for their careful reading and helpful comments to improve this chapter.

References

Althoff, L., Eckert, F., Ganapati, S., & Walsh, C. (2021). *The geography of remote work.* National Bureau of Economic Research Working Paper Series Working Paper 29181. https://www.nber.org/system/files/working_papers/w29181/w29181.pdf

Altmann S., Milsom L., Zillessen H., Blasone R., Gerdon F., Bach R., Kreuter F., Nosenzo D., Toussaert S., & Abeler J. (2020) Acceptability of App-Based Contact Tracing for COVID-19: Cross-Country Survey Study. *JMIR mHealth and uHealth, 8*(8), e19857–v8i8e19857 10.2196/19857

Anand, A., Vessal, S. R., Rathi, K., & Ameen, N. (2021). Show me your mobile and I will tell you who you are: Forecasting consumer compassion and altruism behaviour through smartphone type and usage. *Journal of Retailing and Consumer Services, 63*, 102657. https://doi.org/10.1016/j.jretconser.2021.102657

Anderson, M. (2019). *Mobile Technology and Home Broadband 2019.* Pew Research Center. https://www.pewresearch.org/internet/2019/06/13/mobile-technology-and-home-broadband-2019/pi_2019-06-13_broadband_0-03/

Barsky, A. E. (2019). *Ethics and values in social work: An integrated approach for a comprehensive curriculum* (2nd ed.). Oxford University Press.

Bhanot, D., Singh, T., Verma, S. K., & Sharad, S. (2021). Stigma and discrimination during COVID-19 pandemic. *Frontiers in Public Health, 8.* https://doi.org/10.3389/fpubh.2020.577018

Boorstein, M., Iati, M., & Shin, A. (2021, July 21). Top U.S. Catholic Church official resigns after cellphone data used to track him on Grindr and to gay bars. *The Washington Post.* https://www.washingtonpost.com/religion/2021/07/20/bishop-misconduct-resign-burrill/

Canzian, L., & Musolesi, M. (2015). Trajectories of depression: Unobtrusive monitoring of depressive states by means of smartphone mobility traces analysis. In *UbiComp'15: Proceedings of the 2015 ACM international joint conference on pervasive and ubiquitous computing* (pp. 1293–1304). https://doi.org/10.1145/2750858.2805845

Chow, P. I., Fua, K., Huang, Y., Bonelli, W., Xiong, H., Barnes, L. E., & Teachman, B. A. (2017). Using mobile sensing to test clinical models of depression, social anxiety, state affect, and social isolation among college students. *Journal of Medical Internet Research, 19*(3), e62. https://doi.org/10.2196/jmir.6820

Cox, J. (2021, January 28). More Muslim apps worked with x-mode, which sold data to military contractors. *Motherboard.* https://www.vice.com/en/article/epdkze/muslim-apps-location-data-military-xmode

Dong, E., Du, H., & Gardner, L. (2020). An interactive web-based dashboard to track COVID-19 in real time. *The Lancet Infectious Diseases, 20*(5), 533–534. https://doi.org/10.1016/S1473-3099(20)30120-1

Gasser, U., Ienca, M., Scheibner, J., Sleigh, J., & Vayena, E. (2020). Digital tools against COVID-19: Taxonomy, ethical challenges, and navigation aid. *The Lancet Digital Health.* https://doi.org/10.1016/S2589-7500(20)30137-0

Goh, B. (2020, February 25). China rolls out fresh data collection campaign to combat coronavirus. *Reuters.* https://www.reuters.com/article/us-china-health-data-collection/china-rolls-out-fresh-data-collection-campaign-to-combat-coronavirus-idUSKCN20K0LW

Goldenfein, J., Green, B., & Viljoen, S. (2020, April 17). Privacy Versus Health Is a False Trade-Off. *Jacobin.* https://jacobinmag.com/2020/04/privacy-health-surveillance-coronavirus-pandemictechnology

Gong, J., Huang, Y., Chow, P. I., Fua, K., Gerber, M. S., Teachman, B. A., & Barnes, L. E. (2019). Understanding behavioral dynamics of social anxiety among college students through smartphone sensors. *Information Fusion, 49*, 57–68. https://doi.org/10.1016/j.inffus.2018.09.002

Gupta, R., Bedi, M., Goyal, P., Wadhera, S., & Verma, V. (2020). Analysis of COVID-19 tracking tool in India: Case study of Aarogya Setu mobile application. *Digital Government: Research and Practice, 1*(4), 1–8. https://doi.org/10.1145/3416088

Halbfinger, D. M., Kershner, I., & Bergman, R. (2020, March 16). To track coronavirus, Israel moves to tap secret trove of cellphone data. *The New York Times*. https://www.nytimes.com/2020/03/16/world/middleeast/israel-coronavirus-cellphone-tracking.html

Holmes, A. (2020, March 27). Thousands of spring breakers traveled from one Florida beach to cities across the US. Mapping their phone data shows the importance of social distancing amid the coronavirus outbreak. *Business Insider*. https://www.businessinsider.com/coronavirus-florida-spring-break-location-data-spread-social-distancing-2020-3

Hu, Y., Quigley, B. M., & Taylor, D. (2021). Human mobility data and machine learning reveal geographic differences in alcohol sales and alcohol outlet visits across US states during COVID-19. *PLoS One, 16*(12), e0255757. https://doi.org/10.1371/journal.pone.0255757

Huxhold, W. E., & Craig, W. (2003). Certification and ethics in the GIS profession. *URISA Journal, 15*(1), 51–64. http://dusk.geo.orst.edu/ethics/papers/Huxhold_GISCI_process.pdf

Kamel Boulos, M. N., & Geraghty, E. M. (2020). Geographical tracking and mapping of coronavirus disease COVID-19/severe acute respiratory syndrome coronavirus 2 (SARS-CoV-2) epidemic and associated events around the world: How 21st century GIS technologies are supporting the global fight against outbreaks and epidemics. *International Journal of Health Geographics, 19*(8). https://doi.org/10.1186/s12942-020-00202-8

Kedron, P. & Trgovac, A.B. (2021). Assessing connections and tradeoffs between geospatial data ethics, privacy, and the effectiveness of digital contact tracing technologies. In: Shaw, SL., Sui, D. (eds) Mapping COVID-19 in Space and Time. Human Dynamics in Smart Cities. Springer, Cham. https://doi.org/10.1007/978-3-030-72808-3_7

Keegan, J., & Ng, A. (2021, September 30). There's a multibillion-dollar market for your phone's location data. *The Markup*. https://themarkup.org/privacy/2021/09/30/theres-a-multibillion-dollar-market-for-your-phones-location-data

Kim, J., & Kwan, M.-P. (2021). An examination of people's privacy concerns, perceptions of social benefits, and acceptance of COVID-19 mitigation measures that harness location information: A comparative study of the U.S. and South Korea. *ISPRS International Journal of Geo-Information, 10*(1), 25. https://doi.org/10.3390/ijgi10010025

Kitchin, R. (2020). Civil liberties or public health, or civil liberties and public health? Using surveillance technologies to tackle the spread of COVID-19. *Space and Polity, 24*(3), 362–381. https://doi.org/10.1080/13562576.2020.1770587

Lash, R. R., Moonan, P. K., Byers, B. L., Bonacci, R. A., Bonner, K. E., Donahue, M., Donovan, C. V., Grome, H. N., Janssen, J. M., Magleby, R., McLaughlin, H. P., Miller, J. S., Pratt, C. Q., Steinberg, J., Varela, K., Anschuetz, G. L., Cieslak, P. R., Fialkowski, V., Fleischauer, A. T., et al. (2021). COVID-19 case investigation and contact tracing in the us, 2020. *JAMA Network Open, 4*(6), e2115850. https://doi.org/10.1001/jamanetworkopen.2021.15850

Lee, L., Jones, M., Ridenour, G. S., Bennett, S. J., Majors, A. C., Melito, B. L., & Wilson, M. J. (2016). Comparison of accuracy and precision of GPS-enabled mobile devices. *2016 IEEE International Conference on Computer and Information Technology (CIT)*, 73–82. https://doi.org/10.1109/CIT.2016.9

Loomar, A. (2021, May 13). Rebekah Jones and the consequences of whistleblowing. *WUFT*. https://www.wuft.org/news/2021/05/13/rebekah-jones-and-the-consequences-of-whistleblowing/

Luguri, J., & Strahilevitz, L. J. (2021). Shining a light on dark patterns. *Journal of Legal Analysis, 13*(1), 43–109. https://doi.org/10.1093/jla/laaa006

Mazzei, P. (2020, December 11). A State scientist questioned Florida's virus data. Now her home's been raided. *The New York Times*. https://www.nytimes.com/2020/12/11/us/florida-coronavirus-data-rebekah-jones.html

Mell, P. (2002). Big brother at the door: Balancing national security with privacy under the USA PATRIOT act. *Denver Law Review, 80*(2), 375–427.

Morozov, E. (2013). *To save everything, click here: The folly of technological solutionism*. Public Affairs.

Morrison, S. (2020, March 26). *Like it or not, tech companies can use your phone location data to map social distancing*. https://www.vox.com/recode/2020/3/26/21192653/coronavirus-privacy-social-distancing-data-collection-unacast

Nielsen, N. (2020, March 23). Privacy issues arise as governments track virus. *EU Observer*. https://euobserver.com/coronavirus/147828.

OECD (Organisation for Economic Co-operation and Development). (1980). *OECD guidelines on the protection of privacy and transborder flows of personal data*. http://www.oecd.org/sti/ieconomy/oecdguidelinesontheprotectionofprivacyandtransborderflowsofpersonaldata.html

Popkin, H.A.S. (2020, December 26). Forbes technology awards 2020: geeks step up when governments fail. *Forbes*. https://www.forbes.com/sites/helenpopkin/2020/12/26/forbes-technology-awards-2020-geeks-step-up-when-governments-fail/?sh=361e128c7088

Rout, A., Nitoslawski, S., Ladle, A., & Galpern, P. (2021). Using smartphone-GPS data to understand pedestrian-scale behavior in urban settings: A review of themes and approaches. *Computers, Environment and Urban Systems, 90*, 101705. https://doi.org/10.1016/j.compenvurbsys.2021.101705

Sadowski, J. (2020, April 13). The Authoritarian Trade-Off: Exchanging privacy rights for public health is a false compromise. *Real Life Magazine*.. https://reallifemag.com/the-authoritarian-tradeoff/

Seidl, D. E., Jankowski, P., Clarke, K. C., & Nara, A. (2020). Please enter your home location: Geoprivacy attitudes and personal location masking strategies of internet users. *Annals of the American Association of Geographers, 110*(3), 586–605. https://doi.org/10.1080/24694452.2019.165484

Shoval, N., & Ahas, R. (2016). The use of tracking technologies in tourism research: The first decade. *Tourism Geographies, 18*(5), 587–606. https://doi.org/10.1080/14616688.2016.1214977

Simmons-Duffin, S. (2021, June 3). Why contact tracing couldn't keep up with the U.S. COVID outbreak. *Health Shots by NPR*. https://www.npr.org/sections/health-shots/2021/06/03/1002878557/why-contact-tracing-couldnt-keep-up-with-the-u-s-covid-outbreak

Singer, N., & Choe, S. (2020, March 23). As coronavirus surveillance escalates, personal privacy plummets. *The New York Times*. https://www.nytimes.com/2020/03/23/technology/coronavirus-surveillance-tracking-privacy.html

Spadaro, A. (2020). COVID-19: Testing the limits of human rights. *European Journal of Risk Regulation, 11*(2), 317–325. https://doi.org/10.1017/err.2020.27

Tan, J., Sumpena, E., Zhuo, W., Zhao, Z., Liu, M., & Chan, S. H. G. (2020). IoT geofencing for covid-19 home quarantine enforcement. *IEEE Internet of Things Magazine, 3*(3), 24–29. https://doi.org/10.1109/IOTM.0001.2000097

Tarabay, J. (2021, January 31). Governments tap covid data for other uses, risking backlash. *Bloomberg Businessweek*. https://www.bloomberg.com/news/articles/2021-01-31/governments-exploit-covid-data-for-other-uses-risking-backlash

Taylor, L. (2021). There is an app for that: Technological solutionism as covid-19 policy in the global north. *The New Common*, 209–214. https://doi.org/10.1007/978-3-030-65355-2_30

Taylor, K., & Silver, L. (2019). *Smartphone ownership is growing rapidly around the world, but not always equally*. Pew Research Center. https://www.pewresearch.org/global/2019/02/05/smartphone-ownership-is-growing-rapidly-around-the-world-but-not-always-equally/

Timberg, C., & Harwell, D. (2020, March 19). Government efforts to track virus through phone location data complicated by privacy concerns. *The Washington Post*. https://www.washingtonpost.com/technology/2020/03/19/privacy-coronavirus-phone-data/

Van Ness, L. (2021, March 19). For states' COVID contact tracing apps, privacy tops utility. *Stateline*. https://www.pewtrusts.org/en/research-and-analysis/blogs/stateline/2021/03/19/for-states-covid-contact-tracing-apps-privacy-tops-utility

Vich, G., Marquet, O., & Miralles-Guasch, C. (2017). Suburban commuting and activity spaces: Using smartphone tracking data to understand the spatial extent of travel behaviour. *The Geographical Journal, 183*(4), 426–439. https://doi.org/10.1111/geoj.12220

Zastrow, M. (2020, March 18). South Korea is reporting intimate details of COVID-19 cases: Has it helped? *Nature*. https://doi.org/10.1038/d41586-020-00740-y

Chapter 6
The City and the Pandemic: The Cities' COVID Mitigation Mapping (C2M2) Program

Laura Cline and Melinda Laituri

Personal Story

As a geographer, I have always been curious to see how people live in other places. After 5 years of building professional relationships and friendships with geospatial partners in 16 global cities through our Secondary Cities (2C) Initiative, the abrupt end to travel and the unexpected narrowing of my own geography to the city streets around my home was quite a shock to my system. As I saw many changes in my local community that surprised me, my colleagues and I began to wonder what changes were taking place in the communities we had partnered with for the past few years. In reaching out through our networks, we learned of their innovative local applications of human geography data and the tremendous impacts of the pandemic, and we began to ask what more we could do. We reached out to these international colleagues and partnered to adapt our participatory mapping model to examine COVID's second-order impacts, with a focus on under-examined urban areas. This is our story…*Laura Cline.*

Introduction

In 2020, the rapid spread of the COVID-19 pandemic to cities across the world illuminated the need to understand the distribution of resources and social dynamics in these densely populated heterogeneous areas. Urban geographies are complex and interdimensional, however the data needed to understand these critical socio-economic relationships is infrequently available, particularly in lower- and middle-income countries (LMIC). Such data are necessary to better identify and understand

L. Cline (✉)
US Department of State, Washington, DC, USA

M. Laituri
Colorado State University, Fort Collins, CO, USA

M. Laituri et al. (eds.), *The Geographies of COVID-19*, Global Perspectives on Health Geography, https://doi.org/10.1007/978-3-031-11775-6_6

the second-order impacts of the pandemic. Identifying the spatial outcomes of second-order impacts exposes the locations of inequitable distribution of basic services, newly emerging vulnerable populations, and changing patterns of mobility and migration (Laituri et al., 2021). Decision-makers and communities need this data and additional analysis to identify the second-order social and economic interdependencies that are impacted by first-order public health restrictions. Capturing local knowledge (i.e., who lives where and why?) as human geography data for analysis to mitigate COVID-19 impacts emphasizes the practical need for local geospatial capacity and participatory mapping partnerships among local stakeholders to support community recovery and resilience. Human geography data are inclusive of demographics, critical infrastructure, economies, ethnicities, green space, community and cultural sites, education levels, environmental resources, ecosystem services, and medical facilities. Some of these data are available at the national and state or province scale but are often lacking at local levels (i.e., city, towns and rural villages). The Cities' COVID Mitigation Mapping (C2M2) Program led by the U.S. Department of State's MapGive Initiative from summer 2020 through fall 2021 illustrates the value of participatory mapping partnerships to address the second-order impacts of the COVID-19 pandemic. The geospatial model of data generation and community participation offers lessons applicable to developing urban resilience and sustainability.

Fine Scale Human Geography Data Is Needed to Understand Cities, Especially During the COVID-19 Pandemic

Prior to the emergence of the pandemic in 2020, cities around the world were growing rapidly and the UN predicts that between 2021 and 2050 all of the world's population growth will be in cities (United Nations Department of Social and Economic Affairs, 2019). In LMIC, this growth frequently results in the expansion of informal, unplanned settlements as migrants move to urban areas to flee conflict or seek economic opportunity combined with a continued increase in natural population growth (Perry et al., 2021). Unplanned settlements frequently lack the basic services and social structures which would enable residents to meet public health measures such as washing hands frequently and maintaining social distance to disrupt the spread of COVID-19, while at the same time enabling them to continue to support their families and maintain livelihoods. While cities present challenges due to the prevalence of unplanned development, they also continue to offer opportunities to power local, regional, and national economies, accounting for 70% of the world's wealth prior to the pandemic (Garland, 2015, p1).

All urban geographies are complex and multidimensional. However, cities in LMIC are particularly challenging environments to understand. Expanding informal settlements, migration and movement patterns, or a lack of basic services are difficult to track if data about the fundamental infrastructure is not available.

Mapping infrastructure is key to all functions of daily life: livelihoods, mobility, medical, education, cultural, shopping/markets, religion, recreation, and home (Worldwide Human Geography Working Group, n.d.). Mapping infrastructure is critical for understanding local conditions when data is generated at the appropriate scale. This infrastructure includes not only roads, water supply, and energy networks but also data about the kinds of infrastructure that is a critical part of socio-economic relationships (i.e., education, employment, health).

Human geography data captures local knowledge about these socio-economic aspects, focusing on data that considers people and how they interact within their communities and beyond. Collecting data about waste flows and water management highlights modified hydrology for planning to reduce flooding and water-borne disease. Data can identify impacts of coastal sea level rise to urban transportation and other networks (e.g., energy grids, communication lines) to inform adaptive planning (Perry et al., 2020). During the COVID-19 pandemic, these data illuminate the extent and adequacy of the existing infrastructure, its accessibility, and allow assessment of gaps in serving urban populations (Laituri et al., 2021). Such data enable location-based analysis to site health facilities, COVID-19 testing locations, and vaccination sites. Decision-makers and communities need georeferenced data for analysis to identify the second-order social and economic interdependencies that are impacted by first-order public health measures and mobility restrictions during the pandemic.

Local Partnerships Are Invaluable for Mapping Urban Areas During a Crisis

An avenue to collect human geography data is through participatory mapping that empowers local groups to be included in decision-making about their communities (Crampton & Stewart, 2004). Increasingly geospatial tools and technologies are used in local situations to give communities a voice. This local agency gained through participatory mapping is further enhanced by geospatial capacity building at the local level. Efforts in local participatory mapping combined with geospatial analysis build community networks, support collaborative planning, and enable data-driven decision-making (Crampton & Stewart, 2004).

International communities continue to recognize and emphasize the importance of supporting local agency across humanitarian and development frameworks. When considering the response to humanitarian crises, such as the COVID-19 pandemic, the humanitarian community continues to be guided in part by the Grand Bargain framework. Announced in 2016 at the World Humanitarian Summit, this framework was termed the 'participation revolution' recommending that there should be more emphasis that those who are receiving aid should be making the decisions that will affect their lives (Sanderson, 2019, p 079). In addition, Sustainable Development Goal #11 brings a global focus to making cities and human settlements inclusive, safe, resilient, and sustainable. In 2016, the New Urban Agenda emphasized inclusion, which some jurisdictions interpret as "cities for all" (United Nations Department of Economic and Social Affairs, n.d. & United Nations General Assembly, 2016).

Africa, Asia, and Latin America host large urban populations, which comprised approximately 80% of the world's urban population in 2020 (United Nations Human Settlements Program, 2020). Urbanization and urban populations continue to increase, and such growth is both planned and haphazard. The need for human geography data in rapidly growing cities inspired the creation of a participatory mapping model – the Secondary Cities (2C) Initiative – implemented through the U.S. Department of State's Office of the Geographer. From 2015–2019, the 2C Initiative worked with U.S. Embassy officers to facilitate local partnerships with academics, governments, and NGOs to generate georeferenced human geography data for emergency preparedness, resilience, and sustainability.

The lessons learned during the 2C Initiative are critical in furthering a global network of geospatial experts focused on technical tools and geospatial approaches for decision-making and civic engagement. These lessons included the need to adapt methods created in the developed world to address the different scenarios of LMIC secondary cities. The emphasis on secondary cities was due to the fact that these cities remain under-examined. This lack of data inhibits planning efforts in this type of city which is typically a rapidly growing urban area that provides critical local and regional services. Facilitating local participation and conducting field-based data generation was foundational to establish local partnerships to produce open data, clear and concise printed map products, and accessible web-based tools. Web-based tools were adjusted and optimized for smart phone access when stakeholder communities were primarily reliant on phones for web access vs computers.

During a crisis, local inhabitants provide the initial response in their community. They provide critical information about the status of infrastructure to first responders and can track and monitor community needs for long-term support (UN OCHAb, 2021). Enabling this local agency is critical during the COVID-19 pandemic, as all communities are impacted and need to provide some level of response to raise awareness about local conditions and requirements. Focusing on locally-based geospatial data generation, analysis, and application in urban areas during the COVID-19 pandemic allows for merging the application of geographic methodologies for crises and public health with this greater need for understanding the human geographies of urban areas. Urban geospatial approaches used in or appropriate for LMIC, such as the 2C model, can be adapted to address COVID-19 second-order impacts because of the original focus of furthering human security and development goals. Two decades of development gains to address poverty have eroded during the COVID-19 pandemic, including negative impacts on education, employment, food security, and healthcare (UN OCHAa, 2021). The 2C human geography participatory mapping model is relevant for communities today, as needs for resilience and sustainability are further challenged by the compounding effects of the COVID-19 pandemic restrictions upon development gains and other potential emergencies.

Partnerships to Apply Geospatial Tools to Examine COVID-19 Second-Order Impacts; the Cities' COVID Mitigation Mapping (C2M2) Program

Evan in late spring 2020, it was clear early in the pandemic that the dynamic challenges of urban poverty would be exacerbated by curbs on mobility and other public health restrictions. The U.S. Department of State's MapGive Initiative and its international partners began to adapt its Secondary Cities (2C) model of local participatory mapping to examine the socio-economic problems which could occur as second-order impacts of the COVID-19 pandemic, including the lack of access to basic services, reduced incomes due to closures, and resulting changes in mobility or migration patterns. The MapGive Initiative is a long-term effort of the U.S. Department of State's Humanitarian Information Unit, within the Office of the Geographer and Global Issues, to promote geographic approaches internationally and encourage open mapping and data sharing by remote and local mappers alike. Developing an international applied research program remotely during the pandemic initially relied upon participatory mapping partnerships from the 2C Initiative.

Existing relationships from the 2C Initiative enabled the new Cities' COVID Mitigation Mapping (C2M2) program to identify regional geospatial experts to remotely develop and lead city-level partnerships in mapping and assessing vulnerable communities in LMIC as part of three regional hubs of geospatial excellence, one each in Africa, Asia, and Latin America. These hubs leveraged their existing networks to create 12 city-level C2M2 projects (Fig. 6.1). In Africa, project teams

Fig. 6.1 Cities with C2M2 project activities in 2020–2021. (https://mapgive.state.gov/c2m2/)

examined education and other access issues in vulnerable communities in Bukavu, Democratic Republic of Congo; Nairobi, Kenya; and Pemba, Mozambique. In Asia, project teams examined changing patterns of mobility in Dhaka, Bangladesh; Kathmandu, Nepal; and Ulaanbaatar, Mongolia. In Latin America, project teams examined changing impacts due to increased poverty in Cusco and Lima, Peru; Ouro Preto (Minas Gerais), Brazil; Quito and Santa Cruz (Galapagos), Ecuador; and Santiago, Chile.

Building Partnerships to Understand Communities

Previous experience with the 2C model demonstrated the importance of building partnerships, enhancing geospatial capacity, supporting data creation and analysis, and emphasizing visualization and dissemination to key stakeholders to support data-driven decision-making (Fig. 6.2). The C2M2 program was designed to support data-driven decision-making by local stakeholders developing mitigation strategies for COVID-19 second-order impacts. Initial C2M2 city-level project activities required socializing the project across the community to build partnerships and key stakeholder buy-in for the city-level project design, which allowed for discussions to identify needed datasets that may already exist in the community or could be used as a proxy to answer relevant questions. Local and regional geospatial experts hosted numerous geospatial workshops for local participants and contributors to support data gathering, creation, and analysis; many of these were virtual, but some in-person events occurred if local restrictions allowed and public health requirements could be met. The program leadership offered guidance and technical support when needed. Also, as a key requirement for any participatory mapping program,

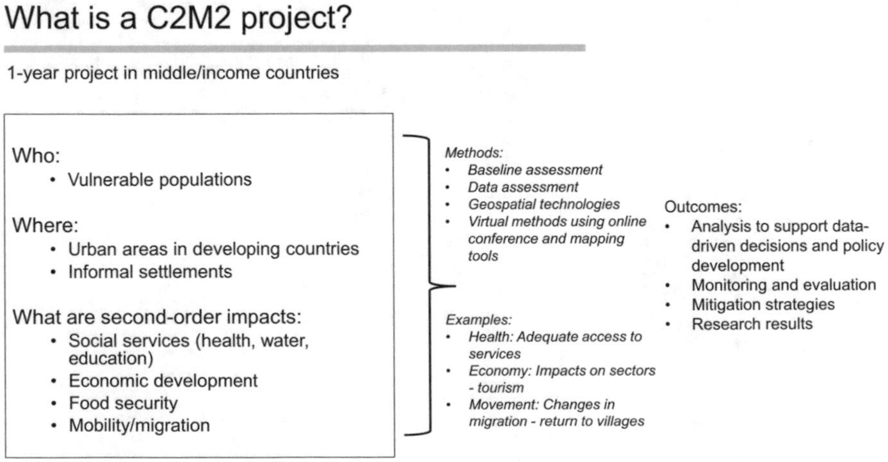

What is a C2M2 project?

1-year project in middle/income countries

Who:
- Vulnerable populations

Where:
- Urban areas in developing countries
- Informal settlements

What are second-order impacts:
- Social services (health, water, education)
- Economic development
- Food security
- Mobility/migration

Methods:
- *Baseline assessment*
- *Data assessment*
- *Geospatial technologies*
- *Virtual methods using online conference and mapping tools*

Examples:
- *Health: Adequate access to services*
- *Economy: Impacts on sectors - tourism*
- *Movement: Changes in migration - return to villages*

Outcomes:
- Analysis to support data-driven decisions and policy development
- Monitoring and evaluation
- Mitigation strategies
- Research results

Fig. 6.2 What is a C2M2 project?

the regional hubs and C2M2 program leadership ensured that all participants were similarly trained on the ethics of participatory mapping by providing a web-based training available to all project participants.

Using Geospatial Tools to Generate Open Data

Foundational to both the C2M2 program and 2C Initiative was the need to examine national and local laws and regulations to determine if there would be any legal or political constraints to mapping a selected city. Some countries have declared their open data policies. Regardless of formal declaration or informal practice, it is the duty of the program and project designers to know if this type of activity will be accepted. Also, if a country has begun to implement its open data policy, then there may be a national or regional metadata standard that is appropriate to use to ensure the data can easily be integrated with other datasets for local decision-makers' use.

Whenever possible, and in agreement with the local communities, the data and products are shared openly, either through open data portals such as OpenStreetMap, the UN OCHA Humanitarian Data eXchange (HDX), or published on websites. Lastly, the international, regional, and local C2M2 project partners came together to identify effective visualization techniques, products, and dissemination methods that took best advantage of their online and in-person reach.

Sharing Results About COVID-19 Second-Order Impacts

One method to share the results of C2M2 analysis included the city-level project partners hosting local in-person and virtual meetings. Both city and regional C2M2 partners also used websites and social media to share meeting announcements, map products, and web-based interactive map portals. International C2M2 partners at the MapGive Initiative and American Association of Geographers (AAG) highlighted announcements across their networks using existing websites and social media accounts. MapGive co-hosted an international symposium as part of an academic partnership with Harvard University's Center for Geographic Analytics, delivering a four half-day virtual event to examine the Geography of COVID-19 Second-Order Impacts, that combined discussions with academics and practitioners on the forefront of examining global impacts on mobility, economics, and gender. The symposium included a day of demonstrations of new geospatial techniques and methodologies being developed to examine these issues. The large audience for the Symposium became so engaged on the topic that many chose to participate in a follow-on virtual networking event 2 months later.

This 18-month C2M2 program demonstrated that a largely virtual international participatory mapping research program can achieve success and overcome a number of challenges when grounded in shared field-based experiences and trusted professional partnerships. Local partnerships were multi-faceted, composed of key community members, municipal decision-makers, and academics or NGOs with technical expertise. Local data created at the appropriate scale and generated according to cultural and ethical guidelines enabled analytic visualization and map production. Field based data generation was conducted using multiple hand-held, mobile devices (i.e., tablets, cell phones, drones). Ensuring the physical safety of any field mappers is essential and can be achieved through coordination with the community and adopting recommended safety measures (i.e., wearing identification badges or outreach to community leaders to raise awareness before the data collection activity). These data informed the geographic tools and methodologies locally appropriate for our project cities in LMIC. Access to these tools through unique licensing relationships with private geospatial industries or the adoption of open source tools is part of the C2M2 approach. Key to the C2M2 program was training to ensure data analysis was conducted based on sound scientific practices and documented through creating replicable workflows. These mapping projects would not have been possible without having technology and infrastructure to connect the program, regional hub, and project partners.

LMIC countries are challenging environments to conduct participatory mapping and geospatial analysis. Connectivity is central to the C2M2 program. Virtual mapping (Esri Story Maps, Google Earth Voyager) and virtual meetings (i.e., Zoom) demonstrated how a sense of community could be built during the pandemic. Sharing stories of the pandemic during meetings solidified relationships and allowed for learning from each other. Adaptations to ensure communication and connectivity were successful included video-recording presentations before meetings, targeting workshops to smaller groups of individuals in a single location, and keeping continuously open lines of communication using email, Slack, and Google folders (to share documents). Identifying virtual common ground across the C2M2 program meant using the appropriate infrastructure for mapping and analysis in a disconnected environment. For example, limiting the use of proprietary software if the licenses will later be prohibitively expensive, and seeking to use existing open data sharing portals when possible, to ensure continued access to and maintenance of the data beyond the end of the program. Partnerships with local geographic professionals, whether in academia or an NGO, help identify existing GIS tools and resources that can be used for a mapping project. The C2M2 projects overcame many challenges, including personal health concerns due to the ongoing pandemic, to deliver quality analytic products for their local decision-makers and their communities more broadly.

Outcomes and Impacts of C2M2

Each regional hub emphasized different aspects of the pandemic due to the myriad geographies reflected by their locations. We asked project partners to focus on second-order impacts most relevant to their project communities (see Chap. 2, Richardson, this volume).

C2M2 Africa Hub

Initially, the COVID-19 impact on African countries followed a gradual trend as opposed to the rapid surges occurring elsewhere. However, variants of the virus have increased COVID-19 cases across the continent, coupled with a slow vaccine rollout that has seen the pandemic surge in selected countries. Africa is an important region to examine and understand how countries have prepared themselves to manage the second-order impacts of the pandemic built upon experiences of contending with other pandemics such as HIV, AIDS, and Ebola. Using the tools and open data from the 2C Initiative, the City Navigator application provides an online, interactive web application to share locations (i.e., health facilities) and information about access to services (i.e., water points) (Mblonglou et al., Chap. 10). The projects in Bukavu, Democratic Republic of Congo; Nairobi, Kenya; and Pemba, Mozambique addressed second-order impacts related to access to social services, such as education, water, and health. The Nairobi project analyzed access to water points and access to technology for remote learning within two informal settlements of the city – Kibera and Mathere. The Bukavu project created the first ever geo-located database of education and healthcare facilities for the city. Pemba project partners also created a geo-referenced database of critical services to identify gaps in services throughout the city.

C2M2 Asia Hub

The Asia Hub guided three city projects: in Dhaka, Bangladesh; Kathmandu, Nepal; and Ulaanbaatar, Mongolia. In Dhaka, the city project studied the migration of economically vulnerable populations to coastal regions. The project examined the changing economic conditions of a sub-district of Dhaka to inform adaptive strategies for employment of those workers most impacted by the pandemic. In Ulaanbaatar, the project worked in ger communities – informal settlements – to assess access to critical health services. The local Asia Hub project partner, Public Lab Mongolia, created a health portal that identifies where health facilities are located in relation to public transportation services for ger communities. Kathmandu, Nepal, focused on the local economy in the form of impacts on tourism and informal employment (Karki et al., Chap. 9). The Asia Hub also created data and analysis methodologies for use throughout the region.

C2M2 Latin America Hub

Latin America experienced critical economic challenges due to the pandemic. The Latin America Hub focused on the spatial analysis and visualization of the emergent properties of poverty from the collapse of the tourism industry in different cities and the resulting impact on crime and domestic violence (Mena et al., Chap. 11). The relationship between economic impacts and social conditions was analyzed using multiple geospatial tools to create heatmaps of cities experiencing increased societal problems. Specifically, projects were undertaken in Quito, Ecuador; Santiago, Chile; and Lima, Peru, to track the impacts of vulnerable populations in urban areas and their access to services and resources to address domestic violence during the time of government lockdowns. The hub has incorporated the results of these studies into a data portal to share models and other decision-making tools for use by stakeholders in the region to mitigate COVID second-order impacts.

Conclusion and Lessons Learned for Geospatial Partnerships

The C2M2 Program supported partnerships to use geospatial data to understand and address second-order impacts of the pandemic (see Chaps. 9, 10, and 11 in this volume). The resulting project deliverables can be divided into three categories: data analyses to create new information about the need to improve location-based decisions for basic services during a disaster; data dashboards and portals to provide access to information for local government and communities; and story maps that document the pandemic and share lessons learned. The creation and dissemination of these products built partnerships between non-governmental organizations and government within our project countries and across borders between non-governmental organizations, industry, and universities.

There are some important lessons learned from this program that can be applied to ongoing efforts to improve local data and address critical issues related to disasters such as the COVID-19 pandemic:

- Short-term programs cannot address long-term problems. The C2M2 program continued to expand a global network based upon the 2C Initiative, but issues of continuity and sustainability remain. Longer-term funding is essential to supporting this network.
- Place-based projects illuminate the importance of physical and human geography and the multiple ways geospatial approaches can be used. Locations, such as urban areas, have specific needs that are varied. Vulnerable populations face unique challenges in both formal and informal urban areas.
- The value of scale-appropriate data. C2M2 projects focused on sub-city locations to examine vulnerable populations that are often overlooked in country-level datasets.
- The need to identify and create georeferenced data at the sub-city scale (i.e., city districts or neighborhoods). Many of the city projects collected data that had never before been created, providing new insights for city planning and decision-making (i.e., locations of all types of healthcare facilities such as clinics, hospitals, pharmacies).

- Partnerships can leverage resources and expand networks. Exploring connections with government offices and international programs can raise awareness of a network and create new connections. Networking with industry and non-governmental organizations can provide additional opportunities for local participants.
- Data generation can be standardized in terms of format, attribute definitions, database management, and data sharing. Geospatial data generation and management can build bridges between projects through sharing such standards.
- Similarly, geospatial data creation and analysis can be applied to many different types of data. However, it is essential to include documentation (metadata, analytical flowcharts, workflows) on processes to ensure replicability and reliability of outcomes and to communicate any limitations of the data.
- Protection of personal information and informed consent. Ethics must underpin surveys and georeferenced data creation. Georeferenced data reveal sensitive information regarding locations and people. Protection of personal information is a necessary part of data generation and sharing. Ensuring participants understood informed consent while protecting anonymity was essential to all projects.
- Visualization (maps, story maps, graphs) of geospatial data can inform decision-making. For example, creating a map of basic services and identifying where there are gaps in services gives a voice to communities and a representation of what is needed to address a problem.
- Sustainability is key to facilitating a network which can build robust databases at a variety of scales (with an emphasis on local-scale data) and continue to maintain and manage data and products. Numerous COVID 19-related websites and databases are already out-of-date and not being maintained.

A key contribution of both the C2M2 Program and 2C Initiative is the creation of a global network of geospatial experts to address the integrated problems associated with vulnerable populations. These programs will continue as participatory mapping partnerships are pursued to further the reach of geospatially-based problem solving efforts. The pandemic exposed the nature of inequalities around the world. This network addresses one aspect of these inequalities by developing support to train and support LMIC partners in the use of geospatial technologies and applications. In the digital data age, access to and an understanding of the uses of geospatial data is essential to data-driven decision-making and place-based solutions.

Acknowledgments This program builds upon a large network of participants that were essential to the implementation of the C2M2 program. The Humanitarian Information Unit in the Office of the Geographer and Global Issues provided management support and oversight – the team included Lee Schwartz, Geographer of the United States, Robert Richardson, Jefferson Science Fellow, Sebastian Viscuso, HIU Intern, Erika Nunez, Communication and Outreach, and Tom Gertin, Technical Support, in addition to us as the Program Director (Laura Cline) and the Principal Investigator (Melinda Laituri). The American Association of Geographers managed budgets and reporting – the team included Candida Mannozzi, Coline Dory, and Julaiti Nilupaer.

Support from the Center for Geographic Analysis in oversight and facilitation of C2M2 student mentoring – Wendy Guan and Kelly Ly. Appreciation to the student volunteers for their creative contributions. Keys to project support, management, and implementation were the C2M2 Regional Hub leads: Gaston Mbonglou, Carlos Mena, and Nama Budhathoki. We also share our appreciation for the team partners who committed time, energy, and hard work to this program during the challenges of the pandemic and the virtual environment.

Disclaimer The views expressed in this report are solely those of the authors and do not represent those of the U.S. government agencies or any of the organizations mentioned. Assumptions made within the analysis are not a reflection of the position of any U.S. government entity.

References

Crampton, J. W., & Stewart, D. J. (2004). Community mapping as a solution to digital equity. In D. G. Janelle, B. Warf, & K. Hansen (Eds.), *WorldMinds: Geographical perspectives on 100 problems* (pp. 523–527). Springer.

Garland, A. M. (Ed.). (2015). *Urban opportunities: Perspectives on climate change, resilience, inclusion, and the informal economy*. Wilson Center.

Laituri, M., Richardson, R. B., Kim, J., Cline, L. V., Viscuso, S., & Schwartz, L. (2021). Examining second-order impacts of COVID-19 in urban areas. *Annals of GIS*. https://doi.org/10.1080/19475683.2021.1954087

Perry, G., Laituri, M., & Cline, L.V. (2020, October 26). Why Secondary Cities Deserve More Attention. *New Security Beat, Wilson Center. https://www.newsecuritybeat.org/2020/10/secondary-cities-deserve-attention/*

Perry, G., Upchurch, C., & Cline, L.V. (2021, July 6). Displacement, Migration, and Urbanization in the 21st Century. *New Security Beat, Wilson Center.* https://www.newsecuritybeat.org/2021/07/displacement-migration-urbanization-21st-century/

Sanderson, D. (2019). *Good practice review 12: Urban humanitarian response*. Overseas Development Institute.

United Nations Department of Economic and Social Affairs. (2019). *World urbanization prospects; the 2018 revision*. United Nations. https://population.un.org/wup/Publications/Files/WUP2018-Report.pdf

United Nations Department of Economic and Social Affairs. (n.d.). *Goal 11: Make cities and human settlements inclusive, safe, resilient and sustainable*. United Nations Publications. *https://sdgs.un.org/goals/goal11*

United Nations General Assembly. (2016). *Resolution Adopted by the General Assembly on 23 December 2016; 71/526. New Urban Agenda.* https://habitat3.org/wp-content/uploads/New-Urban-Agenda-GA-Adopted-68th-Plenary-N1646655-E.pdf

United Nations Human Settlements Programme. (2020). *World Cities Report 2020; The Value of Sustainable Urbanization.* https://unhabitat.org/World%20Cities%20Report%202020

United Nations OCHAa. (UN OCHAa) (2021). *Global Humanitarian Overview 2022; Introduction At A Glance.* https://gho.unocha.org/intro/glance

United Nations OCHAb. (UN OCHAb) (2021). *Global Humanitarian Overview 2022; Local Actors Play a Key Role in Humanitarian Action.* https://gho.unocha.org/delivering-better/local-actors-play-key-role-humanitarian-action

Worldwide Human Geography Data Working Group. (n.d.). *Human Geography 13 Theme Areas.* https://www.wwhgd.org/13themes

Part II
Global Impacts, Local Responses (Overview)

Melinda Laituri

The global impacts of the pandemic have drawn the world closer together where we are experiencing uncertainty with respect to international travel, strains on the supply chain for goods, and dependence upon medical science for finding solutions to the virus. Using case studies and examples of geospatial analyses, Part II examines several places around the world that have experienced the effects of the pandemic in different ways. Each case study uses a geographic lens that highlights the differences and commonalities across space and time where fundamental inequities are exposed and further exacerbated, governmental response is varied, and outcomes remain uncertain. This moment of global collective experience starkly reveals how inequality is ubiquitous and vulnerable populations – those unable to access basic needs – are increasing.

These case studies demonstrate different geospatial methods, models, and analyses that help us to understand the pandemic. Chuluunbaatar and Sumiyasuren (Chap. 7) describe a health data portal that provides information on transportation networks and health facilities services and operating hours in peripheral, informal settlements of Ulaanbaatar, Mongolia. Mblonglou and John (Chap. 11) explain a mobile app – City Navigator – residents are using in multiple cities of sub-Saharan Africa. This app allows participation from citizens to upload information on a variety of city services. This was particularly helpful during the pandemic to find out where and when services were available.

Geospatial modeling and analyses use data to create new information. The intersection of statistical analysis and geospatial visualization is explored by Li et al. (Chap. 13) using demographic data from China of case counts, population density, and transmission. Campbell et al. (Chap. 14) examine the social and spatial patterns where the multiple factors of the pandemic interplay to create a syndemic that extends beyond their case study situated in Aotearoa New Zealand.

M. Laituri
Colorado State University, Fort Collins, CO, USA

The underlying theme of these chapters is an emphasis on vulnerable populations that are found throughout the world. There is no country that has avoided the ravages of the pandemic and attendant issues related to inequity. Two chapters focus specifically on the United States. Huang et al. (Chap. 8) focus on the disparities of demographic and socioeconomic characteristics that drive inequitable outcomes of the pandemic experience in terms of access to resources and vulnerability to the disease. Guhlincozzi and Wallace (Chap. 9) describe how the Latine community is impacted by the pandemic emphasizing immigration and access to health services. The impact of the pandemic due to stay-at-home orders changed patterns of domestic violence. Mena et al. (Chap. 12) track these changing patterns through an examination of domestic violence complaints and create heat maps that identify aggregations of complaints over time in Quito, Ecuador. Yashadhana et al. (Chap. 15) describe the relationship between access to community services for Aboriginal and Torres Strait Islanders where cultural diversity must be prioritized in providing critical health services.

Economic patterns were disrupted during the pandemic and these impacts were also unevenly distributed. Karki et al. (Chap. 10) examined the impact of the pandemic on tourism in Kathmandu, Nepal, where the informal tourist economy was particularly hard hit. Kebirungi and Mwenyango (Chap. 16) describe the impact of government policies on the Boda-Boda drivers of Uganda.

This place-based approach identifies how geospatial analyses and resulting maps depict the pandemic as it ebbs and flows across the globe. Data-driven decision making is needed as we navigate the pandemic and determine ways to address future events to enable local and regional governments in prioritizing limited resources to minimize the long-term consequences of COVID-19. The use of geospatial approaches to data-driven decision making is fraught due to the limitations of the technology, the lack of appropriate data, and access to the specific geospatial platforms. These limitations are juxtaposed with strategies for addressing these issues for improving access and applications in using geospatial approaches. We used examples from around the world to explore the outcomes of the pandemic and how various places adapted and reacted to this global event through spatial analysis.

Chapter 7
Improving Access to Health Services in Mongolia via Open Data During and Beyond the COVID-19 Pandemic

Enkhtungalag Chuluunbaatar, Erdenetsogt Sumiyasuren, Byambatsetseg Lkhagvasuren, and Nyamsuren Tsadmid

Personal Story
The COVID-19 pandemic has been full of new and old challenges for many around the globe. The Public Lab Mongolia team has also been impacted by the pandemic on a personal level where every staff, implementing partners and their families, contracted the virus at some point during the pandemic. We all faced the challenges with health and medical service access during the pandemic while working on precisely this issue in Mongolia. Some of us were stranded abroad, and we learnt to navigate working remotely across several time zones as well as finding ways to conduct fieldwork during lockdowns and social isolation. The insights from our own personal experiences with difficulty in healthcare information and access helped inform the project adapt to the changing social, policy and economic circumstances. We believe that our project in Ulaanbaatar has established a solid foundation for facilitating health service access across the country because of its relevance to our own lives during the pandemic. *Public Lab Mongolia team*

Introduction

Mongolia faces a unique challenge amid the COVID-19 pandemic. The least densely populated country in the world has half of the population living in the rapidly urbanizing capital city of Ulaanbaatar. The centralized health services and inadequate and unequal access to available health services both in urban and rural areas exacerbate the impacts of the pandemic on the population.

The Cities' COVID Mitigation Mapping (C2M2) program is an initiative of the Office of the Geographer and Global Issues at the U.S. Department of State that

E. Chuluunbaatar (✉) · E. Sumiyasuren · B. Lkhagvasuren · N. Tsadmid
Public Lab Mongolia, Ulaanbaatar, Mongolia
e-mail: enkhtungalag.ch@publiclabmongolia.org

M. Laituri et al. (eds.), *The Geographies of COVID-19*, Global Perspectives on Health Geography, https://doi.org/10.1007/978-3-031-11775-6_7

builds partnerships to enhance geospatial capacity, generate data, and share maps to support planning for mitigating COVID-19 second order impacts.

The C2M2 Ulaanbaatar project is focused on reducing the second-order health impacts due to the COVID-19 pandemic for the vulnerable population of the capital city. We worked to create a comprehensive health service information portal of all available health services. The health service information portal would facilitate access to available health services for the city's population during the pandemic. We identified high risk areas and worked to facilitated health service access for the residents of the capital city, more specifically, for the ger area. The ger[1] area is composed of settlements not connected to central infrastructure such as running water, centralized heating, and sewerage. The centralization of health services in the city is primarily located around in the city center, leaving the ger area residents inadequately serviced.

In order to alleviate the existing challenges with inadequate health services, exacerbated by the pandemic, the project worked to provide an easy, safe, and reliable way for the residents of the capital city to find and access health and medical services during the pandemic.

Impacts of COVID-19 on Health Service Access in Mongolia

Mongolia remained COVID-19 free until November of 2020, except for the imported[2] cases. Since the first community outbreak of the COVID-19 virus, the number of cases soared and continues to climb despite the high vaccination rate in the country (https://covid19.who.int/region/wpro/country/mn).

There is very little data, if any, in terms of sizing up the quantifiable impact of change in the level of access to health and medical services during the COVID-19 pandemic. In December 2020, one in three Mongolians who needed health or medical services did not receive it. As of October 2021, one in five people who needed health and medical services did not receive them. The reasons for this inaccessibility to health and medical services include fear of contracting the virus, health and medical facilities being full, and lack of transportation or means of mobility (World Bank, 2021).

We expect that the financial access to health and medical services has decreased due to overall decline in income as well as increase in the price of stable food products (Graceffo, 2022), especially for low-income households. News reports also indicate closures of various essential, non-COVID-19-related health and medical services such as rehabilitative services, urgent care, oncology, and so on (Khukhnokhoi, 2021).

---

[1] *Ger* means "home" in Mongolian and is a traditional Mongolian housing made from circular wooden frame and felt covering.

[2] The imported cases were those who entered Mongolia with COVID-19 infection and were quarantined upon arrival at a government-run isolation facility to prevent community outbreak.

Unfortunately, there is no comprehensive data which documents the changes in the level of access to health and medical services and its impact on the population. Anecdotally, several deaths and numerous stories of challenges in health service access have been reported as a result of inability to access emergency or other medical services due to lockdown measures, physical or geographic distance, and other COVID-19 policies (Tsetseg, 2020).

Examples of Reported Cases of the Impact of Health Service Access

On December 8, 2020, a woman from Arkhangai province died at the Ulaanbaatar city's border due to travel restrictions and being unable to access emergency medical care (TIME, 2020). According to the World Bank Household Survey Round 3, in December 2020, "1 out of 3 who needed medical treatment did not receive services, mainly due to people's concerns of contracting the virus and mobility restrictions". On January 8, 2021, a woman needing post-operative treatment was unable to get the health/medical care she needed at the public hospital due to COVID-19 outbreak at the facility. The family could not afford private medical care. One of the most notable news relating to healthcare access in Mongolia happened on January 21, 2021, where a new mother and her newborn were promptly taken out of the maternity hospital, under what the public considered inhumane conditions, after she was confirmed to be COVID-19 positive, sparking a public outburst (AFP, 2021).

In June of 2021, a COVID-19 positive man dies from complications shortly after being denied medical care and sent home. Ambulance and other health and medical services are unavailable to many during the pandemic. On June 22, 2021, the Minister of Health announced that the country is in deficit of health infrastructure and human resources, especially specialists in intensive and emergency care. Volunteers are called for, in addition to mobilizing resident doctors and final year medical students (Enkhbold, 2021).

Due to increased workload and minimal pay, health workers protested outside the government building on July 6, 2021, demanding additional wages amidst combatting the increasing COVID-19 cases.

Health Service Information Access Behavior

Before we began the project, we conducted an informal interview with randomly selected people of all ages, occupations, and regions to confirm or reject our assumption that people in Mongolia mostly rely on word-of-mouth information sharing when it comes to health service access.

Word-of-mouth inquiry was the most common among interviewees from the outskirts or rural areas or those with less online search experience or skills. We infer from this that many of the rural population, the elderly, and those from low-income or low-education households would rely most on this method of obtaining information about health services. The word-of-mouth inquiry is often used when a person knows someone in the health sector, especially doctors in their social circle. Therefore, not everyone can also take advantage of the word-of-mouth way of getting information about health and medical services.

In-person inquiry refers is where a person would visit the health service site personally by commuting to get the information and possibly getting the service itself. We observed that people who travelled into the city from rural areas for health or medical services or those who recently moved to the city and are still unfamiliar with the city tend to resort to this method of inquiry. This method of inquiry is highly time and energy intensive as one must commute, search for the location, and wait in line in order to get the information on site. We infer that many of the recent migrants in the capital city, living in the fringes of ger area will rely on this inquiry method.

Online inquiries include phone calls, social media, and search engines. Most interviewees expressed that they prefer to talk to someone rather than search for information on websites or social media pages for reasons such as a lack of online research experience and uncertainty about whether the information found is reliable. However, the way to get the phone number for health services is to call an information call center to get the number for the service provider or search online. The call centers may not always have the detailed, or up-to-date contact information for smaller private health or medical service providers. Social media pages are common. However, the amount and reliability of the information is an issue. Search engine results are also sporadic and uncomprehensive and can be out of date.

The informal interviews confirmed our assumption that a comprehensive health service information portal would be highly beneficial.

State of Data on Available Health Services

The information on available health services was not only inaccessible but also, to a certain extent, unavailable. There was no comprehensive, up-to-date database of available health service providers. While we found sets of data on health services in various formats such as Excel, PDF, and lists published on websites, there was no comprehensive dataset to rely on completely.

We worked with the multiple sets of available data to triangulate and develop a final list of health services. The final list contains information such as name, location/address, contact information, opening hours, wheelchair access, services available, and whether the service provider is a private or public entity. However, we found that the information on the health service providers was not accurate, especially regarding location. This prompted the unforeseen need to carry out an extensive level of fieldwork. The fieldwork was completed section by section of the city, where project staff and volunteer mappers went around the assigned administrative units of the city to collect any available health services such as hospitals, pharmacies, dentists, and clinics. The field mappers used the ArcGIS Survey123 tool to collect data, mostly from outside of the Point of Interest, and did not go inside the health services to prevent infection, in addition to wearing masks, using hand sanitizers, and social distancing.

Our interaction with some key stakeholders in the health sector, including the Ministry of Health, revealed this data gap to be the case. A government agency responsible for registration and licensing of health service providers, Medicine and Medical Equipment Registration Agency, sought our support in presenting the locations of available health services to the Ministry of Health of Mongolia. We provided an online link to the collected health service data and also prepared static maps of any available health services by administrative boundaries.

Similarly, the National Emergency Management Agency (NEMA) benefitted from the health service database and the improved OpenStreetMap data developed by the project in COVID-19 response. NEMA uses OpenStreetMap as one of its main geospatial tools. This was especially relevant when we trained local volunteer mappers in 21 provinces to map health and other essential services in all 21 provinces outside Ulaanbaatar, covering the whole country. The data collected are in the process of getting inputted onto OSM for public use.

Challenges with Open Data in Mongolia

The culture of open data has been a significant barrier for Mongolia's development, despite laws and policies describing access to information (Mongolian Law on Information Transparency and Right to Information, 2011).

Inaccessibility and unavailability of more detailed and granular public data present a challenge for all sectors in making informed decisions. For example, we were unable to obtain a consistent, complete, up-to-date, and granular set of socioeconomic status data. Each dataset, we were able to obtain only for certain districts or settlements of the city of Ulaanbaatar, from different years, and at district or large-scale level. Khoroo, or the smallest administrative unit-based data, was lacking consistently. These challenges limited our ability to further our vulnerability assessment.

The vulnerability assessment was developed using existing data on available essential services such as water, health, education, transportation, and grocery stores. It is intended to provide a visual illustration of the communities in Ulaanbaatar with the least access to such services to inform government and non-governmental stakeholders.

Moreover, the use and availability of geospatial data remains limited with key challenges in terms of recency, completeness, and reliability. The overall limited technical capacity at all levels of government entities results in poor or nonexistent geospatial data. More importantly, the lack of technical GIS capacity at the lowest level of public administration is the key barrier because all the data is collected through the primary level of administration where key surveyors are not trained or equipped with the tools and knowledge necessary.

Most government organizations use paid geospatial services such as Mapbox, Google Maps, or ArcGIS or more commonly contract paid map services from a private sector. Our meetings with government stakeholders illustrate the lack of understanding, awareness, and technical capacity for OpenStreetMap or open-source geospatial tools. While geospatial data is becoming more available, there are a few efforts to develop public GIS databases. The efforts are uncoordinated and uncollaborative and lack sustainability planning. For example, two different ministries may commission nearly identical GIS projects to a private service provider without ensuring technical capacity within the government agencies. Such projects soon run out of resources to hire private service providers and maintenance, troubleshooting, and future developments are halted, resulting in outdated, failed platforms.

Description of the Health Service Information Portal

The health service information portal can accommodate nation-wide health service data and search functions from a technical standpoint. The project team was able to collect and verify health and essential services data from 21 provinces outside the capital city of Ulaanbaatar. Inputting the collected and verified data for the provinces outside the capital city would establish the portal as the first-ever national health service search tool that is available (Fig. 7.1). More importantly, extending the product nationally would significantly improve usership and therefore sustainability and enable further development of the portal.

Users in Mongolia can search or browse any available health services at https://www.i-med.mn/. A user can search by using filters including location, service

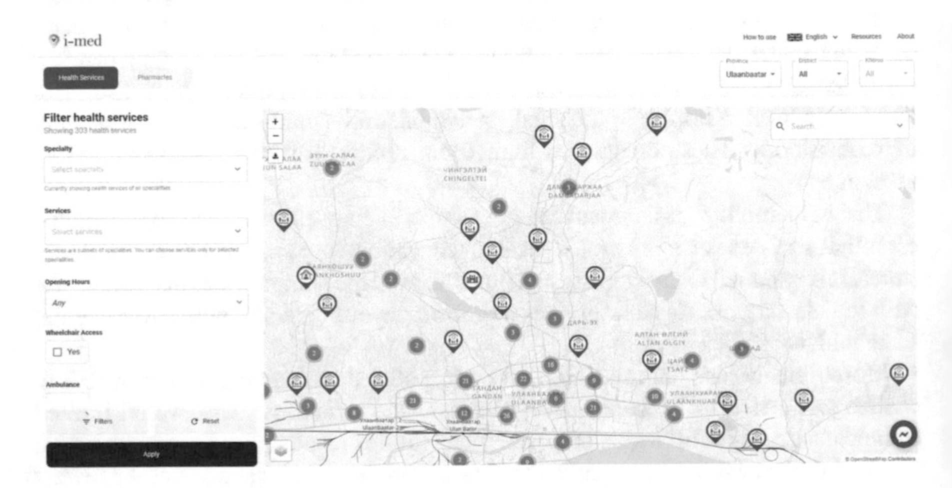

Fig. 7.1 Health service information portal

specialty, a specific service, wheelchair access, opening hours, etc. The browsing function allows the user to use the zoom function on the website to browse available health services near them. The mobile app of the portal, I-Med, is also available for Android phone users with additional functions such as navigation, sharing, and saving a specific point of interest.

The portal is a good foundation to provide information about health services in one place. One of the key functions of the portal also allows users to leave a comment and rate the service they have received from a specific health service provider, be it a hospital, a family clinic, or a dentist. The data collected are still being inputted into OSM as the portal's reference map base. As more data is inputted and more users are using i-med as a reference tool, the quality will continue to improve.

Conclusion

Based on the project results, challenges, and successes, we have compiled a list of recommendations for government and other stakeholders. First, an inter-sectoral stakeholder collaboration is recommended to ensure up-to-date, comprehensive, geo-location data of available health and medical services in the country. Second, building local, technical capacity in GIS and geospatial data management for government stakeholders to maintain and update health and medical service databases. Third, provide timely, useful, easy-to-understand information and resources related to accessing health and medical services on the public-facing information portal where detailed information about available services is accessible.

One of the most important future components of the work carried out during the project is to conduct large-scale awareness and engagement activities for the public, especially those located in areas with higher vulnerability factors. In line with awareness and outreach activities, capacity building, community building and expanding, local representative training workshops, outreach, and engagement activities are recommended to improve and strengthen the existing OpenStreetMap community in Mongolia, but also to promote the culture of open data in the country.

About Public Lab Mongolia

Public Lab Mongolia was founded in 2018 because we saw the need for a voice that advocates for a meaningful collaboration among stakeholders such as civil society, academia, government, and the media. We work to promote the culture of open data so that different actors, leaders, and communities can address the environmental and public health challenges of today and tomorrow by engaging in informed and inclusive discussions. Our mission is to cultivate a healthy environment and resilient communities through open data. We work to foster data-based decision-making and increase youth contribution in achieving sustainable development goals via open data by creating and improving data on environment and environmental health issues.

Acknowledgements The C2M2 Ulaanbaatar project would like to thank the Office of the Geographer and Global Issues at the U.S. Department of State and the global program team. This work is supported by the U.S. Department of State and managed by the Association of American Geographers. We would like to express our appreciation to Dr. Nama Budhathoki, Dr. Melinda Laituri, and Dr. Junghwan Kim for their encouragement, support, and guidance. We also thank the other Asia Hub projects for their valuable insights and shared experiences. We would like to express our deep appreciation for the technical partnership with Kathmandu Living Lab for being an integral part of the project to create the final product of the project. Finally, we would like to thank the colleagues of the Public Lab Mongolia and local partners at the Mongolian National University of Medical Sciences, and Health Development Center for their hard work.

Disclaimer The views and opinions expressed in this article are those of the authors and do not necessarily reflect the official policy or position of any agency of the U.S. government. Assumptions made within the analysis are not a reflection of the position of any U.S. government entity.

References

AFP. (2021). *Mongolians protest at virus curbs after row over mother's treatment*. https://www.france24.com/en/live-news/20210121-mongolians-protest-at-virus-curbs-after-row-over-mother-s-treatment

Enkhbold, S. (2021). *There is a shortage of specialists in intensive care and emergency care*, T.Ulziibayar. https://ikon.mn/n/299g

Graceffo, A. (2022). *Mongolia suffers under China's zero Covid policy*. https://www.lowyinstitute.org/the-interpreter/mongolia-suffers-under-china-s-zero-covid-policy

Khukhnokhoi, K. (2021). *Patients languish as health centers focus on coronavirus*. https://globalpressjournal.com/asia/mongolia/patients-left-behind-health-centers-focus-coronavirus/

Mongolian Law on Information Transparency and Right to Information. (2011). *Mongolian Law on Information Transparency and Right to Information*. https://legalinfo.mn/mn/detail/374

TIME. (2020). *A resident of Arkhangai aimag, stuck behind border checkpoint 22, dies*. http://time.mn/piD.html

Tsetseg, B. (2020). *A resident of Arkhangai died while waiting for an ambulance*. https://news.mn/r/2380850/

World Bank. (2021). *Results of Mongolia COVID-19 household response phone survey (round 3)*. https://thedocs.worldbank.org/en/doc/674291610418865659-0070022021/original/MNGHFphonesurveyR3Final.pdf

Chapter 8
The Inequities in the United States During the COVID-19 Pandemic

Xiao Huang, Siqin Wang, and Xiao Li

Personal Story

As the leading author of this chapter, I am grateful for being invited to contribute to this book. I was born in Wuhan, China, where the initial outbreak of COVID-19 was first declared. My strong ties with Wuhan motivated me to devote my time and direct my research focus to the investigation of the COVID-19 pandemic, hoping to make some contributions to the understanding of the COVID-19-induced societal impact with my data mining expertise. In August 2020, I became an Assistant Professor in the Department of Geosciences at the University of Arkansas. Despite the fact that I had to work from home during the 2020 Fall and 2021 Spring semesters, I was able to maintain my daily routine without worrying about the economic burden. However, not everyone has such a privilege as I do. For many people, being able to work from home is a dream. Vulnerable populations have far fewer economic opportunities, and the jobs they hold do not have a work-from-home option. I started to realize that the long-standing social inequity in the US has disproportionately affected spatially and socially disadvantaged groups and lead to other inequalities in their daily lives. The consequences are probably worse than we have anticipated. *Xiao Huang*

Introduction

> I couldn't help sitting there reflecting about sometimes when you're in the middle of a crisis, like we are now with the coronavirus, it really does … ultimately shine a very bright light on some of the real weaknesses and foibles in our society—Anthony Fauci, Director

X. Huang (✉)
University of Arkansas, Fayetteville, AR, USA
e-mail: xh010@uark.edu

S. Wang
The University of Queensland, St Lucia, QLD, Australia

X. Li
Texas A&M Transportation Institute, Bryan, TX, USA

M. Laituri et al. (eds.), *The Geographies of COVID-19*, Global Perspectives on Health Geography, https://doi.org/10.1007/978-3-031-11775-6_8

of the National Institute of Allergy & Infectious Diseases, White House Coronavirus Task Force.

Major disasters make the poor poorer, if not the rich richer. The COVID-19 pandemic we are facing is not an exception. Amidst the COVID-19 pandemic, inequities and biases have become more visible, as many pieces of evidence reveal that the pandemic affects all segments of the population and is particularly detrimental to members of those social groups in the most vulnerable situations. In this chapter, we aim to summarize relevant studies that reveal, analyze, and discuss the inequities in the United States during the COVID-19 pandemic from three different perspectives: (1) inequity in COVID-19 testing rates, confirmed cases, and mortality, (2) inequity in stay-at-home compliance and short-/long-term recovery, and (3) inequity in vaccination tendency, vaccine allocation, and vaccination rate. These inequities that disfavor the vulnerable populations (a group of people who are less able to respond to, cope with, resist, and/or recover from the impacts of the COVID-19 pandemic) can further compound other disadvantages, such as underlying comorbidities and poor access to high-quality medical care, further causing negative health outcomes for the vulnerable populations. This chapter summarizes evidence that unravels the disparity potentially triggered by sociodemographic inequities and other imposed inequities in terms of infected cases, policy compliance, and vaccination. We provide further suggestions on mitigation strategies to reduce the potential inequity and vulnerabilities induced by future epidemics.

Inequity in COVID-19 Testing Rates, Confirmed Cases, and Mortality

A growing number of studies have demonstrated that the COVID-19 has disproportionate impacts on the population, more profoundly influencing the vulnerable ones. A range of social issues (e.g., discrimination and segregation) and underlying health inequities expose vulnerable populations to a higher risk of infection, developing severe illnesses, and eventually dying from COVID-19. This section summarizes the emerging evidence from existing studies to illustrate the inequities in the COVID-19 testing rates, confirmed cases, and mortality. We also explore the potential causes of these disparities and discuss the suggested interventions.

Testing Rates

COVID-19 testing plays a vital role in preventing viral spread. Equitable and accessible testing can help people determine if they have been infected, leading to the proper action of self-isolation and timely treatment. However, the testing sites were not equally distributed. The disparities of testing rates exist among different population groups and geographic contexts (e.g., communities and counties). For example, Dalva-Baird et al. (2021) conducted a nationwide study to examine the racial and ethnic composition of neighborhoods surrounding COVID-19 testing sites in the US. They compared the current distributions of testing sites with the simulated random distributions. The results demonstrated that testing sites were not equally distributed by race, underrepresenting minority populations, especially the Hispanic residents. The low testing rates among Hispanic residents led to an increased disparity in their mortality. Some studies found that socially vulnerable communities were associated with lower testing rates. For example, Bilal et al. (2021) examined the relationship between communities' social vulnerability levels and testing rates in three US cities – New York, Philadelphia, and Chicago. They modeled the relationship between the test rate and the Social Vulnerability Index (SVI) published by the US Centers for Disease Control and Prevention (CDC) through the analysis of a negative binomial regression. Figure 8.1 presents the CDC SVI (the 2018 version) in the contiguous US at the census tract level. The results from Bilal et al. (2021) showed that a higher level of social vulnerability was associated with 3%, 9%, and 13% lower testing rates in New York, Philadelphia, and Chicago, respectively. In fact, the disparities in testing rates do not just exist at the community or city level; studies found that countries distinct in socioeconomic statuses were significantly associated with different testing rates. Marziali et al. (2021) examined the testing rates of 86 countries with a group of selected socioeconomic

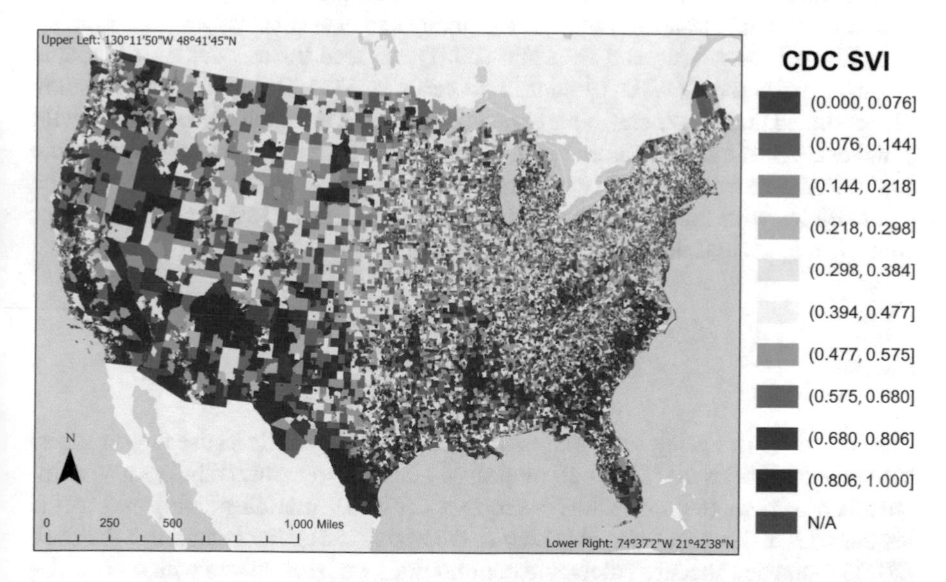

Fig. 8.1 Social Vulnerability Index (SVI) in 2018 at the census tract level in the Contiguous US from the US Centers for Disease Control and Prevention. The variable mapped here is the overall tract summary ranking, i.e., "*RPL_THEMES*" in the dataset. Tracts with higher values (higher levels of vulnerability) are more vulnerable to the pandemic, thus demanding more attention and care from governments and public sectors. SVI values are classified into ten groups via the quantile method

factors, including the Human Development Index (HDI), health expenditure, universal health coverage (UHC), and urban population. They found that testing rates were significantly correlated with the HDI and urban population at the country level; developing and underdeveloped countries with a lower HDI may experience a higher burden of conducting widespread testing, leading to a global inequity in COVID-19 testing.

Confirmed Cases

Existing studies suggested a disproportionate burden of illnesses among racial and ethnic minority groups. For example, the CDC systemically examined the confirmed COVID-19 cases by race and ethnicity. The available data demonstrated that higher rates of confirmed cases were observed from non-White communities. For example, 34% of confirmed cases were Black Americans, who account for only 13% of the total US population. Latinos and Hispanics also showed higher rates of infection (Garg et al., 2020). A similar result was obtained by Mude et al. (2021), who conducted a systematic review by summarizing 72 selected studies to examine the racial disparities in COVID-19 confirmed cases. This study demonstrated that Blacks and Hispanics experienced significantly higher percentages (156% for Blacks and 154% for Hispanics) of COVID-19 confirmed cases compared to Whites. Besides the racial disparities, income inequality can also contribute to the disparity in confirmed cases. Liao and De Maio (2021) assessed the association of income inequality with the COVID-19 confirmed cases in 3141 US counties. This study demonstrated that, in general, a higher level of income inequality is associated with a higher level of infected cases. Meanwhile, residents living in socially vulnerable communities were more likely to get infected. As reported by Bilal et al. (2021), the communities with higher SVI were associated with 22%, 33%, and 27% higher confirmed cases in Chicago, New York, and Philadelphia, respectively.

Mortality

The inequities in testing rates and confirmed cases inevitably caused and exacerbated the disparities in COVID-19 mortality. The aforementioned studies also demonstrated a more severe inequity issue in COVID-19 mortality compared to the testing rates and confirmed cases (Liao & De Maio, 2021). For example, Bilal et al. (2021) found that socially vulnerable communities were associated with 44%, 56%, and 58% higher mortality in Chicago, New York, and Philadelphia; the mortality burden is significantly higher than the testing rates and confirmed cases. A recent report published by the APM Research Lab showed that Pacific Islanders,

Indigenous, Latino, and Black Americans have a much higher death rate—double or more—than White and Asian Americans (APM Research Lab, 2021). Garg et al. (2020) also reported similar mortality disparities that Black and Hispanic Americans experienced substantially higher death rates than White persons. Studies showed that the confirmed COVID-19 patients with previous comorbidities, including cardiovascular disease, hypertension, and diabetes, are more likely to develop severe symptoms or even lead to death. These comorbidities, however, are more prevalent in racial/ethnic minorities and socially vulnerable communities (Raifman & Raifman, 2020). It implies that, even at equal levels of viral exposure, the vulnerable populations could experience higher burdens and more severe consequences from COVID-19.

Discussion and Policy Implications

Health equity is among the primary goals of most governments and public health agencies. However, the apparent inequity in COVID-19 testing sites, confirmed cases, and mortality suggests we are still far from reaching this goal. The CDC selected five social determinants of health that could lead to the vulnerable groups being disproportionately affected by COVID-19, including neighborhood and physical environment, health and healthcare, occupation and job conditions, income and wealth, and education (CDC, 2020). However, some long-standing social issues, such as discrimination, racism, and segregation, significantly influence these social determinants. Note that modern US cities are characterized by strong residential segregation by both ethnicity and income. The racial and residential segregation result in stark differences across communities in multiple factors, such as household crowdedness, access to reliable transportation, assess to quality healthcare, job opportunities and types, the privilege of working from home, quality of living environments, among others (Ellis et al., 2004; Phillips, 2007). These factors, directly or indirectly, exacerbated the transmission of the virus and the severity of illness, exposing the minorities and socially vulnerable communities to higher risks. To stop the spread of COVID-19 and protect vulnerable populations from future pandemics, greater efforts must be made toward promoting health equity. On the one hand, governments and public agencies need to ensure resources (e.g., free testing, affordable treatments, tailored medical, and mental healthcare) are equitably distributed and accessible for all people, helping them maintain physical and psychological health during the pandemic. On the other hand, policymakers, urban planners, and healthcare providers need to consider the negative impacts of segregation and racism when making health equity strategies. Effective solutions and policies are needed to address the structural factors linked to income inequality, racism, and segregation, which could fundamentally minimize the burden of the pandemic and promote population health and health equity.

Inequity in Stay-At-Home Compliance and Short-/Long-Term Recovery

Social distancing has emerged as one of the most widely adopted nonpharmaceutical control measures to curb the viral spread, as it has been proved to effectively reduce physical contact with viral sources, thus decelerating the transmission rate. The US federal and local governments have issued stay-at-home orders to close the majority of nonessential businesses and limit the outdoor activities and large gatherings of residents, leading to a sharp decrease in human mobility and an increase in home-dwelling time (Huang et al., 2020a, 2022). On April 4, 2020, 95% of the US population was under stay-at-home orders (Baek et al., 2020). However, keeping people from unnecessary outdoor activities in rich and/or poor regions, in urban and/or rural areas, and in authoritarian and/or open societies is the ultimate human challenge (Van Rooij et al., 2020). These mitigation measures implemented in the US, with different temporal coverage, effectiveness, and stringency, have a voluntary nature (Rose-Redwood et al., 2020). People with different socioeconomic statuses and demographic backgrounds tend to respond differently to the implementation and lifting of restrictions, which exaggerates the discrepancies in stay-at-home compliance and short-/long-term recovery. This section summarizes the studies that reveal the disparity in stay-at-home compliance and recovery and explore the potential factors that lead to such disparity. We also discuss what can be suggested for better policymaking.

Stay-At-Home Compliance

Efforts have been made to investigate the disparate responses to stay-at-home orders. Many pieces of evidence at various scales have been found, pointing to the social construct largely determined by demographic and socioeconomic variables (Huang et al., 2020b, 2022; Chiou & Tucker, 2020). Taking the Atlanta-Sandy Springs-Roswell metropolitan statistical area as a study case and utilizing 45 million digital devices in the US, Huang et al. (2020b) explored how people with demographic/socioeconomic backgrounds respond differently to the stay-at-home order. Their study suggested that demographic/socioeconomic variables can explain the disparity in home-dwelling duration in response to the stay-at-home order. Vulnerable populations are less likely to follow the order to stay at home, pointing to the extensive gaps in the effectiveness of social distancing measures between vulnerable populations and others. An extended study of the twelve most-populated metropolitan statistical areas in the US investigated, from a modeling perspective, the contribution of selected demographic/socioeconomic variables to the increase in home-dwelling time when stay-at-home orders were effective (Huang et al., 2022); their results indicated that median household income and percentage of high income are the two most important variables in predicting the increase in home-dwelling duration, which proved the dominance of economic status in driving the adherence to stay-at-home orders. Similarly, Chiou and Tucker (2020) revealed that high-income earners generally spent more time at home during stay-at-home orders and stated that access to high-speed Internet potentially plays a vital role. Besides the above efforts, other studies have linked racial/ethnic composition, risk awareness, educational attainment, and political affiliations to the disparate stay-at-home policy adherence.

Short-/Long-Term Recovery

Numerous efforts have been made to explore the disparity in recovery that presumably results from the disparity in sociodemographic settings. Leveraging multi-source mobility datasets from Google, Apple, Descartes Labs, and Twitter, Huang et al. (2021) investigated how people reduced their travels during the mobility-restricting period and how mobility recovered after the reopening at the US county level. Their results revealed a great disparity in mobility dynamics in the recovery phase, as the poor counties tended to gain earlier and greater upward momentum in mobility following the lifting of stay-at-home orders than the rich counties. Such a rapid, short-term recovery following the lifting of stay-at-home orders in poor communities can be explained by their employment types with limited options to remote working and their economic burdens that need to be addressed, which potentially caused unequal exposure to COVID-19 that disproportionately affects vulnerable groups (Huang et al., 2021). As for long-term recovery, even though it is difficult to decide (at the time of writing) when the post-pandemic recovery really starts, numerous endeavors have been made to investigate whether the degree of long-term recovery differs in communities with varying demographic and/or socioeconomic characteristics. One notable effort was by Saenz and Sparks (2020), who provided a broad and comprehensive overview of the inequities in job losses and recovery using data from the US monthly Current Population Survey microdata from the Integrated Public Microdata Series (IPUMS – CPS). Their results suggested that, by June 2020, Blacks and Latinos had about 12% fewer jobs than they held in February, compared with 7.5% job losses among Whites. Meanwhile, women have experienced higher likelihoods of unemployment than men during the pandemic, with women of color and Latina immigrant women having the most elevated jobless rates.

Discussion and Policy Implications

The disparate exposure to vulnerable populations resulting from the stay-at-home compliance and rapid bounces following the lifting of orders can further compound other disadvantages, such as underlying comorbidities, limited access to COVID-19 testing centers, and poor access to and low utilization of high-quality healthcare, further causing negative health outcomes for the vulnerable populations. Mitigation measures at all levels need to recognize and account for the disparity in policy compliance and recovery due to these inequities, keeping the public value of social equity at the forefront of actions to support stronger, more effective readiness for future epidemics. As for the disparity in long-term recovery, attention should be paid to the differences in vulnerability across sectors, which could create disproportionate risk levels for lower-income workers, minority business owners, and business owners with less educational attainment. Protecting small businesses owned by vulnerable populations from permanent closure is of great importance, given the many roles they play in the economy as employers, engines of entrepreneurship, economic multipliers, and community hubs (Dua et al., 2020). It is imperative to reduce immediate health effects and ensure equitable allocation of healthcare resources and the proper allocation of financial resources, such as subsidies, for more vulnerable populations.

Inequity in Vaccine Tendency, Vaccination Allocation, and Vaccination Rate

Apart from the inequity described in the preceding two dimensions, inequity also exists in vaccination tendency and vaccine allocation, which collectively lead to inequity in vaccination rates. Current estimates suggest that 60–90% of the US population will need to be vaccinated to reach herd immunity in order to effectively interrupt virus transmission (Bartsch et al., 2020). However, vaccination rates vary across population groups and geographic contexts, in tandem with the inequity in vaccine tendency amongst people with different demographic and socioeconomic statuses and the disparity of vaccination allocation across urban and rural spaces. We illustrate below how the vaccination tendency, vaccine allocation, and vaccination rate are intertwined and speculate on the mechanisms behind the inequity in these three aspects based on the findings and observations from existing studies.

Vaccine Tendency

Prior research has unveiled that demographic and socioeconomic factors, such as age, sex, race/ethnicity, income, and education, are correlated to vaccine tendency (Szilagyi et al., 2021). Demographic and socioeconomic factors are understood to be dimensions of social positioning, reflecting life chances and experiences, subsequently influencing the attitude, perception, and assessment of risk in the context of vaccination (Duan et al., 2021). Current studies commonly observed that vaccine hesitancy varied by race and ethnicity. For example, Black/African Americans have reported the highest vaccine hesitancy and the lowest vaccine confidence (Duan et al., 2021). The high level of their vaccine hesitancy would be concerning of vaccine safety on their own but also combined with the worrisome evidence that COVID-19 is exacting a disproportionate financial burden on racial minorities. In addition, racism within the medical establishment has been deeply rooted in the history of racist exploitation and past instances of medical experimentation, possibly triggering the distrust of the medical care system that may be more prevalent among racial and ethnic minority populations (Willis et al., 2021). The inequity in vaccine hesitancy by race and ethnicity could be further exacerbated by other factors, including age, education, employment, financial status, and health conditions. A certain social group with the privilege of working from home (e.g., Internet-based workers) or with private medical insurance may have a lower intention of getting vaccinated than the essential workers who must work on-site with more risks of virus infection (Huang et al., 2022). Age and education may also affect the receiving of information about vaccination disseminated by the government and other authorities, given the various access to Internet, public and social media among people at different ages and educational levels. The existing inequity in individual health risks may also influence vaccine hesitancy among the US non-Hispanic Black, Hispanic, and American Indian/Alaskan Natives (Willis et al., 2021), as certain types of vaccines (e.g., Moderna and Pfizer) may bring larger risks to people with the onset of chronic conditions, such as cardiovascular disease, pulmonary disease, and diabetes. Taken together, the safety concern, improper dissemination, and misleading perception of vaccines potentially relate to the inequity in the vaccine that should be addressed essentially by vaccination authorities and governments as the priority.

Vaccine Allocation

The equitable allocation of vaccines means that all areas should have equitable access to vaccines regardless of their socioeconomic status and topographic barriers. However, it was widely noted that rural/regional areas and socioeconomically disadvantaged areas have more difficulties in accessing vaccines (Bachireddy et al., 2021). People of color (e.g., Black and Latinx communities) were observed to be less likely to have private medical insurance, sufficient access to private transportation, and healthcare facilities. These disadvantages make it difficult for these communities to get vaccinated and strengthen the disproportionate impact of the pandemic on these groups (Szilagyi et al., 2021). In addition, the distribution of vaccines was imbalanced across urban and rural spaces, possibly due to the availability of physicians and nurses that have the capability to give vaccines. Although the US government dispatched mobile vaccine clinics in remote and rural areas, they may be insufficient in provision to cover the population living in such areas, particularly when most vaccines need more than one dose. The imbalanced distribution of vaccines may also be related to the effectiveness and stringency of vaccination policies (Agarwal et al., 2021). Some states (e.g., California and Connecticut) that started vaccination uptake at an earlier stage or implemented vaccination policies more restrictively have obviously higher vaccination rates (Hughes et al., 2021). With the principles of equity, ethics, and effectiveness, the strategies for the equitable allocation of COVID-19 vaccines should be prioritized to offer an overarching guideline for vaccine deployment to assist policymakers in governments at various levels.

Vaccination Rate

Taking the preceding mechanisms to explain the inequity in vaccine tendency and allocation, it would be easier to understand the inequity in vaccination rates that need to be analyzed in the context of the dynamic process of infection, policy implications, and vaccine tendency and allocation (Agarwal et al., 2021). The reasons for the inequitable vaccination rates are multifactorial. People who have been infected by COVID-19 or residing in high-risk areas may have a stronger intention to get vaccinated. Areas and regions (e.g., California, Connecticut, and Florida) where vaccination policies were implemented effectively and stringently may have vaccine rollout and uptake more quickly, leading to a higher vaccination rate (Hughes et al., 2021). Moreover, the demographic and socioeconomic profiles of communities and their locations matter. Racial and ethnic minorities or communities in a lower socioeconomic status may be more likely to have barriers to obtaining sufficient access to transportation, insurance, and health facilities where they can get vaccines. Such barriers were also exposed to those living in rural and remote areas. Therefore, policy implications can be given to address the current issues in vaccine tendency and allocation to boost vaccination rates, with recommended initiatives detailed as below.

Discussion and Policy Implications

To diminish the concern and hesitancy about vaccination, governments and health authorities should provide proper education and dissemination of vaccination to all populations, correcting the misinformation and false perception of vaccine safety. Meanwhile, clinicians, physicians, and other healthcare professionals should administer vaccines, make better use of the vaccines already at their disposal, and strongly recommend vaccines to all patients. Empowered and visible community/political leaders need to set an example. Multiple parties (e.g., non-profit organizations, faith-based entities, and community leaders trusted by racial and ethnic minorities) should partner up to establish trust and identify the best ways to meet the healthcare needs of disproportionately affected populations. To overcome the barriers in vaccine allocation, it is critical to pair the rising interest in vaccination with improved access to vaccines. Legislation that provides financial support to local communities, including critical access hospitals and mobile clinics in remote/rural areas, may provide continued medical services and attract physicians to work and distribute vaccines to such areas. Furthermore, vaccine outreach to vulnerable populations will need to continue and even intensify these efforts. It is recommended to create hotline supports and telephone-based appointment systems to help those who cannot access the Internet. Taking on board the initiatives in the preceding two aspects, vaccination rates are expected to increase more equally to better achieve the threshold of vaccination coverage that would need to prevent the virus from spreading in the long term.

References

Agarwal, R., Dugas, M., Ramaprasad, J., Luo, J., Li, G., & Gao, G. G. (2021). Socioeconomic privilege and political ideology are associated with racial disparity in COVID-19 vaccination. *Proceedings of the National Academy of Sciences of the United States of America, 118*(33), e2107873118.

APM Research Lab. (2021). *The color of coronavirus: COVID-19 deaths by race and ethnicity in the U.S.* Retrieved October 5, 2021, from https://www.apmresearchlab.org/covid/deaths-by-race

Bachireddy, C., Dar, M., & Chen, C. (2021). Medicaid and COVID-19 vaccination—Translating equitable allocation into equitable administration. In *JAMA health forum* (Vol. 2, No. 2, p. e210114). American Medical Association.

Baek, C., McCrory, P. B., Messer, T., & Mui, P. (2020). Unemployment effects of stay-at-home orders: Evidence from high frequency claims data. *The Review of Economics and Statistics*. https://doi.org/10.1162/rest_a_00996

Bartsch, S. M., Ferguson, M. C., McKinnell, J. A., O'shea, K. J., Wedlock, P. T., Siegmund, S. S., & Lee, B. Y. (2020). The potential health care costs and resource use associated with COVID-19 in the United States: A simulation estimate of the direct medical costs and health care resource use associated with COVID-19 infections in the United States. *Health Affairs, 39*(6), 927–935.

Bilal, U., Tabb, L. P., Barber, S., & Diez Roux, A. V. (2021). Spatial inequities in COVID-19 testing, positivity, confirmed cases, and mortality in 3 US cities: An ecological study. *Annals of Internal Medicine, 174*(7), 936–944.

Centers for Disease Control and Prevention (CDC). (2020). *Introduction to COVID-19 racial and ethnic health disparities*. Retrieved October 5, 2021, from https://www.cdc.gov/coronavirus/2019-ncov/community/health-equity/racial-ethnic-disparities/index.html

Chiou, L., & Tucker, C. (2020). *Social distancing, internet access and inequality*. Working Paper 26982. National Bureau of Economic Research. https://doi.org/10.3386/w26982

Dalva-Baird, N. P., Alobuia, W. M., Bendavid, E., & Bhattacharya, J. (2021). Racial and ethnic inequities in the early distribution of US COVID-19 testing sites and mortality. *European Journal of Clinical Investigation, 51*(11), e13669. https://doi.org/10.1111/eci.13669

Dua, A., Ellingrud, K., Mahajan, D., & Silberg, J. (2020). *Which small businesses are most vulnerable to COVID-19–and when*. McKinsey & Company. Retrieved October 5, 2021, from https://www.mckinsey.com/featured-insights/americas/which-small-businesses-are-most-vulnerable-to-covid-19-and-when

Duan, Y., Shi, J., Wang, Z., Zhou, S., Jin, Y., & Zheng, Z. J. (2021). Disparities in COVID-19 vaccination among low-, middle-, and high-income countries: The mediating role of vaccination policy. *Vaccine, 9*(8), 905.

Ellis, M., Wright, R., & Parks, V. (2004). Work together, live apart? Geographies of racial and ethnic segregation at home and at work. *Annals of the Association of American Geographers, 94*(3), 620–637.

Garg, S., Kim, L., Whitaker, M., O'Halloran, A., Cummings, C., Holstein, R., et al. (2020). Hospitalization rates and characteristics of patients hospitalized with laboratory-confirmed coronavirus disease 2019—COVID-NET, 14 States, March 1–30, 2020. *Morbidity and Mortality Weekly Report, 69*(15), 458.

Huang, X., Li, Z., Jiang, Y., Li, X., & Porter, D. (2020a). Twitter reveals human mobility dynamics during the COVID-19 pandemic. *PLoS One, 15*(11), e0241957.

Huang, X., Li, Z., Lu, J., Wang, S., Wei, H., & Chen, B. (2020b). Time-series clustering for home dwell time during COVID-19: What can we learn from it? *ISPRS International Journal of Geo-Information, 9*(11), 675.

Huang, X., Li, Z., Jiang, Y., Ye, X., Deng, C., Zhang, J., & Li, X. (2021). The characteristics of multi-source mobility datasets and how they reveal the luxury nature of social distancing in the US during the COVID-19 pandemic. *International Journal of Digital Earth, 14*(4), 424–442.

Huang, X., Lu, J., Gao, S., Wang, S., Liu, Z., & Wei, H. (2022). Staying at home is a privilege: Evidence from fine-grained mobile phone location data in the United States during the COVID-19 pandemic. *Annals of the American Association of Geographers, 112*(1), 286–305.

Hughes, M. M., Wang, A., Grossman, M. K., Pun, E., Whiteman, A., Deng, L., et al. (2021). County-level COVID-19 vaccination coverage and social vulnerability—United States, December 14, 2020–March 1, 2021. *Morbidity and Mortality Weekly Report, 70*(12), 431.

Liao, T. F., & De Maio, F. (2021). Association of social and economic inequality with coronavirus disease 2019 incidence and mortality across US counties. *JAMA Network Open, 4*(1), e2034578.

Marziali, M. E., Hogg, R. S., Oduwole, O. A., & Card, K. G. (2021). Predictors of COVID-19 testing rates: A cross-country comparison. *International Journal of Infectious Diseases, 104*, 370–372.

Mude, W., Oguoma, V. M., Nyanhanda, T., Mwanri, L., & Njue, C. (2021). Racial disparities in COVID-19 pandemic cases, hospitalizations, and deaths: A systematic review and meta-analysis. *Journal of Global Health, 11*, 05015. https://doi.org/10.7189/jogh.11.05015

Phillips, D. (2007). Ethnic and racial segregation: A critical perspective. *Geography Compass, 1*(5), 1138–1159.

Raifman, M. A., & Raifman, J. R. (2020). Disparities in the population at risk of severe illness from COVID-19 by race/ethnicity and income. *American Journal of Preventive Medicine, 59*(1), 137–139.

Rose-Redwood, R., Kitchin, R., Apostolopoulou, E., Rickards, L., Blackman, T., Crampton, J., et al. (2020). Geographies of the COVID-19 pandemic. *Dialogues in Human Geography, 10*(2), 97–106.

Saenz, R., & Sparks, C. (2020). *The inequities of job loss and recovery amid the COVID-19 pandemic*. University of New Hampshire. Retrieved October 15, 2021, from https://carsey.unh.edu/publication/inequities-job-loss-recovery-amid-COVID-pandemic

Szilagyi, P. G., Thomas, K., Shah, M. D., Vizueta, N., Cui, Y., Vangala, S., & Kapteyn, A. (2021). Likelihood of COVID-19 vaccination by subgroups across the US: Post-election trends and disparities. *Human Vaccines & Immunotherapeutics, 17*(10), 3262–3267.

Van Rooij, B., de Bruijn, A. L., Reinders Folmer, C., Kooistra, E. B., Kuiper, M. E., Brownlee, M., et al. (2020). Compliance with COVID-19 mitigation measures in the United States. *PsyArXiv Preprints*. https://doi.org/10.31234/osf.io/qymu3

Willis, D. E., Andersen, J. A., Bryant-Moore, K., Selig, J. P., Long, C. R., Felix, H. C., et al. (2021). COVID-19 vaccine hesitancy: Race/ethnicity, trust, and fear. *Clinical and Translational Science, 14*(6), 2200–2207. https://doi.org/10.1111/cts.13077

Chapter 9
The Latine Community and COVID-19: Nuances, Experiences, and Data

Aída Guhlincozzi and Deshira Wallace

Personal Stories

I stood under a tent in the rain-soaked parking lot to collect information from food pantry visitors, taking dozens of English and Spanish ID cards, some plastic, some fragile paper. I needed to quickly gather crucial information from every participating household. While dodging raindrops, the paper form I scribbled information on slowly became saturated. Pursuing brevity, I used shorthand for nearby towns and dashes for no information. Through an oft-used KN-95 mask, I shouted over the rain in Spanish and English, "How many? "Cuántos personas?" straining to hear the shouted answers through many accents and rain hammering on the tent above me, contorting words. Tres became seis, and street names I drove on daily became unrecognizable. I tried my best to transcribe it all. By closing time, I had four soggy pages of attendee information, each record listed with questionable legibility. I walked inside, submitted the records, and breathed relievedly, seeing the tacos al pastor awaiting volunteers. I forgot about the records written in my rain-addled shorthand and enjoyed my meal. *Aída Guhlincozzi*

The COVID-19 pandemic meant balancing either—in fear—avoiding healthcare systems or—in need—going to healthcare systems. I faced this difficult choice when accompanying a family member to the emergency department. The room was crowded, people waiting to be seen for over 10 h, including us. In the triage room, the nurse spoke mostly to me, a common action when clinicians hear my family member's accent. I tried redirecting them to speak directly to my family member to no avail. When the nurse entered the information, I noticed the demographics section. Race: Black/

A. Guhlincozzi (✉)
University of Missouri, Columbia, MO, USA
e-mail: argvfz@missouri.edu

D. Wallace
University of North Carolina at Chapel Hill, Chapel Hill, NC, USA

M. Laituri et al. (eds.), *The Geographies of COVID-19*, Global Perspectives on Health Geography, https://doi.org/10.1007/978-3-031-11775-6_9

African American, Ethnicity: Not Hispanic/Latino. That was incorrect, but a consistent experience of poor demographic data collection anytime my family interacted with any healthcare system. I wanted to correct the nurse, "Black AND Latino", but we had more pressing matters at hand. Still, I knew the implications of this mistake—notably, inaccurate reports on healthcare utilization. Another misunderstanding is who is being affected by COVID-19 when supposedly "self-reported" demographics are inaccurately observer-reported. This incident clarified the nuance of reported race and ethnicity data and who is and is not made visible during data collection. *Deshira Wallace*

Introduction

Collecting accurate health data is always challenging, but doing so on historically marginalized populations can be especially difficult. Fine-tuning our data collection methods, including our questions, can improve our data quality and lead to better solutions. When looking at the issues of a group that is racialized or othered by structural racism, questions frequently focus on what a person's self-reported race or ethnicity is. Yet, responses to these questions are broad and can be interpreted in various ways depending on context and lived experiences.

Further, researchers should interrogate why they need to know race and ethnicity. Race should not be used as a proxy for the effects of systemic racism or suiting a false notion of biological race (Lett et al., 2022). Questioning how we discuss the "Latine"[1] community and their COVID-19 experience is central to this chapter because too often, only white Latines are conceptualized as "Hispanic or Latino."

For example, Cecilia Márquez (2019) applies the racialized and geographic realities of Latines within the context of legalized racial segregation (i.e., Jim Crow) in the contemporary US South. *Juan Crow*, appeared in 2008 as a response to the racist, anti-immigrant laws affecting Latines in the South. Márquez critiques the term *Juan Crow* for erasing the historical and contemporary presence of Afro-Latines in the South, who were subjugated to Jim Crow – a US version of apartheid – as racialized Black people experiencing Black-specific oppression in the past and continuing to be impacted in the present. This framing of Juan Crow as an immigrant and *mestizo* issue reifies the idea that Black and Latine peoples are categorically "separate groups living under discrete systems of racial control" and that Latines across different geographies in the U.S. are always "new arrivals" or forever-immigrants (Márquez, 2019). Laws constructed to define racialized space by the state went beyond Jim Crow, they included racial covenants, which were written in property deeds to explicitly not sell homes to non-white people for the preservation of white

[1] Of note, there is no preferred term for people whose ancestors or have immigrated from Latin America, but in this publication, we choose to use the gender inclusive term, Latine.

spaces (Gibbons, 2018). Additionally, "redlining" which legitimized the practice of maintained racial segregation through real estate practices and government programs restricting access to home loans and purchasing for non-white, but specifically for Black individuals (Gibbons, 2018).

Yet, as the law constructs racialized stratification through geography in the United States, it also does so in health and well-being. This was experienced during the COVID-19 pandemic, which has unevenly impacted the US population through the laws and policies passed to mitigate the pandemic's effects (Asad & Clair, 2018). The law's role in emergency response can contribute to the inequities experienced by minoritized communities, immigrants, and those living in congregated spaces (Parmet et al., 2021). Joseph (2017) found a similar pattern in the healthcare-seeking behaviors of various Latine groups in Boston. Dependent on whether they were racialized as white, Black, or "Latino" was the likelihood of being profiled on immigration and healthcare treatment status (Joseph, 2017). This is only one example of the type of nuance critical to capturing health and healthcare access research, especially as it relates to COVID-19 and the varied Latine experience of the pandemic.

In this chapter, we briefly discuss who in the Latine population is counted, and how, and why this leads to flawed data collection on Latine health and healthcare experiences. We summarize the available COVID-19 data and the societal structures contributing to COVID-19 disparities. To support new data collection on this population and speak more broadly to the use of race and ethnicity in health and healthcare data, we provide a framework for potential ways of asking about the Latine population and conclude with future directions.

Latine Community: Who "Counts" and Why?

We begin our exploration of data collection with the foundational practice of defining the population. According to the US construction of Latines, "Hispanic" or "Latino" refers to a geopolitical grouping of "Cuban, Mexican, Puerto Rican, South or Central American, or other Spanish culture or origin *regardless of race*" (Census, 2021).

The US Census is important framing in the discussion of public health because socially constructed ethnoracial groups in the United States have been legitimized by the Census since the late 1700s (Zuberi, 2001). The Census currently uses national origin (or place of ancestral migration), tribal affiliation and membership, and physical characteristics (i.e., skin color) to categorize race. The model of counting individuals for population estimates based on demographic criteria can change over time and place due to sociopolitical contexts. One example is how nations can change in size and territory, which can lead to differences in national identity. Yet, how public health conducts population-level and community-level studies is closely tied to this Census operationalization, even if it does not reflect the complexities of race and ethnicity. Instead, the questions posed in the Census are, "Is this person of

Hispanic, Latino, or Spanish origin?" and "What is this person's race?" (US Census, 2010, 2011). These questions are not always simple to respond to for Latines. The complexities of each question are explored briefly below.

First, the "ethnicity" question ties the nation-state as a descriptor for the ethnic group. This assumption of ethnicity as rooted in nation-states extends the settler-colonial project of Latin American countries and flattens the Indigenous identities within these countries (Cardenas, 2018). According to US Census records, the Latine "ethnic" group, makes up approximately 18.5% of the population; however, the question remains if current conceptualization and enumeration is reflective of who the purported Latine populations are.

To address this issue, we explore race as the next domain. Clara Rodriguez (2000) noted that at the time, the conceptualization of race for the typical Latine person was the equivalent to "la raza", which is a continuum and reflective of racial mixture, thereby having no categorically distinct races. Rodriguez argues that it is this disconnect between *raza* as operationalized in Latin America and *race* as operationalized in the United States that causes this confusion. For example, in the 1990 Census, 43.5% of the self-identified "Hispanic" population elected "other race" when asked to fill in their self-identified race. This was similar in 1980 (~40%), 42.2% in 2000, and 36.7% in 2010 (US Census, 2001). Note that "other race" often came in second to self-identified "white" race for the Latine population, which increased over this same timeframe. This does provoke the notion that while Latines hypothetically ascribe to a non-racial ideal, they will still endorse whiteness over other ethnoracial options. This perceived confusion about Latine ethnoracial positioning appears often during these Census reports, with one 2010 Census headline printing that over 75% of Puerto Ricans on the Island of Puerto Rico checked "white" for race, leading to think pieces of Latines, or in this case Puerto Ricans, understanding whiteness differently (US Census, 2010). Whereas results from the 2020 Census – in which only 17% of Puerto Ricans on the island endorsed white – only resulted in more dialogue about Puerto Ricans' self-identity and their racialized experiences (Jensen et al., 2021; Ortiz-Blanes, 2021). However, although endorsement of white-only labels has shifted over time, what was evident in the data was the consistent action of respondents positioning themselves away from endorsing Black or Indigenous self-identities, which are supposedly part of Latine culture (López Oro, 2021).

These examples can be interpreted as forms of racialized, geographic, and contextual-specific fluidities in race, and these also demonstrate the challenges of relying on self-identified race questions when the aim is not enumeration but to address structural determinants of health. Further, the argument of Latines not having a set race and being fluid is perpetuated in public health, which has stunted the critical data collection of ethnoracial-focused demographics.

The questions "what is the person's race?" or the alternative, "what is your race?" can be interpreted varyingly. Wendy Roth (2016) summarized the multidimensionality of measuring the concept of "race" as including racial identity, self-classification, observed race, reflected race, phenotype, and racial ancestry. Most public health studies operationalize race and ethnicity questions only as self-classification questions separated from time, the state, and power. However, according to Charles Mills (1997), what self-classification and similar types of questions get at are how the individual places themselves within the state-constructed racialized hierarchy. This hierarchy in the United States and across the Americas designates white as a class privileged with full personhood and non-white as individuals without full personhood. Thus, according to US Census patterns from the 1980s to today, when Latines across geographies select "white" or "other" as their self-classified race in a Census they are consenting to their own position in the racial order of being white or approximating whiteness rather than ascribe to Blackness or Indigeneity given the sociopolitical implications of aligning with these racialized groups.

Considering the social, political, and historical realities of the United States, the question of who are Latines is dependent on time, geography, and how we consider "race" in both law and practice. The current operationalization of Latine using standardized surveys attempts to simplify and flatten Latine populations as a homogenous group based on geography and language(s). However, public health and social science more broadly will continue to miss the nuances within this group particularly when addressing structural factors that impact health and well-being if we continue to aim to simplify demographic questions without interrogating context. The COVID-19 pandemic demonstrated issues with data collection and with the categorization of ethnoracial groups. Particularly for Latines living across the United States, the COVID-19 data did not highlight the full story of who was affected and how.

COVID-19 and Data on Latine Population

As we consider these critical perspectives of the data we collect regarding Latine populations, how do we understand their experiences of COVID-19? Data from the Center for Disease Control and Prevention (CDC) show trends that the pandemic differentially affected racial and ethnic groups. At the peak of reported cases in January 2021, the highest number of incident cases per 100,000 of the population was among reported "Hispanic" individuals (CDC, 2021). However, it is unclear if the reports were self-reported and/or observer-reported race and ethnicity. At the start of the year, approximately 465.7 cases per 100,000 people were reported among "Hispanics." This was closely followed by 418.5 per 100,000 cases among American Indigenous groups. However, there is an issue with the percentage of cases that reported race in general. The proportion of the data linked to reported racial or ethnic groups were 62.42%, indicating the potential for error or bias, in the case estimates. These data are also in flux, as case numbers are updated with reports from various regions in the United States.

As of November 2020, Latines accounted for one-third of all confirmed COVID-19 cases, where race/ethnicity data are collected (Rodriguez-Diaz et al., 2020). Latines account for one in five confirmed COVID-19-related deaths

(Rodriguez-Diaz et al., 2020). A study of national transmission rates of Latines in the United States found that COVID-19 diagnoses were greater in counties that had a significantly greater share of Latines than the national average (>17.8%) compared to counties with fewer proportions of Latines. Greater cases of COVID-19 cases and deaths were in the Midwestern Latine counties, and greater cases of COVID-19 in the Northeastern counties. County level data were collected from the US Census Bureau ACS and COVID-19 cases and deaths were from USAFacts (Rodriguez-Diaz et al., 2020). Rodriguez-Diaz et al. (2020), a scoping review published in 2021, evaluated 37 articles that focused on the impact of the COVID-19 pandemic on the living conditions of Latines in the United States. The articles were specifically evaluated using social determinants of health framework. Key risk factors were working an "essential" job, living in an area with a high population density of Latine and Black people, overcrowded living conditions, limited English proficiency, and barriers to execute recommended preventive behaviors.

Beyond actual exposure to COVID-19, and in relation to essential work, Latines represent a disproportionate number of workers in the food industry sector, and in service areas such as construction and housekeeping (Hawkins, 2020). These and similar jobs that could not be transitioned to a "work-from-home" environment, were affected earlier on in the pandemic when physical distancing measures were in place and organizations/companies had to lay off or furlough employees, causing extreme financial strain particularly for financially precarious individuals and families (Salgado de Snyder et al., 2021).

Studies out of New York City (NYC) serve as examples of the use of data to demonstrate health disparities by race and geography. A study by Renelus et al. (2021) examined racial disparities in COVID-19 related hospitalizations and in-hospital mortality. The authors used electronic health record (EHR) data from a single hospital center for hospitalization data and used demographic data from the U.S. Census Bureau and the New York City Department of Health to calculate rates of hospitalization. Ultimately the authors found that Black people were more likely than white people to require hospitalization for COVID-19, and Latine people were more likely to experience in-hospital mortality compared to white people. In another NYC-based study on COVID-19 positivity, hospitalization, and mortality, the study authors used electronic health records from one NYC healthcare system and Census tract data to geocode patients on neighborhood socioeconomic status (Ogedegbe et al., 2020). The results of the study were that Black and Latine patients were more likely to have a positive COVID test than white patients but were less likely to be hospitalized in this healthcare system. Race and ethnicity were the predictor variables, and this study coded self-reported ethnicity as Hispanic regardless of race. If ethnicity was missing, self-reported race was assigned and for patient data with multiple entries for race, the entry with the most information was chosen. Although Ogedegbe et al. (2020) used EHR data and purported to focus on self-reported race and ethnicity data, these data may not have only been self-reported. Rather, these data could include observer-reported data, which could account for differences at different clinical encounters. These data also reference to the paradoxical standardization of Latines having both all races and no race at all.

Lessons learned from these studies in relation to how Latines were affected by COVID include the strict dichotomization of Latines as separate from other ethnoracial groups. Latines are often described as multicultural, which can include multiracial; however, that does not mean that all Latines experience the same racialized experiences, particularly in an environment such as New York City, which is the hub of Black Latine migration from the Caribbean and Central America. Contextualizing COVID-19 data from New York City brings into question why intersectional methods were not used to account for the intersectional identities of a multitude of ethnoracial groups, including Latines. Intersectional methods such as those recommended by Lett et al. (2022) would include incorporating additional data from other sources to contextualize the data and questioning what experiences study participants would have with racism and other biases. Further, the primary question these and similar studies evoke is, do studies that enumerate Latines using one dimension or "race" improve our understanding of health inequities? Or are they simplifying the experiences of Latines to the point of providing incomplete information about health risks and outcomes?

Improving Public Health Data and Experiences of Marginalized Communities

In the discussion of affected race and ethnic groups, the CDC notes there is potential for error in the reported race and ethnicity of COVID-19 cases. In the work of Salgado de Snyder et al. (2021), they describe the only unifying element of "Latinos" is the shared Spanish language. However, we know this ignores the French and Portuguese colonization of Latin American countries. This also entirely erases people from Indigenous nations in Latin America still fighting for their land and resource rights. Still, the geographic-rooted identity of Latines is just as diverse as the languages spoken and racial identities. With 33 nation-states in Latin America, and even more villages, townships, and other unique scales of local identity, there are many ways to diversify and examine the experiences of Latines. Failing to do so leads to gaps in knowledge, coarse details in our data, and an inability to fine-tune health and healthcare access interventions.

McLafferty et al. (2021) present the three P's of data: perspective, purpose, and privacy. They encourage researchers to ask what the biases are embedded in the data creation and analysis, who is served by the data and results collected, and the analyses conducted, and ask how the data, results, and related analyses impact the privacy the data is collected from.

While we already know some information about Latines, we do not know the data quality. Just as the stories this chapter opened with suggested, data collection contains errors affected by many factors. These include the environments the data were collected in, the humans doing the collection, the tools being used to collect the data, and the humans the data are being collected from. Incorporating these factors into the data collection process can help researchers address these biases, and better understand the nuances of the population being studied.

Data collection and analysis requires interrogating our methods and biases as researchers as well (Gieseking, 2018; Graham, 1999). Considering how perspective, purpose, and privacy can impact the data collection, a broad set of questions when designing a study with data collection may include the following:

- What do I want to know?
- What information will help me answer that question?
- Why do I think this information will help me answer my question?
- How do I want to retrieve this information?
- How do I ensure that the data collected will not cause harm?
- Who is the referent group and why?
- Where is the referent group located and how would their placement potentially change their experiences from others?

Looking for information on Latines in the United States based on self-reported characteristics is a flawed measure. Racially, Latines in the United States have modified their racial identification to suit their safety, depending on their geographic placement, social status, and socio-economic structure of the societal context they are located within. Further, when we are asking about Latines' health and wellbeing as researchers, what exactly are we asking? Who is in the cohort we are envisioning? Are they white, Black, or the often-fluctuating concept of "brown"? Are they bilingual, Spanish-speaking, Portuguese-speaking, or speaking another language? What is their immigration status? What is their educational background? The potential answers to these questions alone are numerous. The potential additional questions to include similarly so.

When using data collected by others, where efforts on asking questions around race and ethnicity are limited or constrained, questions to ask may be:

- What other information has been collected and can data from national censuses, local government structures, and other organizations be consulted to contextualize the information? (Lett et al., 2022)
- How do these other sets of information provide details on the groups I am interested in asking about?
- How are these other sets of information able to speak to the structures that impact health and healthcare access for the group I am seeking to learn about?

We cannot always control the quality of the data available for our research. We can use more nuanced and critical frameworks when evaluating that data. For example, many Latin American countries do not collect data for Black or "Afrodescendent" people in their censuses. However, shifting discourse has created opportunities for data collection incorporating Black people and their experiences. By 2010, 16 countries in Latin America *had* collected this information at one point in time. Using data compiled critically for the United Nations Economic Commission for Latin America and the Caribbean "Social Panorama of Latin America 2016" Report,[2] this map of Afrodescendent people in Latin America was created (ECLAC, 2017)

[2]Population data is compiled from the censuses of only 16 Latin American countries spanning several years, with some population numbers estimated based on population growth. For more details, please see the Social Panorama of Latin America 2016, by the ECLAC, on page 222.

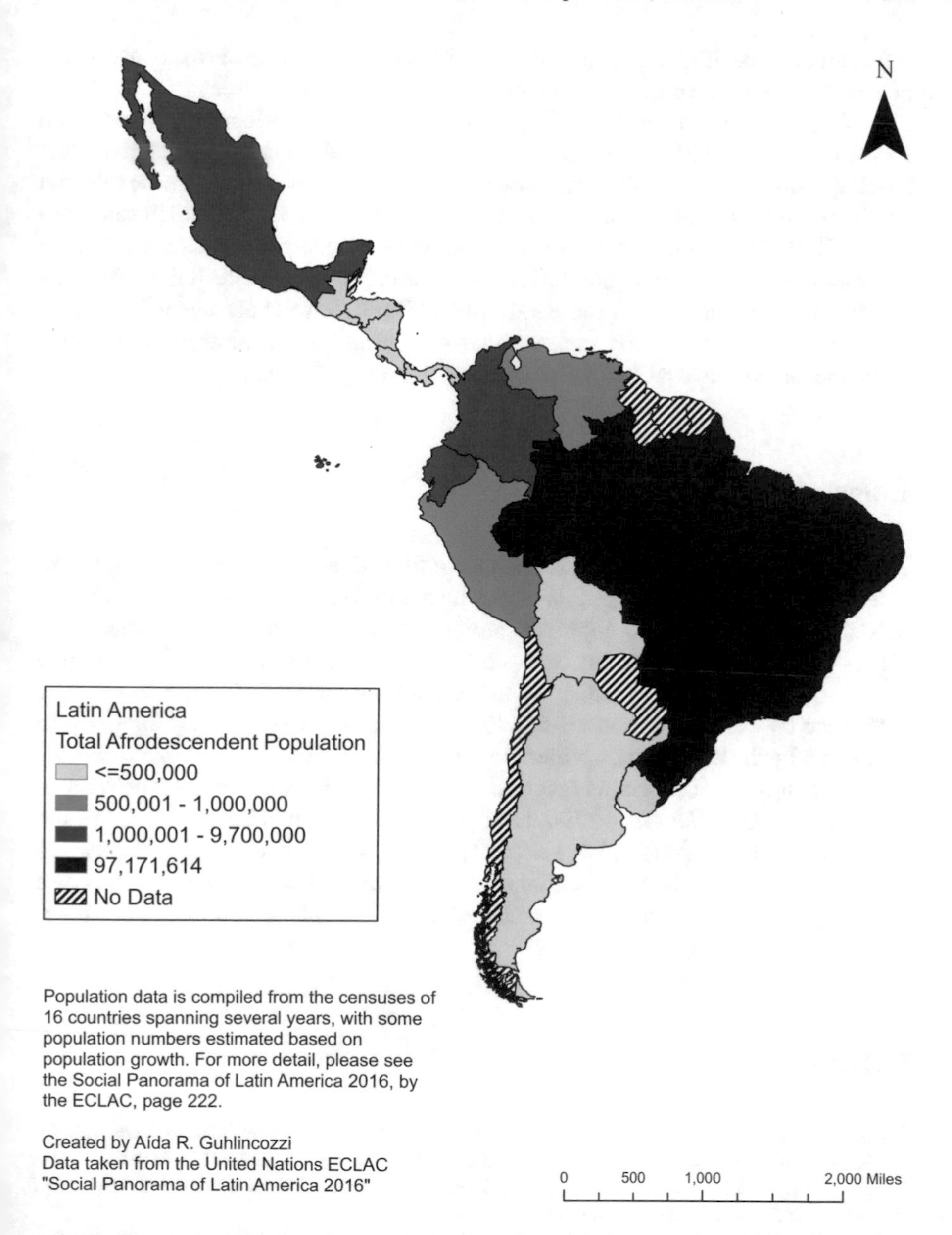

Fig. 9.1 Latin American countries by total population identified as Afrodescent

(Fig. 9.1). While limited in data, and ranging in timeframe, providing details of potential errors allows researchers to make better choices for analysis.

Other data limitations also exist. The Uncertain Geographic Context Problem (UGCoP), first named in 2012, is how geographic units used in analysis (such as census tracts) are not drawn to match the actual, socially-relevant geographic context. This uncertainty in the geographic context can lead to less accurate findings on environmental health effects, such as assuming good air quality because of where a person lives rather than where they work and spend most of their day (Kwan, 2012). The Neighborhood Effect Averaging Problem (NEAP) considers health research focusing on the residential neighborhood of the population under study, given that mobility can mean a significant amount of time is spent outside the residential neighborhood and thus have a significant impact on health effects from the environment (Kwan, 2018). Geographic research seeks more fine-grained data and collection methods for these problems and more attentively addresses the unique contexts and impacts on health given these variations. Similar efforts are needed in health research on Latines, and the varied impacts on health and healthcare access systemic racism can have on Black, Indigenous, and white Latines.

Conclusion

Between 2019 and 2020, there has been a significant push towards applying health equity in public health research. Whether or not this shift is short-lived or leads to needed change in how we approach public health, one major issue remains the same—our data collection tools have stayed the same since the late 1700s, despite technological advances. Whether we are examining health disparities to record a difference between groups or striving to shift towards a health equity approach to examine why the differences are there in the first place, we must interrogate how we are collecting data. If collecting race and ethnicity data has not significantly changed since the first U.S. Census in 1790, how do we work towards improving the health and quality of life of people? For Latines specifically, continuing to place Latines as a homogenous group linked to an imagined, singular cultural lived experience is a disservice to the diverse and rich narratives that individuals in different Latine communities bring with them.

References

Asad, A. L., & Clair, M. (2018). Racialized legal status as a social determinant of health. *Social Science and Medicine, 199*, 19–28. https://doi.org/10.1016/j.socscimed.2017.03.010

Cardenas, M. E. (2018). *Constituting central American-Americans: Transnational identities and the politics of dislocation*. Rutgers University Press.

CDC. (2021). *COVID data tracker*. https://covid.cdc.gov/covid-data-tracker/#datatracker-home. December 13, 2021.

Economic Commission for Latin America and the Caribbean (ECLAC). (2017). *Social panorama of latin america, 2016 (LC/PUB.2017/12-P)*. Santiago.

Gibbons, A. (2018). *City of segregation: 100 years of struggle for housing in Los Angeles*. Verso Books.

Gieseking, J. J. (2018). Focus: critical data, critical technology. *The Professional Geographer, 70*(1), 150–156.

Graham, E. (1999). Breaking out: The opportunities and challenges of multi-method research in population geography. *The Professional Geographer, 51*(1), 76–89. https://doi.org/10.1111/0033-0124.00147

Hawkins, D. (2020). Differential occupational risk for COVID-19 and other infection exposure according to race and ethnicity. *American Journal of Industrial Medicine, 63*(9), 817–820.

Jensen, E., Jones, N., Rabe, M., Pratt, B., Medina, L., Orozco, K., & Spell, L. (2021). The chance that two people chosen at random are of different race or ethnicity groups has increased since 2010. US Census. https://www.census.gov/library/stories/2021/08/2020-united-states-population-more-racially-ethnically-diverse-than-2010.html

Joseph, T. D. (2017). Falling through the coverage cracks: How documentation status minimizes immigrants' access to health care. *Journal of Health Politics, Policy and Law, 42*(5), 961–984. https://doi.org/10.1215/03616878-3940495

Kwan, M.-P. (2012). The uncertain geographic context problem. *Annals of the Association of American Geographers, 102*(5), 958–968.

Kwan, M. P. (2018). The neighborhood effect averaging problem (NEAP): An elusive confounder of the neighborhood effect. *International Journal of Environmental Research and Public Health, 15*(9), 1841. https://doi.org/10.3390/ijerph15091841

Lett, E., Asabor, E., Beltrán, S., Cannon, A. M., & Arah, O. A. (2022). Conceptualizing, contextualizing, and operationalizing race in quantitative health sciences research. *Annals of Family Medicine, 20*(2), 157–163. https://doi.org/10.1370/afm.2792.Annals

Márquez, C. (2019). Juan crow and the erasure of blackness in the Latina/o South. *Labor, 16*(3), 79–85.

McLafferty, S. L., Guhlincozzi, A., & Winata, F. (2021). Counting COVID: Quantitative geographical approaches to COVID-19. In *COVID-19 and similar futures* (pp. 409–416). Springer. https://doi.org/10.1007/978-3-030-70179-6_54

Mills, C. (1997). *The racial contract*. Cornell University Press.

Ogedegbe, G., Ravenell, J., Adhikari, S., Butler, M., Cook, T., Francois, F., et al. (2020). Assessment of racial/ethnic disparities in hospitalization and mortality in patients with COVID-19 in New York City. *JAMA Network Open, 3*(12), e2026881.

Oro, P. J. (2021). Refashioning afro-latinidad: Garifuna new yorkers in diaspora. In A. Y. Ramos-Zayas & M. M. Rúa (Eds.), *Critical diálogos in latina and latino studies*. New York University Press.

Ortiz-Blanes, S. (2021). *In 2000 most Puerto Ricans identified as white. In 2020, few did. Why that number dropped.* Miami Herald. https://www.miamiherald.com/news/nation-world/world/americas/article254982442.html

Parmet, W. E., Burris, S., Gable, L., de Guia, S., Levin, D. E., & Terry, N. P. (2021). COVID-19: The promise and failure of law in an inequitable nation. *American Journal of Public Health, 111*(1), 47–49.

Renelus, B. D., Khoury, N. C., Chandrasekaran, K., et al. (2021). Racial disparities in COVID-19 hospitalization and in-hospital mortality at the height of the New York City pandemic. *Journal of Racial and Ethnic Health Disparities, 8*, 1161–1167. https://doi.org/10.1007/s40615-020-00872-x

Rodriguez, C. E. (2000). *Changing race: Latinos, the census, and the history of ethnicity in the United States* (Vol. 41). NYU Press.

Rodriguez-Diaz, C. E., Guilamo-Ramos, V., Mena, L., Hall, E., Honermann, B., Crowley, J. S., Baral, S., Prado, G. J., Marzan-Rodriguez, M., Beyrer, C., Sullivan, P. S., & Millett, G. A. (2020). Risk for COVID-19 infection and death among Latinos in the United States: Examining heterogeneity in transmission dynamics. *Annals of Epidemiology, 52*(2020), 46–53, e2. https://doi.org/10.1016/j.annepidem.2020.07.007

Roth, W. (2016). The multiple dimensions of race. *Ethnic and Racial Studies, 39*(8), 1310–1338. https://doi.org/10.1080/01419870.2016.1140793

Salgado de Snyder, V. N., McDaniel, M., Padilla, A. M., & Parra-Medina, D. (2021). Impact of COVID-19 on Latinos: A social determinants of health model and scoping review of the literature. *Hispanic Journal of Behavioral Sciences, 43*(3), 174–203. https://doi.org/10.1177/07399863211041214

US Census. (2010). *2010 census puerto rico profile*. Retrieved from: https://www2.census.gov/geo/maps/dc10_thematic/2010_Profile/2010_Profile_Map_Puerto_Rico.pdf

US Census Bureau. (2001). *Overview of race and hispanic origin – Census 2000 brief. Table 10. Hispanic and not hispanic population by race for the United States: 2000*, 11 pages. Retrieved from: www.census.gov/prod/2001pubs.cenbr01-1.pdf

US Census Bureau. (2011). *Overview of race and hispanic origin. 2010–2010 census briefs. Table 2. Population by hispanic of latino origin and race for the United Status: 2010*, 24 pages. Retrieved from: www.census.gov/content/dam/Census/library/publications/2011/dec/c2010br-02.pdf

US Census Bureau. (August 12, 2021). Retrieved from: https://www.census.gov/library/stories/2021/08/2020-united-states-population-more-racially-ethnically-diverse-than-2010.html

Zuberi, T. (2001). *Thicker than blood: How racial statistics lie*. University of Minnesota Press.

Chapter 10
An Overview of the Impact of COVID-19 on Nepal's International Tourism Industry

Asmod Karki, Nama Raj Budhathoki, and Deepak Raj Joshi

Personal Story

May 2021 was one of the worst months in Nepal from the COVID perspective. There was an unprecedented stress on the country's limited health facilities: hospital beds were completely occupied, intensive care units (ICUs) were full of patients, ventilators were out of stock, and oxygen supply was in severe shortage. Despite all my efforts to limit mobility and keep social distance, I got COVID positive on the second of May 2021. Naturally, I was anxious. I was also equally fearful about the possibility of infection to other members of my family. Fortunately, all of them tested negative. That gave me a huge relief. However, I still needed to spend 14 days in isolation before I could spend time with the family. I used to get a bit happy every morning as I could see people walking in my neighborhood and even hear them talking to each other from the window of my room. Passing time was relatively easier during the day. But I used to get worried when the day used to begin to convert to evening and then night. A couple of questions used to frequently come and disturb my sleep: What if my situation deteriorates in the night? How will I get to the hospital? Which hospital should I go to get medical care? I knew these questions were equally valid during the day as well, but somehow, I found them more worrisome during the night. I felt nights were longer than they usually used to be. Although I defeated the virus without any medical support, those 2 weeks were strange times. *Nama Budhathoki*

A. Karki
Kathmandu Living Labs, Chundevi, Nepal

N. R. Budhathoki (✉)
Kathmandu Living Labs, Chundevi, Nepal

Humanitarian OpenStreetMap Team (HOT), Washington DC, USA
e-mail: nama.budhathoki@kathmandulivinglabs.org

D. R. Joshi
World Tourism Network, Honolulu, HI, USA

M. Laituri et al. (eds.), *The Geographies of COVID-19*, Global Perspectives on Health Geography, https://doi.org/10.1007/978-3-031-11775-6_10

Introduction

Tourism is a major contributor to Nepal's economy. In 2019, the sector contributed more than seven percent to Nepal's GDP (WTTC, 2020). In addition to being a significant contributor to the economy, the tourism industry is also one of the major job providers. It is the fourth largest industry by employment of total formal jobs in the country (Prasain, 2021).

The surge of the Western tourists in Nepal began in the late 1960s and 1970s, where Western visitors discovered a pristine land with "natural" beauty (Lal, 2017). To capitalize on the potential of the tourism industry, the government of Nepal launched several campaigns to promote Nepal as an attractive tourist destination. For instance, in 1998, it launched the Visit Nepal 1998 campaign, which aimed to increase the tourist flow and length of stay in Nepal. Almost half a million tourists visited Nepal during the campaign year (Shakya, 2017).

The tourism industry has suffered several setbacks. From 1996 to 2006, Nepal's civil war impacted tourism due to the conflict. As the internal politics of Nepal transitioned from a deadly civil conflict to a peaceful resolution towards the mid-2000s, the tourism industry showed signs of recovery. The World Bank (2022) data shows that the country received 0.8 million international visitors in 2011. Increased tourism continued into the mid-2000s, until the massive earthquake hit Nepal in 2015. The megaquake of 2015 had a devastating impact on the tourism industry of Nepal. The quake severely damaged the cultural center of Kathmandu, a UNESCO World Heritage site and key tourist attraction.

However, the industry bounced back from these shocks in the succeeding years. Nepal's GDP grew by 6.94 percent in 2016–2017. This was the fastest growth Nepal had seen after the mid-1990s. This impressive growth can be partly attributed to the growth in the hotel and restaurant sector, a large subsector of the tourism industry, as it contributed for more than seven percent of the total GDP growth in both in 2016/17 and in 2017/18.

The upward trajectory of the growth in the tourism industry and in Nepal's economy buoyed the tourism stakeholders, i.e., the businessowners, workers, and the government. As a result, the government, in collaboration with the private sector and other major stakeholders in the tourism industry, announced the Visit Nepal Year (VNY) 2020 campaign in mid-2018. One of the main objectives of the campaign was to attract 2 million tourists to Nepal in the year 2020 (Mali, 2020). The preparations for the VNY 2020 began with the government appointing a prominent businessperson as a Campaign Coordinator to promote the campaign.

COVID-19, Lockdowns, and Cancellation of VNY 2020

Just as the VNY 2020 began and a year full of tourist activities in Nepal was anticipated, COVID-19 started spreading around the world. In the third week of January 2020, Nepal reported its first COVID-19 case. As the rate of infections accelerated in the subsequent months, the government started taking precautionary measures nationwide to curb the rapid rise in infection. Beginning March 2020, the government gradually started imposing travel restrictions. In the first week of March, the government issued a travel advisory against all non-essential travel. By late March, the government cancelled all international flights and Nepal entered its first lockdown on March 24, 2020.

Tourism, which relies on the mobility of people, was severely impacted due to the travel restrictions and the subsequent lockdown. The cancellation of international flights and the lockdown meant that the tourists already in the country had limited mobility and those who wished to arrive in Nepal had to hold their plans. The change in the number of international tourist arrivals between 2020 and 2021 provides a proxy to understand the impact. The year-on-year arrivals of international tourists declined by 34 percent in 2021 compared to 2020 (Department of Immigration, 2021, 2022).

As there were signals of the travel restrictions and lockdown prolonging beyond a few weeks or months, the government decided to cancel the much anticipated and planned VNY 2020 in March 2020 (Nepali Sansar, 2020). This was a major blow to the tourism sector's entrepreneurs and workers, as the businesses had planned and prepared to absorb the increased flow of tourists.

Prolonged Lockdown and Travel Restrictions

The daily reported COVID cases amounted to less than a thousand in the first few months of the lockdown. This could be either due to the low rate of infection due to the complete lockdown imposed all over the country or it could have been due to low rates of testing during the period. In fact, until July 2020, the average daily reported cases hovered below thousand per day (Fig. 10.1).

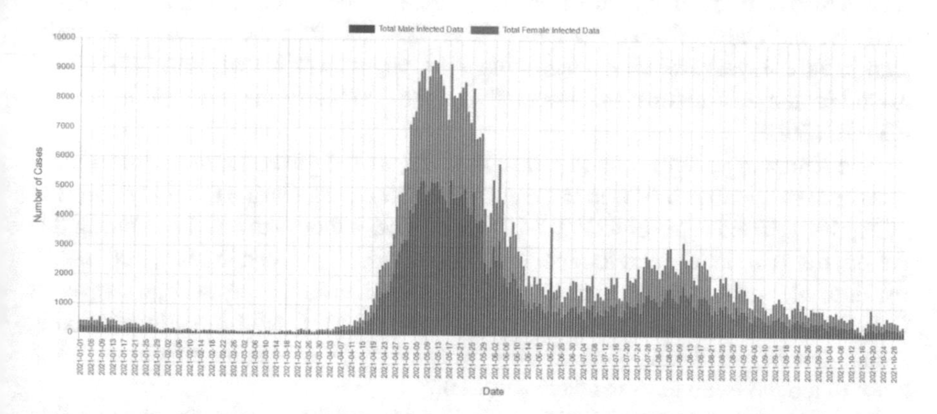

Fig. 10.1 Daily reported cases of COVID-19 in Nepal between January 23 and December 31, 2020. (Source: Covid-19 Dashboard, 2021)

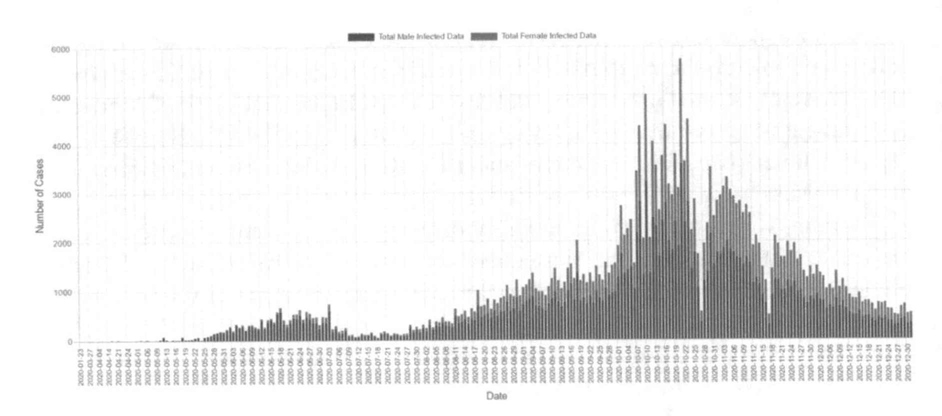

Fig. 10.2 Daily reported cases of COVID-19 in Nepal between January 1 and October 31, 2021. (Source: Covid-19 Dashboard, 2021)

The lockdown checked the reported rate of infection while at the same time largely halting economic activities in the country. This lockdown had an overarching impact on the overall economy, not only the tourism industry. As concerns over the state of the economy grew and the reported rate of infection did not show signs of acceleration, the government ended the first country-wide lockdown on July 21, 2020—almost 4 months after it was first imposed (Nepali Times, 2020).

When the lockdown ended, travel restrictions were still in place. Recreational places such as banquet halls, gymnasiums, and exhibition sites were still barred from opening. While hotels and restaurants were open, restrictions were imposed on their opening hours. The government announced that chartered international flights would resume from mid-August 2020. However, the government put a cap on the total number of travelers to 800 per day as travel restrictions were still in place in several countries around the world (CCMC, 2020).

As mobility increased and economic activities resumed, the infection rate also started to rise. In 2020, the infection rate started to accelerate in August, peaked in October, and started decreasing from November onwards. As a result, the initial plan to ease travel restrictions beginning mid-August was halted. The government extended the restrictions till September 2020 to curb the rise in infections (CCMC, 2020).

The situation in 2021 was like that in 2020, as the government-imposed lockdown and restrictions followed the rise of infection rates. During the first quarter of 2021, the reported cases of COVID-19 were below a thousand per day. During this period, the government eased travel restrictions, which resulted in increased mobility and high-risk activity for COVID-19 transmission. The second wave of COVID-19 hit Nepal during the second quarter of 2021, with infections peaking around late May and early June. During this period, the total daily infection peaked close to ten thousand on some days (Fig. 10.2).

The government-imposed lockdown mirroring the 2020 response, as the infections peaked in mid-2021. The authorities announced a week-long lockdown beginning April 29, 2021. What began as a week-long lockdown extended for months until August 2021 (Business Standard India, 2021). The prolonged lockdowns and restrictions had a devastating impact on the overall economy. The tourism industry was also severely impacted as the industry was heavily reliant on the mobility of international tourists.

Impact on the Tourism Industry

While a detailed study analyzing the impact of COVID-19 on Nepal's tourism industry has yet to be carried out, the next two sections will consolidate the *economic and non-economic impact to provide a holistic understanding* of the secondary impact of COVID-19 on Nepal's tourism industry.

Economic Impact

The overall economic impact of COVID-19 on Nepal's economy and on the tourism industry by extension is difficult to assess as the majority of the country's workforce is involved in the informal economy (ILO, 2021), and the data for the informal sectors are either non-existent or limited. The available studies and estimates, therefore, should mostly be seen as proxies. Shivakoti (2021) has penned one of the few studies that strives to estimate such impact. In the study, she estimates that the tourism sector was likely to lose "85 thousand direct jobs and 36 thousand indirect jobs (p. 9)." In the same paper, she references estimates from several government bodies. She notes that the Nepal Rastra Bank, the central regulatory bank in Nepal, estimated that the expected tourist arrival in Nepal to fall by 31.6 percent.

In 2019, pre-pandemic year, Nepal received 1.2 million IVAs (international visitor arrivals) and USD 48.00 per day per tourist expenditure. The average length of stay was approximately 13 days. In 2020, Nepal experienced 81% decrease in IVAs (Ministry of Culture, Tourism and Civil Aviation, 2020).

Tourism has employed more than 1 million people (directly and indirectly) in Nepal, which is nearly 7% of total jobs (WTTC, 2020). It signals the magnitude of impact on tourism created by COVID-19 pandemic. While the exact numbers are difficult to estimate, such figures could be taken as a proxy for the devastating secondary impact of COVID-19. A study conducted by Kathmandu Living Labs (KLL) in 2021 also presents a similar picture. KLL—with support of the State Department and American Geographers Association—conducted a survey of tourism businesses and workers to assess the secondary impact of COVID-19 (C2M2 Kathmandu Portal, 2021).[1] The survey was distributed through a network of tourism partners

[1] The portal can be accessed here: http://tourismincovid.klldev.org/

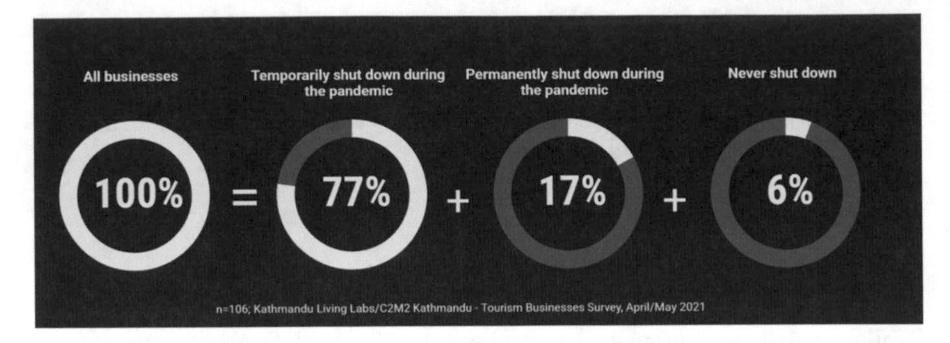

Fig. 10.3 Survey findings from the Kathmandu Living Labs (KLL). (Source: Kathmandu Living Labs, 2021)

such as the Hotel Association of Nepal (HAN), Thamel Tourism Development Council (TTDC), Joint Tourism Trade Union Forum (JOTUFF), Union of Trekking Travels Rafting Workers Nepal (UNITRAV), and Trekking Agencies' Association of Nepal (TAN).[2]

The survey, which included more than a hundred businesses and 2 hundred workers, found that more than nine out of ten tourism businesses shut down during the pandemic (Fig. 10.3). Similarly, more than three out of five businesses had their revenue completely stop during the pandemic. As a result of the reduced revenue, four out of five businesses could not cover their operating costs.

Such figures, if not already alarming in themselves, look worrisome when combined with the economic impact of COVID-19 on tourism sector workers. Most of the workers faced challenges sustaining their livelihoods. For instance, around four out of five workers surveyed in the KLL's study mentioned they had difficulty in paying rents. Similarly, more than ninety percent of the tourisms sector workers mentioned they had outstanding debts (C2M2 Kathmandu Portal, 2021).

Non-economic Impact

Besides the direct economic impact, there are also several non-economic impacts of COVID-19 on Nepal's tourism industry. However, such non-economic impact is difficult to assess given the relative lack of data measuring such impact. The studies have aimed to capture some of the non-economic impact, which provide us a glimpse of the wider effect. For instance, Shivakoti (2021) notes that issues related to mental health such as stress and fear due to financial insecurity had also risen during the period. Similar findings can be observed in the KLL's study. The study sought to assess the psychological impact due to COVID-19. Ninety-five percent of

[2] The business and workforce survey questionnaires can be downloaded from the landing page of http://tourismincovid.klldev.org/

the respondents mentioned that the pandemic has had some form of psychological effect. Similarly, eighty-seven percent of the tourism workforce in the survey mentioned that they had sought psychological support to deal with the pandemic. Given the stigma surrounding mental health in Nepali society, such high self-reported figures of people seeking mental health support are noteworthy.

Expectations from the Government

The response to the severe impact of COVID-19 on Nepal's tourism has been multifold. Among the many stakeholders in the tourism sector, such as media, business owners, workforce, tourists, and think-tanks, this section focuses on and presents expectations of two major stakeholders, i.e., business owners and workforce. Businesses and the workforce are the ones most impacted by COVID-19. Business owners' expectations are included with the assumption that it will provide an organizational perspective on the impact of COVID-19 as business owners' interests lie in sustaining business ventures. Similarly, including the expectations from the workforce provides an understanding of the expectations from a large segment of the tourism sector service providers, many of whom are employed as workers. Understanding the expectations of businesses and workforce provides useful insights for reviving the sector.

Businesses

The tourism sector's businesses were largely shut during the lockdown period in 2020 and 2021, following the government's orders. During the lockdown period, many tourism businesses were shut down as their sources of revenue were completely or partially halted. This dire situation among businesses engendered expectations from the government. While there are no formal studies consolidating the response from the businesses, available evidence from interactions with the business owners and media reports suggest that they had significant expectations from the government. Initially, their expectations were related to receiving relief packages for businesses and workers. Over the course of time, they demanded that the government make provisions including exemption of taxes and fines (Shrestha, 2020).

Not all the demands pertained to economic relief. Some of the demands from businesses related to easing the flow of tourists. They demanded that the government make Polymerase Chain Reaction (PCR) testing available in the airport. Similarly, they also demanded the government to implement a "visa on-arrival" system that would ease international tourists' arrival in Nepal (RSS, 2021).

Workforce

As the businesses were largely shut, the livelihood of the workforce was also severely impacted. While businesses largely looked to the government to come up with relief measures and regulatory responses to help alleviate the impact, the response from the workforce was diverse. Like businesses, the workforce, through labor associations, demanded that relief packages be provided to withstand the economic impact of COVID-19.[3] It was also interesting to note that the tourism workers diversified their livelihood sources as their primary income through tourism activities halted. Informal conversations with labor union leaders suggest that many formal and informal workers adapted secondary professions such as ridesharing, investing in the stock market, and agriculture. These findings, however, should be further corroborated through larger studies as it is still unknown how the workforce coped with a significant economic and psychological shock due to COVID-19 imposed lockdown and travel restrictions.

Government's Response

The Nepal government's response was largely reactive in response to the secondary impact or economic impacts of COVID-19. Initially, the focus of the government was on mitigating the first-order impact, i.e., curbing the infection rate and providing health services to those affected by COVID-19. The government used lockdown as the primary means to restrict people's movement to minimize the infection rate.

Such a unidimensional response was, however, inadequate to address the complexity of challenges brought forth by COVID-19. While the restrictions and lockdown helped curb infection rate temporarily, such measures were not without significant economic and social costs in a developing country such as Nepal, where around 5 million people (around 18 percent of the total population) are still under multidimensional poverty (UNDP, 2021). In the tourism industry, the government lost a significant tax income (equivalent to Rs. 34 billion between March and June 2020), businesses lost significant revenues, and a large majority of the workers had difficulty sustaining livelihoods (Ministry of Culture, Tourism and Civil Aviation, 2020).

To address the secondary impact of COVID-19, the government introduced a variety of policy measures. One of the significant measures was the devolution of authority to impose lockdown. In late 2020, the federal government devolved the authority to impose lockdown to local governments. For the tourism industry, this meant that authorities in major tourism clusters (i.e., Kathmandu, Pokhara, and Chitwan) could independently decide on policy and program measures. Similarly, various government agencies and offices introduced safety measures to ensure the

[3] Based on conversations with labor union leaders on January 10, 2021.

gradual reopening of the tourism industry. The Ministry of Tourism issued guidelines to be adapted across all tourism sectors (Ulak, 2021). Similarly, the Ministry of Foreign Affairs also issued guidelines "affecting the control and preventive measures" against the "new variant" of COVID-19 (Ministry of Foreign Affairs, 2021).

One of the most significant policy responses from the government was introduced in the budget for the fiscal year 2078/79 (2021/22). The budget has a range of provisions to help businesses and the workforce recover from the COVID-19-related shock. For instance, the budget has allocated a 50 billion rupees fund as a stimulus package for Nepali businesses, including the tourism businesses (NRB, 2021). Similarly, the government has sought to promote domestic tourism through the budget, considering the scenario that international travel restrictions might continue in tourism source countries. As part of the domestic tourism promotion campaign, the government has allocated a 10-day paid leave to all government employees (Online Khabar, 2021).

The budget has also allocated a host of resources targeting the workforce's resilience and recovery. The Prime Minister's Employment Program, with a budget of NPR 12 billion, is one such program that is intended to absorb the unemployed and semi-employed workforce into the workforce. Similarly, the government has allocated NPR 1.4 billion to upskill and reskill the workers so that they can diversify and upgrade their skillsets in the event of another economic shock.

While the government's responses might not have been as holistic and as targeted as demanded by the tourism sector's businesses and workers, the government has introduced a series of policies and programs targeting businesses and workers to mitigate the secondary impact of COVID-19.

Recommendations

The second and third waves of COVID-19 and the spread of new variants such as Omnicorn continue to create threats both for lives and livelihoods. In this context, the government should give special focus to the survival and revival of the tourism industry. Some key recommendations are as follows:

- The implications of crises increasingly require the government to engage with the private sector to improve crisis preparedness, management, and recovery plans.
- Crisis preparedness must focus on building trust-based coalitions, assessing readiness, and developing emergency action plans as well as enhancing education.
- The effective management of a crisis requires the rapid activation of emergency plans as well as quick, accurate, and transparent communication.
- The government should ensure a speedy recovery with full transparency. Helping the sector regain its confidence and preparing it to attract more of the international visitors to Nepal are crucial.

- The government should allocate financial resources for technical assistance and revival of the Tourism Recovery Task Force, including supporting the team to gain alternative sources of revenue and using social media marketing and communication tools for the tourism sector.
- Support the implementation of several potential pipeline ecotourism and infrastructure projects such as the Khaptad Tourism Project and Ski Resort Project in Mustang to provide employment opportunities for tourism workers in destination areas and aid post-pandemic market recovery.[4]
- Support the sustainable development of nature-based, ecotourism or adventure-related tourism destinations.
- Support local governments' upgrading of ecotourism facilities, management of control of COVID-19 transmission risks, and destination development.

Conclusion

The secondary impact of COVID-19 on Nepal's economy, particularly in the tourism sector, was significant given the contribution of the industry to Nepal's overall GDP. The tourism businesses were severely impacted due to travel lockdowns and restrictions in Nepal and in major tourist source countries such as China, India, and the United States. This resulted in a large percentage of businesses losing their revenues with many of them permanently closed. The workers were also highly vulnerable to the secondary impact. Many of the workers resorted to taking loans and are in debt to sustain their livelihoods. The psychological impact due to lockdown induced social seclusion and financial insecurity.

Both the businesses and the workers had significant expectations from the government for relief measures. As a response, the government of Nepal has introduced a series of policy measures to address the demands of businesses and the workforce. However, introducing the measures in themselves might not be adequate. What remains to be seen is how those policies and programs are implemented to address the needs of the stakeholders, such as the business owners and workers who are impacted the most. The resilience of the people of Nepal to withstand major socio-economic shocks has been demonstrated repeatedly in the last couple of decades, e.g., concluding a decade old civil war in 2005/06 and recovering after the 2015 mega-quake. We need to wait for some time to see how Nepal's tourism sector fares this crisis. If the sector revives, it can provide one more example for Nepal's resilience history. In addition, it would be valuable to anyone interested in learning how a developing country responds to the secondary impact of COVID-19.

[4] A long list of major tourism projects could be found in the Investment Board Nepal's site: https://ibn.gov.np/project-bank/

Acknowledgements This book chapter would not have been possible without the exceptional work by a number of people at Kathmandu Living Labs (KLL). We are grateful, in particular, to:

1. Sazal Sthapit
2. Arogya Koirala
3. Aishworya Shrestha
4. Roshan Poudel
5. Manoj Thapa

Our special thanks to Dr. Melinda Laituri for her constant encouragement to write this chapter as well as for her constructive feedback on our earlier draft. This research was possible thanks to the Cities' COVID Mitigation Mapping (C2M2) Program, developed by the U.S. Department of State's Humanitarian Information Unit and the American Association of Geographers (AAG).

References

Business Standard India. (2021, August 3). *Covid-19 lockdown extended in kathmandu valley.* Retrieved from https://www.business-standard.com/article/international/nepal-covid-19-lockdown-extended-in-kathmandu-valley-till-aug-11-121080301817_1.html

C2M2 Kathmandu Portal. (2021). *COVID-19 and its impacts on nepalese tourism.* Retrieved from http://tourismincovid.klldev.org/

CCMC. (2020). *CCMC's 11th board meeting.* Crisis Management Coordination Center. Retrieved from https://ccmc.gov.np/key_decisions/key%20decision%202077.05.29.pdf

COVID19-Dashboard. (2021). Retrieved from https://covid19.mohp.gov.np/

Department of Immigration. (2021). *Arrival departure data from January to December 2020.* Retrieved from https://www.immigration.gov.np/public/upload/e66443e81e8cc9c4fa5c099a1fb1bb87/files/Data_jan_Dec_2020(1).pdf

Department of Immigration. (2022). *Arrival and departure record of 2021.* Retrieved from https://www.immigration.gov.np/page/arrival-departure-report?page=1

ILO. (2021). *Informal economy in Nepal.* Retrieved from https://www.ilo.org/kathmandu/areasofwork/informal-economy/lang%2D%2Den/index.htm

Lal, A. (2017, August 28). From the shangri-la to a hippie paradise. *The Record.* Retrieved from https://www.recordnepal.com/from-the-shangri-la-to-a-hippie-paradise

Mali, D. S. (2020, January 7). Making visit Nepal 2020 a success. *The Himalayan Times.* Retrieved from https://thehimalayantimes.com/opinion/making-visit-nepal-2020-a-success

Ministry of Culture, Tourism and Civil Aviation. (2020). *Press Release.* Retrieved from https://www.tourism.gov.np/files/Press%20RELEASE%20FILE%20PDF/Covid_19_WorkProgress_77_3_19.pdf

Ministry of Foreign Affairs. (2021, June 23). *Guidelines to be followed by Nepali and foreign nationals travelling to Nepal.* Retrieved from https://mofa.gov.np/wp-content/uploads/2021/06/Guidelines-to-be-followed-by-Nepali-and-foreign-nationals-travelling-to-Nepal-2021.pdf

Nepali Sansar. (2020). *COVID-19 threat: Nepal government cancels 'visit Nepal 2020.* Retrieved from https://www.nepalisansar.com/tourism/covid-19-threat-nepal-cancels-visit-nepal-2020/

Nepali Times. (2020). *Nepal ends COVID-19 lockdown.* Retrieved from https://www.nepalitimes.com/latest/nepal-ends-covid-19-lockdown/

NRB News. (2021). A publication of central bank of Nepal. *NRB Monetary Policy for 2020–21. 41*(1).

Online Khabar. (2021, December 28). *Government announcement to give staff 10-day travel leave yet to be implemented.* Retrieved from https://english.onlinekhabar.com/travel-leave-not-yet.html

Prasain, S. (2021, June 17). Tourism is Nepal's fourth largest industry by employment study. *The Kathmandu Post*. Retrieved from https://kathmandupost.com/money/2021/06/17/tourism-is-nepal-s-fourth-largest-industry-by-employment-study

RSS. (2021). *Tourism entrepreneurs demand COVID vaccine within 30 days*. Rastriya Samachar Samiti. Retrieved from https://thehimalayantimes.com/nepal/tourism-entrepreneurs-demand-covid-vaccine-within-30-days

Shakya, A. (2017, October 11). A vision for visit Nepal 2020. *New Business Age*. Retrieved from http://www.newbusinessage.com/MagazineArticles/view/1943

Shivakoti, A. (2021). Impact of COVID-19 on tourism in Nepal. *The Gaze: Journal of Tourism and Hospitality, 12*(1), 1–22.

Shrestha, P. M. (2020, December 20). *Covid-19 affected businesses to protest demanding relief and rehabilitation package*. Retrieved from https://kathmandupost.com/money/2020/12/20/covid-19-affected-businesses-to-protest-demanding-relief-and-rehabilitation-package

The World Bank. (2022). *International tourism and number of arrivals*. The World Bank Group. Retrieved from https://data.worldbank.org/indicator/ST.INT.ARVL?locations=NP

Ulak, N. (2021). COVID-19 pandemic and its impact on tourism industry in Nepal. *Journal of Tourism & Adventure, 3*(1), 50–75. https://doi.org/10.3126/jota.v3i1.31356

UNDP. (2021). Nepal multidimensional poverty index 2021. *UNDP in Nepal*. Retrieved from https://www.np.undp.org/content/nepal/en/home/library/poverty/Nepal-MPI-2021.html

WTTC. (2020). *Economic impact report*. World Travel and Tourism Council. Retrieved from https://wttc.org/Research/Economic-Impact

Chapter 11
Data and Dashboards for Measuring the Social Impact of COVID-19 in African Cities

Gaston Mbonglou and Ranjit John

Personal Story

As a consultant, I used to travel to Europe and Africa every 2–3 months. With the onset of COVID-19, everything changed! The impact of the pandemic and news about the casualties across Europe in 2020 forced me to reconsider my travel plans. With my work on the C2M2 project, I had good insights on preventive measures and vaccination in most European and African countries where I normally travel to. Toward Q4 of 2021, I finally decided to travel to Germany and then to Cameroon for long awaited family visits. Of course, I got my vaccination and my booster shot and was ready to go.

On November 19th, I landed in Berlin and was set to go to Dresden on November 21 to visit my family, go shopping and do some sightseeing. Unfortunately, the government in Saxony (where Dresden is located) declared a state of emergency that week to curb the spread of the omicron variant. Implemented measures included mask mandates everywhere, proof of vaccination to go shopping, limited business hours, closing of cultural sites, and more. Everything seemed like the early days of COVID again.

What this really thought me was the fact that this pandemic may be here for a while, and we just need to be Agile while we travel to quickly adjust to local realities. Of course, I did not want to stop my trip in Germany. In fact, I used that experience to better prepare myself for the unexpected while on the way to Cameroon, including making sure I had my vaccination card handy, I saved a digital copy of my vaccination and yellow fever cards, had enough masks, packed few bottles of gels. Welcome to our new world! *Gaston Mbonglou*

G. Mbonglou (✉)
USA and Africa Sourcing and Growth Advisors, King of Prussia, PA, USA
e-mail: gaston.mbonglou@uasgadvisors.com

R. John
Hawkai Data, San Jose, CA, USA

M. Laituri et al. (eds.), *The Geographies of COVID-19*, Global Perspectives on Health Geography, https://doi.org/10.1007/978-3-031-11775-6_11

Introduction

The COVID-19 pandemic has had a profound impact on our world. It has disrupted economies, created new educational and health-care challenges, and affected the livelihoods of millions. The economic, social, and cultural changes induced by the COVID-19 pandemic will have significant long-term effects. The sooner we understand and quantify the impacts, the better placed we are in addressing them in the future. The ability to measure the social impact of COVID-19 is critical to establishing future best practices that are sustainable and that reinforce urban and regional resiliency in managing pandemics, natural disasters, conflicts, and climate-change related events.

One of the challenges in measuring impact is identifying appropriate globally agreed upon social metrics that can be used to track progress (or lack thereof), identify gaps, drive actionable insights, and at the same time provide a standard way to benchmark and compare status across regions and nations. Under the Secondary Cities (2C) project,[1] a field-based initiative of the Office of the Geographer, data was collected to map and track a city's resiliency and emergency preparedness. The focus was on urban jurisdictions (not the capital city in the respective country) that perform vital governance, logistical, and production functions at a subnational or sub-metropolitan region level within a system of cities in a country. The 2C project built partnerships in tier-2 cities to create local geospatial data skills, enhance understanding through data and mapping, and enable data-driven decision-making. The 2C project categorized health and educational facilities in these cities and showed that a facility-based data collection model could be used to create insights into a city's readiness in managing emergencies.

The United Nations Sustainable Development Goals (UN SDG)[2] provides a shared aspirational blueprint for achieving economic prosperity for all nations and a sustainable future for the planet. The goals recognize that ensuring economic prosperity and preserving the environment starts with removing inequities, creating opportunities, and making sure that basic services like food, health, and education are available to all.

This chapter details our experiences with using a facility-based data collection process and supplementing it with census data to do a baseline analysis for several of the UN SDGs in a few cities in Africa – Bukavu, Democratic Republic of Congo; Kericho, Kenya; and Pemba, Mozambique as case studies for the use of analytics and data dashboards during the pandemic for lower- and middle-income countries (LMIC) to demonstrate how geospatial tools, technologies, and applications can be used for decision support. These baseline metrics allow for the tracking of local developmental goals and can also be used to study second-order impacts of pandemics like COVID-19. It summarizes the best practices on how data can be

[1] Secondary Cities (2C), U.S. Department of State, https://secondarycities.state.gov/

[2] https://sdgs.un.org/goals

collected and structured leveraging standards and open-data principles to track, monitor, and create real-time actionable dashboards and community-based services.

Measuring Impact

To measure social impacts, we had to identify data and the appropriate metrics that can track the desired social outcomes and benchmark current status. The UN SDG lists specific targets that should be met by the year 2030. Some of these are macro-level targets that are measured and reported by governments and typically require government-level policies to achieve targets. For UN SDG 3 (Health for All), targets include reducing the global maternal mortality ratio, ending preventable deaths of newborns and children, ending epidemics of AIDS, tuberculosis, malaria, water-borne diseases, and other communicable diseases, etc. Others are micro-level targets that specifically target the individual, one of which is – *Achieve universal health coverage, including financial risk protection, access to quality essential health-care services and access to safe, effective, quality, and affordable medicines and vaccines for all.*[3]

How do we benchmark status and progress (or lack thereof) on these micro-level targets that may require individual or household enumeration? How do we measure the causal impact of COVID-19 towards these targets? Many countries measure the progress towards these goals during their national census. A national census is a considerable and expensive undertaking and is typically done about once a decade. The data for in-between years is extrapolated and may not accurately account for the dissimilar impact of pandemics and other events on different communities and regions. There is a need for more up-to-date local data for policy makers to make timely, data-driven, and well-informed decisions.

One way the SDG 3 target can be measured is by using a statistical sampling of the population during the national census. Another way that the target can be measured is by using a facility-based survey and periodically checking whether the availability of doctors, nurses, beds, medicines, vaccines, etc. meets the needs of the local population (GBD 2015 SDG Collaborators, 2016).

Facility-Based Data

Facility-based data collection involves identifying the specific type of facility of interest, visiting the facility, and collecting data on service availability, service capacity, and service readiness. Once this data was available and linked with the

[3] https://sdgs.un.org/goals/goal3

Table 11.1 UN SDG targets, categories, and data

UN SDG	Facility type	Service availability	Service capacity	Service readiness
SDG 3 Good Health and Well-Being	Dentist, diagnostic lab, hospital, medical clinic, pharmacy, veterinary care	Services offered, hours of service	Number of beds, number of doctors, number of nurses	Power reliability, primary water source
SDG 4 Education for All	Primary school, Secondary school, high school, vocational, university	Online options	Number of classrooms, number of teachers, number of students	Power reliability, primary water source

population and demographic data from the census, interactive dashboards were created to:

- Identify service and resource gaps, where to introduce new services and which services to scale
- Establish a baseline for service availability, capacity, and service readiness
- Monitor service availability, capacity, and service readiness over time and analyze trends and causal factors
- Prepare and provide transparency in policy-making for location-based services.

To establish a baseline for SDG 3 (Health for All) and SDG 4 (Education for All), data about the presence and distribution of health service and educational facilities and the capacity and availability of services was collected. For health facilities, data was collected about health services offered, and capacity in terms of beds, doctors, nurses, etc. For educational facilities, questions were asked about online options, and the number of classrooms, teachers, students, etc.

Table 11.1 illustrates the facilities that were surveyed and the survey questions to collect data for service availability, service capacity, and service readiness.

Social Metrics (Healthcare, Education, Access to Water)

To measure progress on the SDGs, metrics that are key indicators of the desired outcomes need to be identified. For a facility, these metrics should benchmark service availability, service capacity, and service readiness. We used the data and metrics to answer the following questions:

- Is the service accessible? This is a measure of the proximity of the service to the local population.
- Does the service have adequate capacity for the served population? For the health category, does the facility have enough beds, ventilators, vaccines, etc., to meet the needs of the local population?
- What is the quality of service? This is harder to measure using a single metric and may have to be derived from multiple streams of data or by a weighted

Table 11.2 UN SDG outcomes, indicators, and targets

UN SDG	Desired outcome	Key indicator	Desired target
SDG 3	Improve access to medical care	Medical facilities per 100,000 population	4 facilities per 100,000
SDG 3	Improve care capacity	Doctors per 1000 population, beds per 1000 population	1 Doctor per 1000, 3 nurses per 1000, 5 beds per 1000
SDG 5	Gender equality among doctors, nurses, teachers	Female-to-male ratio among doctors	Match local gender mix
SDG 6	Access to safe water	Distribution of safe and unsafe water sources	Safe access to water for all

average of multiple indexes. An erratic power supply might limit hours of service. A poor infrastructure leading to an unreliable supply chain will affect service readiness. To get a measure of service readiness, we collected data on a facility's primary power source and its access to water and sanitation.

One other aspect that was considered was data and metric granularity. Higher data granularity provides more information but must be weighed against data collection costs and quality. A health facility might report data on the number of beds available for patients. The number of beds could be made more granular by specifying the type of beds – ICU beds, ventilator beds, oxygen beds, VIP beds, and general beds. This additional level of detail provides more information and can be useful in providing readiness metrics for specific events, but it requires the willingness of surveyed facilities to provide this level of detail and methods to ensure the accuracy of the data that they provide.

The metrics were defined based on the SDG targets. The metrics determine the data model that drives the analytics and dashboards, which in-turn determines the data that is collected, and the framing of the questions in the survey forms. The table below illustrates target metrics that are key indicators for the desired outcomes. The World Health Organization (WHO) and the US Agency for International Aid (USAID) specify some of the minimum desired target metrics for meeting SDG outcomes (Table 11.2).[4,5]

Data Model

A data model is an abstract that determines the structure of data and how data is collected, stored, and exchanged. The data associated with a particular facility or point of interest, was grouped into three types – base, static, and dynamic. The type of data determines how it is collected and at what frequency.

[4] https://www.who.int/data/gho/data/indicators

[5] https://data.worldbank.org/indicator/

Base data attributes are common to all points of interest. These typically include name, location, and contact information like the name of the facility, address, zip code, latitude/longitude, and phone numbers. This facility-based data is public information that can be used to locate and communicate with the facility.

The base data attributes were collected using KoBoToolbox.[6] KoBoToolbox is an open source tool for field data collection that is adapted for challenging environments, i.e., locations with limited to no Internet connectivity. With KoBoToolbox, we created forms for each type of facility. A team of data collectors visited each facility and recorded information that was then uploaded to the KoBoToolbox servers. KoBoToolbox provides tools for checking the consistency and completeness of the data collected (Fig. 11.1). Data was reviewed, cleaned and exported as CSV data that was then ingested into the Hawkai Data platform.[7]

Static data attributes are data that are mostly common to all points of interest with values that hardly change. These include the facility's operating hours, primary water source, power source, etc. The information collected about the primary water and power sources was used to benchmark status for SDG 6 (Clean Water and Sanitation) and SDG 7 (Affordable and Clean Energy).

Dynamic data attributes are data that are associated with a specific category and facility type and that change over time. These may include current inventory of goods, available capacity of services, number of employees, etc. Data was also collected on the gender mix of employees, which was used to benchmark status for SDG 5 (Gender Equity).

Dynamic data needs to be collected and updated at periodic intervals. Sending data collectors to each of the points of interest to record data is expensive. Instead,

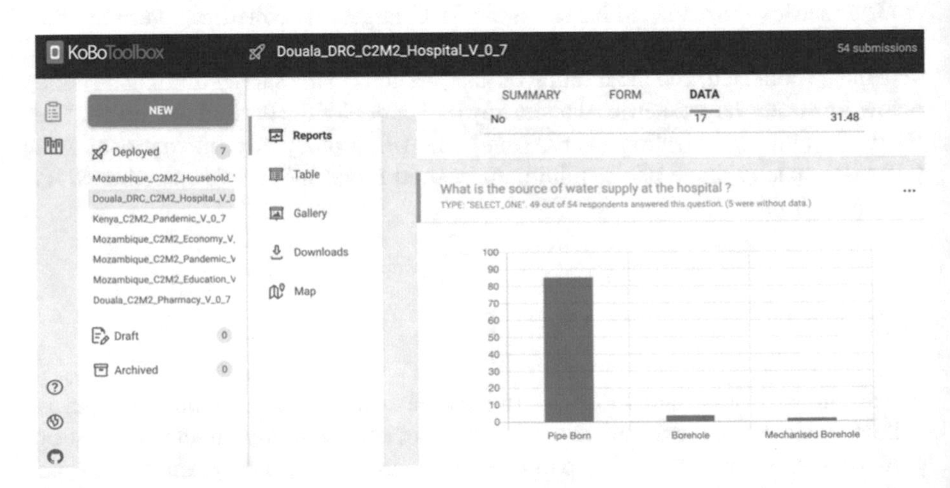

Fig. 11.1 KoBoToolbox provides tools to review data integrity

[6] Tool is available at https://www.kobotoolbox.org/

[7] Platform is available at https://hawkai.net

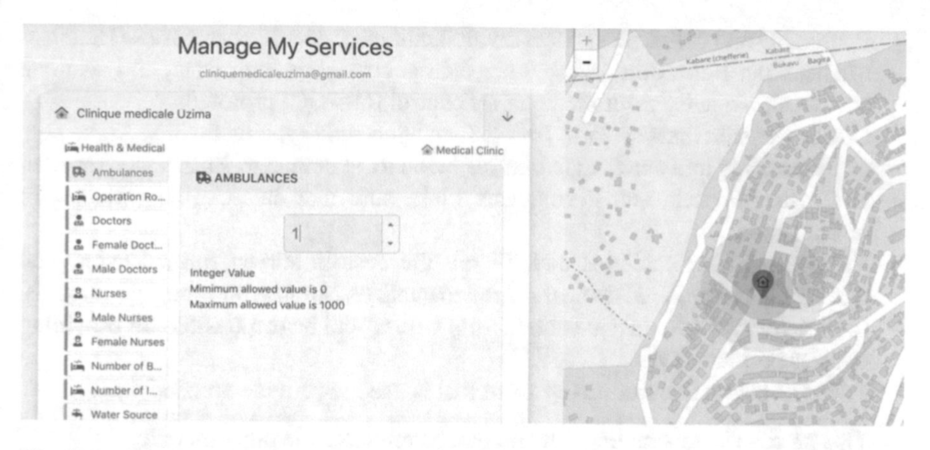

Fig. 11.2 Form for self-serve update of data from a health facility

we identified a contact person at each of the facilities who took ownership for updating the data. The person was either an employee of the facility or was someone with the agency and empowerment to record the dynamic data about a facility on a periodic basis.

The Manage My Service form was created in the Hawkai Data platform to update the static and dynamic attributes for a specific facility (Fig. 11.2).

Accurate data collection is key to maintaining the integrity of the metrics and the dashboards. A formal data collection process ensures that the data that is collected is consistent and accurate. This process also requires tools to collect and record the different types of data attributes.

Data Collection Process

Different methodologies exist for conducting surveys and collecting data to benchmark and measure the desired outcomes. The choice of methodology depends on context, availability of resources with the requisite data expertise, and willingness of the facilities to share information. The following was the process developed for the 2C and C2M2 projects to collect data:

- Focus on specific SDGs and identify the corresponding category and type of facility that needs to be enumerated. C2M2's focus was on SDG 3 and SDG 4. Identify key desired outcomes.
- Recognize the interlinked nature of the SDGs. Framing the survey questions appropriately and collecting data for SDG 3 and SDG 4 could provide key performance indicators for SDG 5 (Gender Equality), SDG 6 (Clean Water and Sanitation), and SDG 7 (Affordable and Clean Energy).

- Enumerate and collect basic data on all facilities of specific categories (Health & Medical and Education). Make sure data is consistent, complete, and accurate and develop quality assurance/quality control (QA/QC) protocols.
- Create a custom survey form for each category and type of facility. Make sure that the survey tool can handle queries in the local language. Forms were created in English, French, and Portuguese. Make sure that data can be collected in offline mode.
- Visit each of the facilities and fill out the custom survey form. Review and cleanse data as required using the defined quality assurance and control protocols.
- Identify/recruit a contact at each facility who would be responsible for providing updates periodically.
- Train the contact on the survey form that is used to provide updated data.

This process is general enough and can be replicated in any other city.

Analytics and Dashboards

Live dashboards use up-to-date data to create visualizations that provide actionable insights.[8] The following lists some of the possible dashboards that can be made available using the data collected for the Health & Medical and Education SDG categories.

- Trend charts for the availability of hospital beds, doctors, patients, and medicines over time by region.
- General health indexes – patients by population, chronic medications by population. These indexes can be used to track the progress of UN SDG 3 and also for checking the secondary effects of COVID.
- Epidemic preparedness – availability of vaccines.

Sample Dashboards for Access to Care in Bukavu

Figure 11.3 shows the summary data for the health category for Bukavu, Democratic Republic of Congo (DRC). The city of Bukavu was selected for this project given its strategic position and role in the DRC. Bukavu is in the eastern region of the Democratic Republic of the Congo at the extreme south-western edge of Lake Kivu. It is west of Cyan Gugu in Rwanda and separated from it by the outlet of the Ruzizi River. It is the capital of the South Kivu province, and as of 2012, it had an estimated population of 806,940. In the 1990s, the city received thousands of refugees

[8] https://hawkaidata.net/apps/a/portal.html

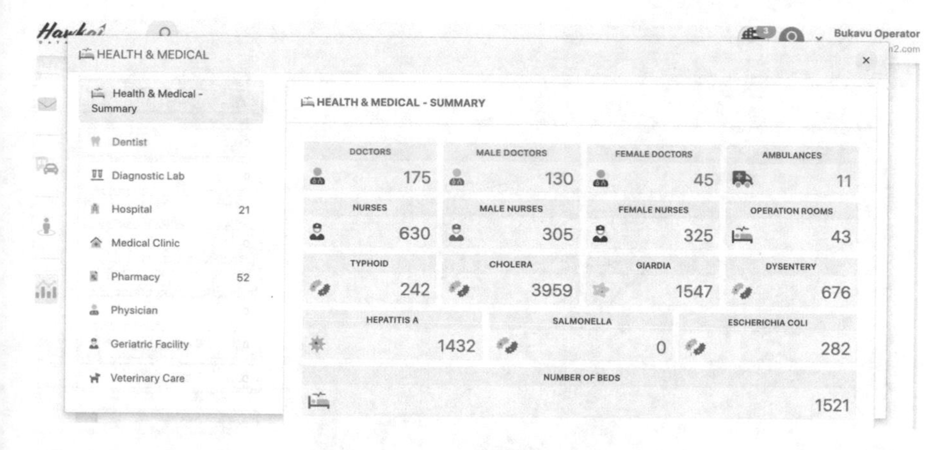

Fig. 11.3 Dashboard summarizing the Health & Medical category in Bukavu, DRC

fleeing ethnic violence in neighboring countries, Burundi, and Rwanda, and is now a cultural melting pot.

Bukavu is a commercial and industrial hub as well as a tourist destination. It is one of the most densely populated cities in the DRC and home to a major university with the resources required to successfully execute the C2M2 project.

Since the beginning of the COVID-19 pandemic in 2020, business activities have suffered a sharp decline for most of the population. To reduce and hopefully contain the spread of the coronavirus disease, authorities in the Democratic Republic of Congo established a quarantine in Bukavu starting May 27, 2020. While these measures significantly impacted both the citizens' social life and the city's economy, they did help reduce the spread of the disease. By August 2020, Bukavu was no longer the epicenter of the disease, and this project, launched in October 2020, was a well-timed opportunity to assess the second-order impacts of COVID-19 and provide recommendations on how to mitigate them going further.

Specific categories can be browsed visually, and service availability and gaps can be determined (Fig. 11.4).

Figure 11.5 shows the gender breakdown by occupation in the health and education categories in the city of Bukavu, DRC.

Sample Dashboards for Access to Water in Nairobi

Figure 11.6 maps and grades the access to water in the division of Kibera, Nairobi, Kenya. Nairobi was the epicenter of Kenya's COVID-19 outbreak during the global pandemic. The Kenyan government imposed a lockdown in March 2020, which included the prohibition of in-country travel. This resulted in massive economic damage. Some trade was allowed to continue, but most individuals were negatively impacted. Schools were also ordered shut. While the country removed most of this

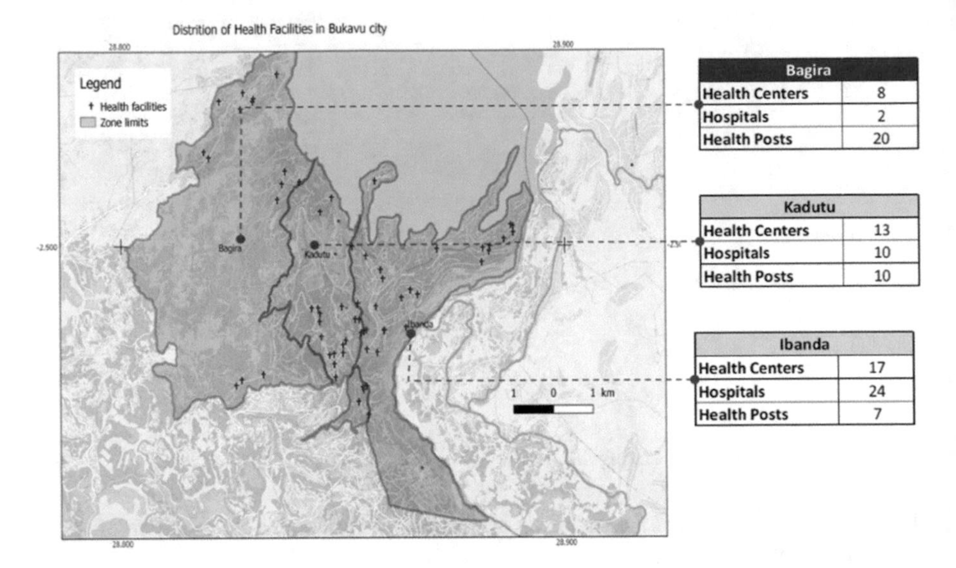

Fig. 11.4 Availability of healthcare service centers in Bukavu, DRC

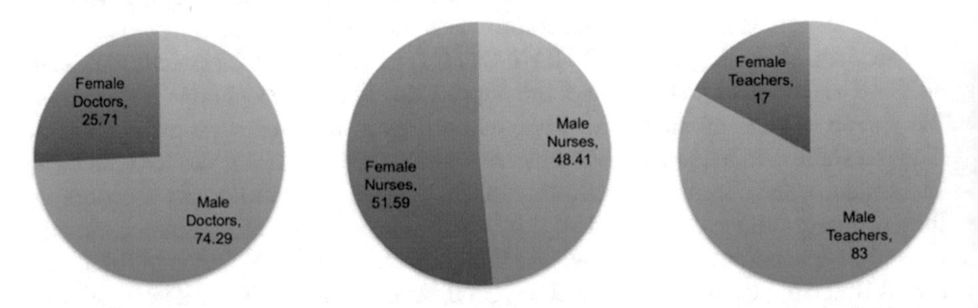

Fig. 11.5 Gender splits across occupations in Bukavu, DRC

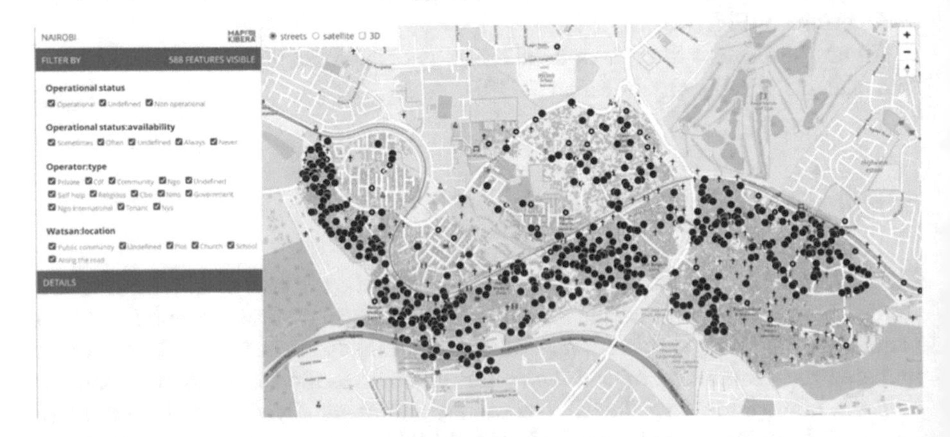

Fig. 11.6 Access to water (mapped water access points) in Kibera, Nairobi, Kenya; GroundTruth Initiative

lockdown during the latter half of 2020, on March 26, 2021, a new lockdown was imposed in response to the spike in COVID-19 cases throughout the country, with emphasis on the five counties with the highest caseloads including Nairobi. Those hit hardest by the economic toll of these lockdowns were people living in the informal settlements of Nairobi, such as Kibera and Mathare, where access to water became a major challenge.

These trends and metrics provide a quantifiable way for a city to measure and justify the impact of its spending on local initiatives while also providing a way for regions to respond effectively during disasters, drive economic growth, and increase resilience, gender equity, and social capital.

Community Services

A facility registry for a city provides data and insights to local officials and the tools to measure, manage, and respond to the needs of the local populations. Collected personal data will meet defined security and privacy requirements. This same data can be used to create services for city residents who can access them on their mobile phones. The Cities Navigator app[9] (Fig. 11.7) makes it easy for users to find facilities of interest, check resource availability, and communicate with the facility. These capabilities illustrate the importance of geospatial data to facilitate access to care in under-examined locations. Using built-in dashboards, officials can track application usage and access metrics on service quality that will enable them to initiate policies to improve the quality of life of the local population. Sample metrics include, i.e., location of the nearest hospital, number of doctors in the city, number of beds, number of ICU beds, etc. This responsive web-based application can be accessed using any browser. No installation is required.

Such community-based services will bridge the digital divide and reduce the impact of such pandemics on vulnerable populations with a smartphone. Local data skills, data quality, data integrity, and up-to-date data will determine the success of the apps and their continued usage. Citizen adoption of these apps and services will increase social equity and reduce the impact of any future pandemics, disasters, conflicts, and climate-change events. The role of local governments is critical for the success of such applications in low- and middle-income countries as it provides legitimation for the application. Some of the ways this could be achieved include using the app by local officials to communicate with the population, providing incentives for public services to continuously update the application to ensure accurate information, making sure city planning and emergency management services adopt the application, and using collected data for planning purposes. Below are some of the lessons learned in Bukavu that illustrate the impact of such applications and the benefits for the population and local governments:

---

[9] App is available at http://citiesnavigator.com/lp/index-e.html

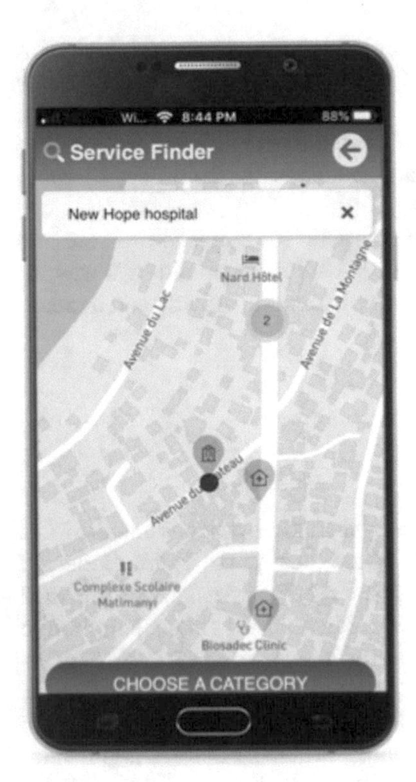

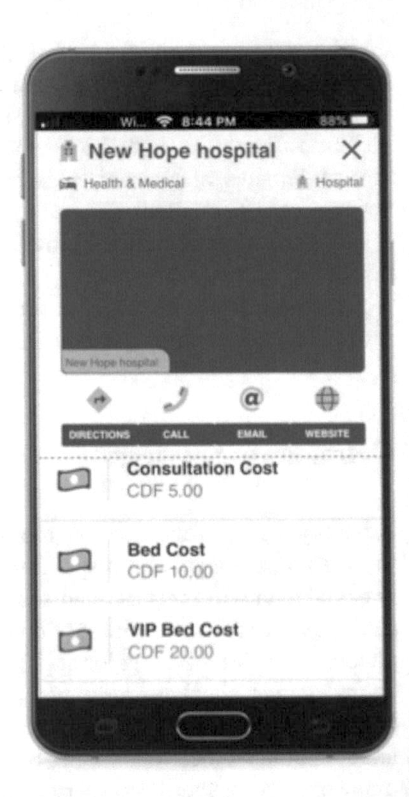

Fig. 11.7 The Cities Navigator app enables community services using facility data

- The project enabled the city to create the first ever geo-localized database of education and healthcare facilities.
- The ability to feed collected data into the application and use it for decision support and communication is of paramount importance and was welcomed by local authorities and stakeholders who attended the final project workshop. As a result, the regional health department expressed interest in using the application for planning purposes.
- Collecting dynamic data in an unusual context of repetitive lockdowns and movement restrictions was a big challenge for the team. This project allowed the team to develop adaptive strategies to approach and communicate with the respondents to complete data collection.
- Working with education and healthcare facilities to gather geospatial data was a challenging undertaking. In this context, the collaboration with the state services in charge of education and health was very beneficial for smooth project execution.

Conclusion

A facility-based data collection model was used to gather data in cities in Africa on service availability, service capacity, and service readiness. The data collected from the health and educational facilities was used to benchmark the region's SDG 3 and SDG 4 status. The interlinked nature of the SDGs allowed us to use the data from these facilities to get insights into SDG 5 (Gender Equality), SDG 6 (Clean Water and Sanitation), and SDG 7 (Affordable and Clean Energy). We show that the facility-based data model offers a low-cost alternative to household enumeration to do a baseline analysis of SDG status. This data, when collected on a continuous basis, will allow cities and regions to track SDG progress, identify gaps, and gain insights into a region's readiness to manage and mitigate the impact of disasters.

Disasters are inherently discriminatory (Belkhir & Charlemaine, 2007). The economically, socially, and even informationally disadvantaged communities are disproportionately impacted. We show that, if data is collected on a periodic basis, the quality and integrity of data are maintained, and this information is accessible to local communities, it can lead to more social equity during disasters.

Acknowledgments This research was possible thanks to the Cities' COVID Mitigation Mapping (C2M2) Program, developed by the US Department of State's Humanitarian Information Unit and the American Association of Geographers (AAG). We would like to acknowledge the following organizations for their support in collecting and analyzing the data:

- Université Officielle de Bukavu, Faculté de Sciences, Democratic Republic of Congo (DRC)
- GroundTruth Initiative, Nairobi, Kenya
- Catholic University of Mozambique, Faculty of Tourism, Pemba, Mozambique
- Digital Divide Data, Kenya

References

Belkhir, J. A., & Charlemaine, C. (2007). Race, gender and class lessons from hurricane katrina. *Race, Gender & Class, 14*(1/2), 120–152. http://www.jstor.org/stable/41675200

GBD 2015 SDG Collaborators. (2016). Measuring the health-related sustainable development goals in 188 countries: A baseline analysis from the global burden of disease study 2015. *Lancet (London, England), 388*(10053), 1813–1850. https://doi.org/10.1016/S0140-6736(16)31467-2

Chapter 12
COVID-19 and Domestic Violence Complaints in Quito, Ecuador: Temporal and Spatial Patterns and Drivers

Carlos F. Mena, Byron Lozada, Patricia Martinez, Fatima L. Benitez, Carolina Sampedro, and María B. Zapata

Personal Story
During the COVID-19 pandemic, Ecuador was hit very hard, especially at the beginning of the crisis. The negative cascading effects are still developing and present in all Ecuadorian families. In my nuclear family, one of the secondary effects of the pandemics is related to education: for me, the challenge to teach online university classes and for my daughters the challenge to attend high school and socialize with their peers using the Internet. We might never know the effect of COVID-19 on education, at home or in Ecuador, but we know that this effect will be huge in impoverished families with poor access to the Internet. We need to push for better education, at all levels, after this crisis, for a better world. *Carlos Mena*

Introduction

There has been an alarming increase in domestic violence during the COVID-19 pandemic, from 2020 to 2021, compared to previous years (Anurudran et al., 2020; Boserup et al., 2020; Leslie & Wilson, 2020), especially during different stages of confinement. COVID-19 and the increase in domestic violence have been defined as a double pandemic (Bettinger-Lopez & Bro, 2020). Piquero et al. (2021) in a meta-analysis of 18 studies across different countries find an increase in domestic violence between pre- and post-COVID-19 lockdown periods, which differ in time and intensity. The increase in domestic violence can be considered a second-order effect of the COVID-19 pandemic.

C. F. Mena (✉) · B. Lozada · P. Martinez · F. L. Benitez · C. Sampedro · M. B. Zapata
Institute of Geography, Universidad San Francisco de Quito, Quito, Ecuador
e-mail: cmena@usfq.edu.ec

M. Laituri et al. (eds.), *The Geographies of COVID-19*, Global Perspectives on Health Geography, https://doi.org/10.1007/978-3-031-11775-6_12

Domestic violence, or intrafamily violence, refers to violence that occurs within a home and is exercised by one member of the family against others (Chamorro, 2008). Domestic violence, including the use of physical force, harassment, intimidation, and provocation, occur within a family and can take different forms: physical, psychological, sexual violence and the murder of women as an extreme result of gender violence (femicide). Domestic violence exerted upon different types of victims include violence against women, violence against men, child abuse, child-parent violence, and elder abuse (Gelb & Palley, 2009).

There are several triggers or factors that can explain domestic violence, for women, minors, and the elderly. In Latin America, among these factors, the lack of economic resources in families, educational level, employment situation, marginalization, access to prosecutors and complaint units, as well as adverse conditions for the adequate care of members of the family (Wilson, 2014; Heaton & Forste, 2008; Flake & Forste, 2006). Thus, the current economic situation, as a result of the COVID-19 pandemic, appears as a factor that generates more domestic violence (Leslie & Wilson, 2020), especially in households where there has been stress-inducing income loss (Silverio-Murillo et al., 2020).

The response of the legal system to domestic violence is one of the most important institutional or formal answers for victims. Unfortunately, in Latin America, the administration of justice is full of obstacles for help-seeking victims, including inadequate laws and incorrect application of the specific legislation, excess bureaucracy and long and inefficient procedures, little privacy for women who report the offense, lack of specialized personnel, and slowness to face emergencies (Sagot, 2005). In Ecuador, according to the public defendant, COVID-19 has created additional barriers for victims of domestic violence who cannot move or visit complaint units, leading to a significantly lower number of formal complaints than the number of actual crimes (Defensoría del Pueblo del Ecuador, 2020).

Within this context, the objectives of this study are twofold: (a) to explore the temporal and spatial patterns of formal complaints on domestic violence across the city before, during, and after the confinement caused by the COVID-19 pandemic and (b) to determine the demographic and socioeconomic factors that explain the reporting of domestic violence during these periods. More specifically, we want to explore if the lockdown affected domestic violence reporting.

This chapter uses a combination of geospatial tools and statistical analysis. Both approaches provide an insight into the importance of spatial relationships to explain intra-family violence, specifically during the period of COVID-19 pandemic. This chapter uses a set of georeferenced data on violence and makes use of tools to understand the degree of spatial autocorrelation, and geographical and temporal trends. This chapter shows the value of geographic data and spatial data analysis by adding spatial-explicit results to potential solutions to the challenges of reporting domestic violence.

Study Area: Quito City

This study focuses on the city of Quito, the capital of Ecuador. Quito is located at 2850 meters above sea level in a valley surrounded by the Andes. Quito is the largest city in Ecuador with a metropolitan area of about 4217 km^2 and 2.7 million people (INEC, 2022). According to the Municipality of Quito (Distrito Metropolitano de Quito, 2022), 51.3% of the population is female and 27.4% of the population under 14 years old. Additionally, in 2021, 12.5% of the population was unemployed and 20.7% had informal jobs (Fig. 12.1).

Poor areas of Quito, which account for 8.2% of the population, are mainly located in the north-east and south-west of the city (Wei et al., 2015) and are considered vulnerable to external shocks including the second-order effects of the COVID-19 pandemic. An econometric analysis shows how distance to the main source of employment, density of public transportation, education, healthcare centers, and ethnicity influence the probability of being poor or not in Quito (Olarte, 2019). Quito is also a very unequal city as different processes of urbanization have produced new geographies of socio-spatial segregation, which include repeated discrimination against migrants, LGBTI groups, and women (Carrión & Pinto, 2019).

Because of COVID-19, families in Quito have been forced to create new modes of coexistence, and new modes of violence have also emerged in households or have increased in families that had previously presented violent forms of relationships (Medina et al., 2020). In Quito, like in many other cities in Ecuador, services and legal remedies to deal with domestic violence exist; however, they are unevenly distributed across the city and fragmented in terms of availability, including economic empowerment, expansion of services to diverse groups of victims, increased access to information, and trauma-informed training of staff (Sabina & Figueroa, 2019).

Data and Methods

This analysis is based on exploratory spatial methods and regression analysis, which uses the following databases:

(a) Individual formal complaints of domestic violence occurred in Quito, between 2018 and 2021, presented to the General Prosecution Office of Ecuador (Fiscalia General del Estado, 2022). This database includes information about the type of violence, date, and geographic coordinates of events. The database contained data for 9933 individual cases.
(b) The 2010 Ecuadorian Population and Housing Census (INEC, 2010), which includes information about the magnitude, structure, growth, and distribution of the population and its economic, social, and demographic characteristics.

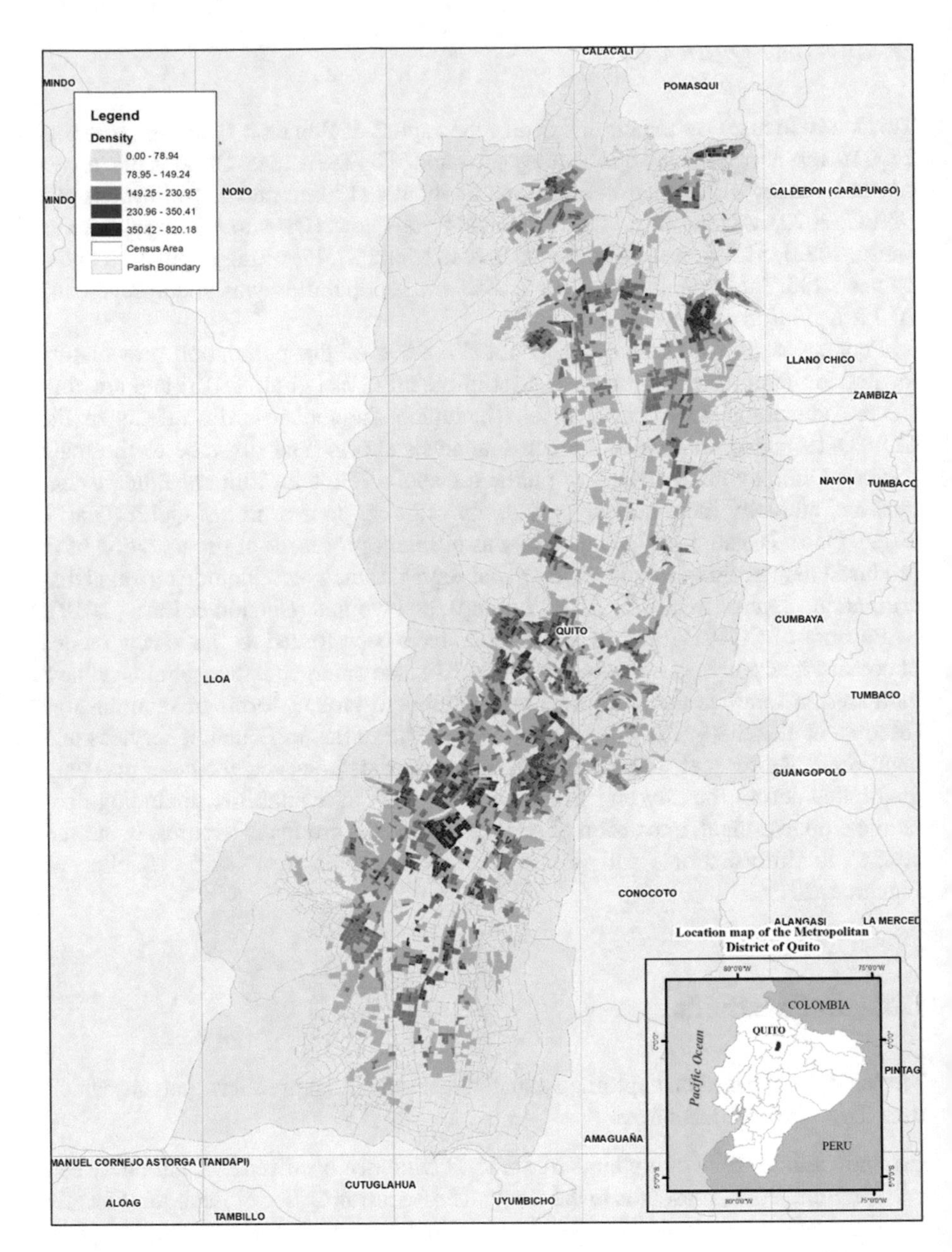

Fig. 12.1 Map of Quito, with census sectors with population density. (Census sectors are the units of analysis in this research)

(c) The population projection at the cantonal level for the years 2010–2020 prepared by the National Institute of Statistics and Censuses (INEC).
(d) The 2014 Social Registry database, a cadaster containing social, economic, and demographic information at the individual and family level; that allows national public institutions to identify their target populations, in order to better focus policy efforts on groups living in poverty.

In terms of spatial resolution, individual records from the FGE were grouped into political division called "parishes" and "census sectors", which is the unit of observation in this analysis. A census sector contains a defined number of census blocks. Our study covers 8264 census sectors. Similarly, the violence database from FGE was grouped according to their type: domestic, psychological, sexual violence, and femicide. In this study, we explore only domestic violence. All individual data has been aggregated at the census sector level to guarantee the privacy of the individual victims. Finally, different types of mobility restrictions were assigned to each event, including complete confinement due to COVID-19 (March 2020–August 2020).

Geographical Analysis

To better understand the changes in spatial patterns of complaints to the legal system during the COVID-19 pandemic, two methods for exploration of spatial dependence were chosen: Moran's Index and Getis-Ord's I (Fig. 12.2). Moran's I (Moran, 1948) is a widely used correlation coefficient that measures general spatial autocorrelation of the data. Moran's I range from −1 to 1 and is positive if the data is clustered, negative if it is dispersed, and close to zero if it is random. The Getis-Ord Index (Ord & Getis, 1995) is another measurement of spatial autocorrelation and measures the concentration of high or low values in a fixed study area. In some

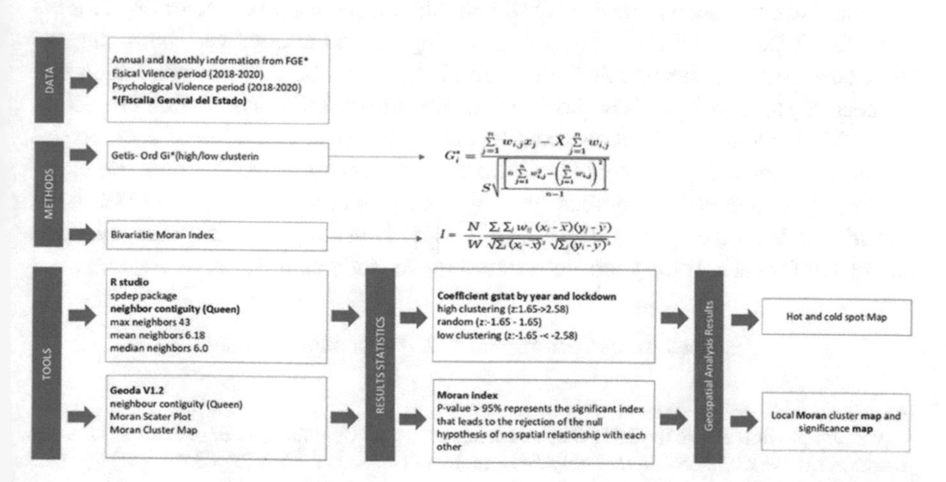

Fig. 12.2 Scheme of the process followed to obtain the spatial correlation and clustering indices

cases, the Getis statistic is dependent to geographical scale, including the size of the study area and the degree of detail (Fan & Myint, 2014).

In Ecuador, a "traffic light" system was used to communicate restrictions in mobility during 2020. "Green" was used to illustrate no restrictions to mobility and was applied between September and December of 2020, "yellow" illustrated some restrictions to mobility during the period of June to August 2020, and "red" illustrated complete lockdown during the period of March to May of 2020. In our analysis, different periods in 2018, 2019, and 2020 were compared to explore differences due to mobility restrictions (Comité de Operaciones de Emergencia Nacional, 2020).

Statistical Analysis

This analysis seeks to understand the drivers of the formal complaints of domestic violence to the competent entity and if there were some changes related to different restrictions in mobility. An Ordinary Least Squares (OLS) regression model was built to explore the direction and statistical significance of the relationship between the number of domestic violence cases in each census sector (dependent variable) and of independent variables that included:

Lockdown Indicates the months when there was a complete lockdown or mobility restrictions (March to August 2020).

Demographic Variables Include percentage of people (age, gender) and number of households per census sector. One of the hypotheses explored is that the levels of domestic violence in each census sector are related to the size of the population in each sector. Given that the only reliable and available source of information, the 2010 Population Census, does not show the current size of the population, it was necessary to carry out a projection of the population at the census sector level for the years 2018, 2019, and 2020. After considering several alternatives (including the use of the 2001 Census to calculate an annual growth rate),[1] it was concluded that the best way to carry out this projection was by using the INEC Population Projection 2010–2020 available only at cantonal level.[2] First, the participation of each census sector in the total population of the Quito canton was calculated. Second, the population projections at the cantonal level for the years 2018, 2019, and 2020 were considered. With these data, the estimated population at the census sector level was calculated for each of the years considered in the study as follows:

$$\text{Estimated Population}_{\text{iy}} = \text{INEC Proyection}_{\text{y}} \times \text{Share}_{\text{i}}$$

[1] It was not possible to use the 2001 Census because changes were made in the census sectors due to administrative decisions, so that only 5457 sectors of a total of 8264 (66.03 percent) coincide between the bases of the FGE and the 2001 Census.

[2] The political-administrative division of Ecuador comprises, from highest to lowest hierarchy, provinces (24), cantons (221), parishes (1499), and census sectors.

where (i) represents each of the census sectors and (y) the years considered.

Distance to the Closest Crime Reporting Facility A variable was constructed that reflects the average distance from the place where domestic violence was committed in each census sector to the closest reporting unit. Subsequently, an average of the distances of crimes in each census sector was obtained for each month and year from 2018 to 2020. It is important to note that only the distances of 9933 individual cases (crimes) could be calculated out of a total of about 12,500 cases, because of errors in the collection of coordinates.

Urban This is a variable that indicates if the census block is urban or rural, according to the definition of the National Census.

Education Variables Include, (a) years of schooling or average years of study of people aged 24 or over and (b) the percentage of children aged 6–12 year old who report non-attendance to primary school.

Poverty and Vulnerability These indicators are constructed according to the Unsatisfied Basic Needs (UBN) methodology.[3] The simple indicators selected to measure the deprivations of the population were two composite indicators of poverty by UBN were constructed at the census sector level: (i) percentage of individuals/households in poverty by UBN (one or more unsatisfied basic needs) and (ii) percentage of individuals/households living in extreme poverty by UBN (two or more unsatisfied basic needs). The unsatisfied basic needs are: (a) percentage of individuals/households in a situation of economic dependence (insufficient economic capacity); (b) percentage of individuals/households with children 6–12 years of age who do not attend school; (c) percentage of individuals/households whose dwelling is built with materials considered deficient; (d) percentage of individuals/households with poor access to basic services; (e) percentage of individuals/households in conditions of critical overcrowding.

Employment Percentage of people over 15 years of age in employment or subemployment.

Economic Dependency Indicates the percentage of individuals, children, and elderly, in a situation of economic dependency for each census sector.

At the end of this data management process, a final database with more than 70 different variables at the census sector level was obtained, which was used to perform the regression analysis. In this analysis, we present the most important variables.

[3]The UBN methodology is a multidimensional poverty measure developed in the 1980s by the Economic Commission for Latin America and the Caribbean (ECLAC). This methodology seeks to determine, with the help of some simple indicators, if the basic needs of the population are covered. Groups that do not reach a fixed minimum threshold (i.e., one or more unsatisfied basic needs) are classified as poor (Battiston et al., 2013).

Results

Temporal and Spatial Patterns of Domestic Violence Complaints

There is considerable decrease in the level of formal complaints of domestic violence during the months of mobility restrictions, i.e., confinement, during March and April of 2020 in Quito, compared to previous years (Fig. 12.3). In May 2020, the number of complaints was considerably higher coinciding with a period of more mobility.

Spatial patterns are similar between the 3 years (2018, 2019, 2020). In 2020, there is a decrease in cases in the "modern or commercial center" of the city and an increase in intensity in the southern, poorer, areas (Fig. 12.4). This is an indication that domestic violence cases moved from areas of employment to areas of residence.

In terms of spatial autocorrelation of domestic violence complaints, Tables 12.1 and 12.2 shows the results for Moran's I and Geary's C, respectively. In this case, traffic light colors were used to communicate restrictions of mobility in the city. Results of Moran's I indicate statistically significant and positive spatial autocorrelation, only for "yellow level" of mobility restrictions, when we compare 2020–2018 and 2019. At the local level, using Geary's C, there is a mix between random and clustered patterns for the 3 years. Important to note that during March and April of 2020, the periods of more restrictions, the spatial pattern is random.

Figure 12.5 shows the spatial representation of the results of the Getis-Ord Index.

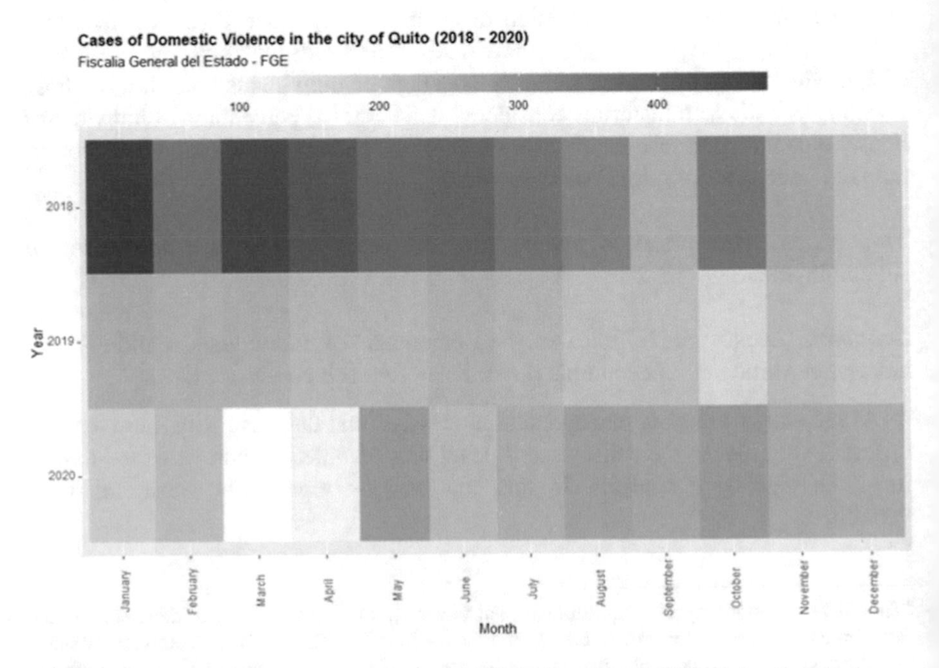

Fig. 12.3 Heatmap that illustrate temporal pattern, March and April 2020, Quito, Ecuador

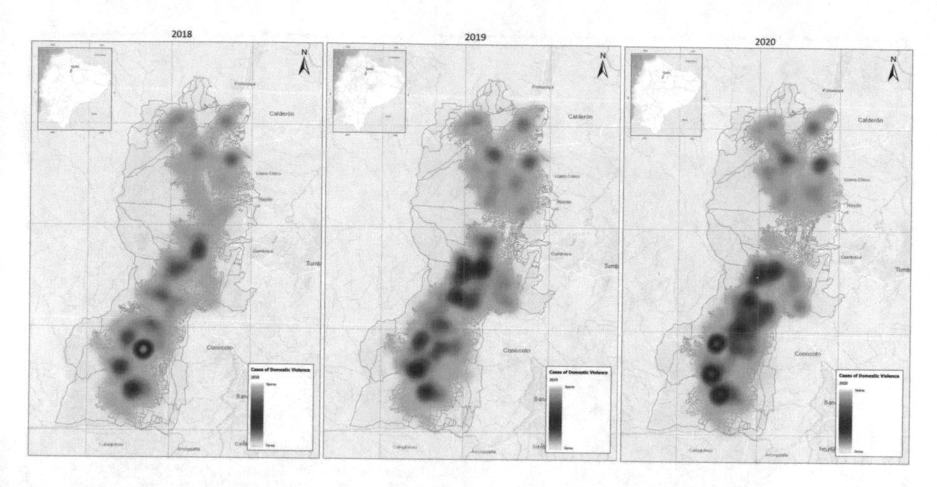

Fig. 12.4 Spatial patterns of the location of domestic violence reported to the General Prosecutor Office, 2020

Table 12.1 Results of Moran's I comparison across different periods in 2018, 2019, and 2020; *refers to restricted mobility?

Period of comparison	Mobility restriction period	Moran's I
2020–2018	Red	1.5309
	Yellow	3.40*
	Green	1.50
2020–2019	Red	1.48
	Yellow	2.3562*
	Green	0.32

Table 12.2 Results of the Getis-Ord Index to explain different patterns of clustering across different months

2018		2019		2020	
January	Clusters	January	Clusters	January	Clusters
February	Clusters	February	Random	February	Clusters
March	Clusters	March	Random	March	Random
April	Clusters	April	Clusters	April	Random
May	Clusters	May	Clusters	May	Clusters
June	Random	June	Random	June	Clusters
July	Random	July	Clusters	July	Clusters
August	Clusters	August	Clusters	August	Clusters
September	Clusters	September	Clusters	September	Random
October	Random	October	Random	October	Random
November	Clusters	November	Clusters	November	Random
December	Clusters	December	Random	December	Random
Total	Clusters	Total	Clusters	Total	Clusters

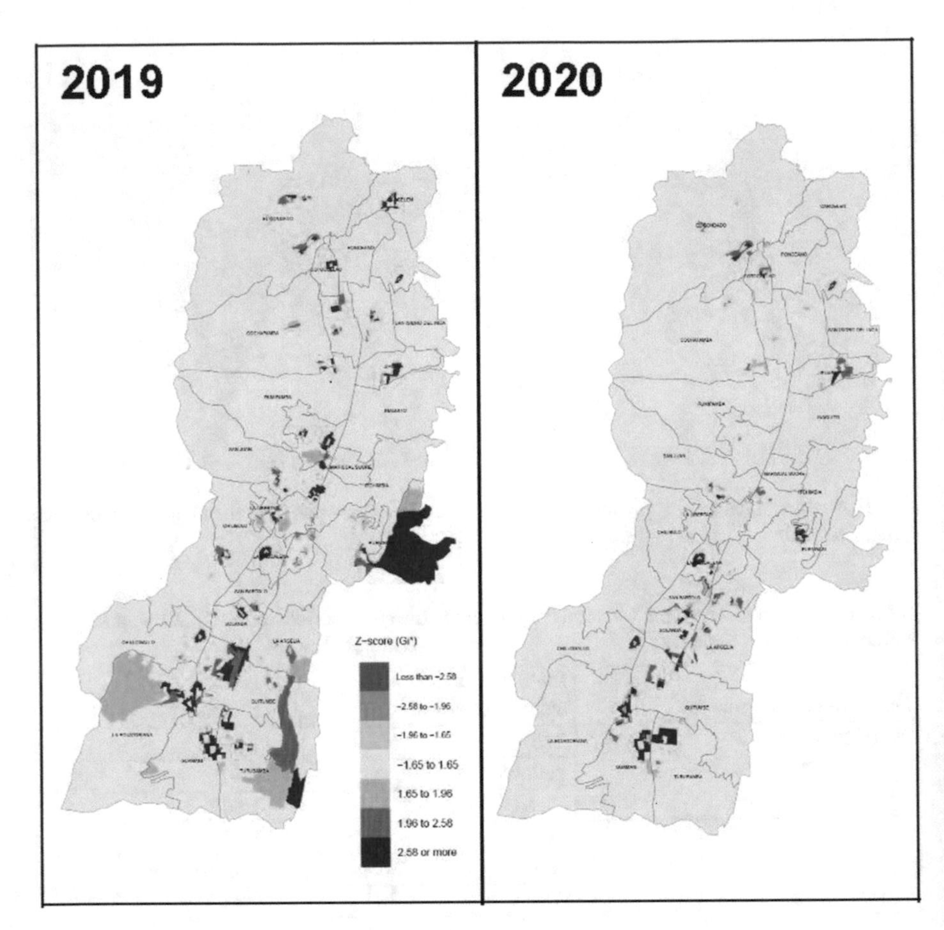

Fig. 12.5 Results of the Getis-Ord Index. Reddish colors indicate concentrations of high values of domestic violence in 2019 and 2020

Drivers of Domestic Violence Complaints

Domestic violence and how victims decide to present complaints to the judicial system is complex with different underlying causes. In this exploratory analysis, we try to understand if the lockdown period or different mobility restrictions played a role in the number of complaints.

Table 12.3 shows the result of the OLS model. The model is a combination of selected variables, which were chosen based on domestic violence theory and the degree of multicollinearity. In general, we find that demographic variables are not statistically significant, while economic and geographic variables are important.

The lockdown variable, restrictions to mobility, is negative and statistically significant to the number of complaints (−0.1982), which indicates that the lockdown affected the capacity of victims to seek legal solutions to violence. Distance to

Table 12.3 Results of the Ordinary Least Squares (OLS) model

Variables	Coefficient	SE
Lockdown	−0.1982	−0.06222***
Population	0.00002	−0.00029
Number of households	−0.00013	−0.0011
Share of women	−0.45512	−0.53685
Distance to the complaint unit	−0.06959	−0.02358***
Urban	−0.03044	−0.16037
Years of schooling	0.06551	−0.01419
School absenteeism	2.36377	−1.25083**
Employment	1.01896	−0.34246***
Economic dependence	1.59331	−1.1543*
Deficient housing materials	0.89382	−0.93042
No access to basic services	1.04665	−0.89468
Overcrowding	1.52544	−0.95386*
Poverty NBI	−0.88592	−0.87416
Extreme poverty NBI	−1.71428	−1.22036*
Constant	0.4844	−0.74508
Year dummies	✓	
Month dummies	✓	
Parish dummies	✓	
Number of observations	7618	
R-squared	0.0671	

Note: All specifications include a constant
***Significant at the 1 percent level (*P*-value <0.01); **(*P*-value <0.05); *(*P*-value <0.1)

complaint unit is also statistically significant and negative (−0.06959), which indicates that there is a geographic factor that mediates whether someone will submit a complaint. The average distance to the closest complaint units in each of the census sectors has a negative effect on the number of complaints. The greater the distance, the lower the number of reported cases. Logically, given the lack or inefficiency of alternative non-face-to-face reporting mechanisms, physical distance from reporting units turns out to be a factor that reduces the possibility of people reporting this type of crime. This also reflects the fact that people find it difficult to get around and, therefore, the more remote reporting units are, the less likely it is that people will bear the transportation and opportunity costs that they must incur to report crimes.

How economic factors, including poverty, affect the level of domestic violence complaints is also very important to understand, especially in very unequal cities like Quito. Employment is a variable that is statistically significant and negatively related to the number of complaints. Census sectors with larger proportions of households with employment have a lower number of complaints, which indicates that income stability linked to employment (or the opposite, stress related to unemployment) affects domestic violence and therefore the number of complaints (Staggs

& Riger, 2005). A sector with a higher level of employment means that it has more women working and, therefore, empowered to make the decision to report to the authorities when they suffer some type of abuse.

Economic dependence (1.59331), no school attendance (2.36377), overcrowding (1.52544), and extreme poverty (−1.71428) all are statistically significant but have positive and negative relationship with the number of domestic violence complaints. Less extreme poverty leads to more complaints, which might indicate that domestic violence complaints are linked to a certain level of wealth. No school attendance for children and overcrowding are also indicators of economic stress and the census sectors that with more extreme poverty generate less domestic violence processes.

The economic dependency indicator has a positive effect (1.59331) on domestic violence. Therefore, the higher the percentage of families with more than three economically dependent persons, the higher the number of reported cases.

The indicator of school absenteeism has a positive effect of a considerable magnitude (2.36377) on domestic violence in general. This means that, as the percentage of children who do not attend school regularly increases, the greater the number of cases of violence that are reported in the census sectors. Without a doubt, school absenteeism is an important indicator of the poverty situation in which many families live in the city of Quito. Not having sufficient resources, many households find it necessary to force their children to work so that they can contribute financially to meet the immediate needs of the family, instead of enrolling them and sending them to school classes (Psacharopoulos, 1997). This serious problem is a trigger for the intergenerational transmission of poverty; since children do not acquire the education and human capital, they require to overcome this situation in the future and end up trapped in what is known as a "poverty trap" (Aldaz-Carroll & Morán, 2001). Furthermore, school absenteeism appears to increase the probability of the occurrence of domestic violence significantly. This could be due to the fact that the mechanisms used by parents to force their children to work and make some financial contribution to the home often involve the use of violence, especially psychological violence (Pasquier-Doumer & Brandon, 2015).

Third, overcrowding inside the home also has a positive and significant effect (1.52544) on the number of reported cases of violence. It seems logical to think that the greater the number of people who share the same space, the greater the friction between them that can lead to the commission of these crimes. The magnitude of the coefficient is quite significant, reflecting that overcrowding is one of the factors that most influences the emergence of violence within the home. In general, three of the five simple indicators of poverty (economic dependency, truancy, and overcrowding) have a positive effect on total cases of domestic violence. This evidence confirms that the vulnerability and precarious economic situation of families explain a large part of the violent crimes committed within them. There are many factors that could explain this relationship, among which we can mention are the abuse of power of those who financially support the rest, the mechanisms that are used to force the other members to contribute financially and the frictions that are generated by coexistence in tight spaces.

However, it is important to mention that poverty due to UBN does not have a significant effect on reported cases of violence; while extreme poverty has, surprisingly, a statistically significant negative effect (−1.71428). The absence of poverty effects could be due to the fact that all of these impacts are being driven by the simple indicators of economic dependency, school absenteeism, and overcrowding. In other words, the poverty level of families increases domestic violence; however, it does so only through these channels, so that once included in the regression the percentage of poor is no longer significant to explain the levels of violence.

The fact that extreme poverty reduces cases of violence might seem somewhat more surprising. However, it must be considered that the cases of violence considered are only those that are reported. Therefore, it seems logical to think that, when the poverty situation of families is extreme, they have greater difficulties and less motivation to report the abuses of which they are victims. This would explain that as the percentage of households living in extreme poverty within a sector increases, the lower the number of violent cases that are reported. There are dimensions of poverty that, when present, increase the number of cases of violence, but when these dimensions are combined or are present at the same time, the situation of poverty is so serious that families stop reporting crimes.

Finally, the fixed effects of the year, month, and parish were controlled. The estimates show that the year 2019 presented a statistically significant reduction (−0.14359) in the total cases of violence compared to the other years. In addition, the months of March, April, and May 2021 have positive effects on domestic violence. The occurrence of these crimes increases during this particular time period. The positive effects are 0.1266, 0.1346, and 0.1351, respectively, for each of these months. Lastly, belonging to a particular parish has no effect on overall levels of domestic violence.

Conclusions and Recommendations

Domestic violence can take different forms and respond to complex processes. This is an exploratory analysis that tries to understand different temporal and spatial patterns of domestic violence through the complaints that victims submitted to the legal system during the COVID-19 pandemic.

As many reports suggest, domestic violence is a secondary effect of the COVID-19 pandemic. We can point out that our preliminary analysis in Quito shows that restrictions to mobility and lockdown have affected the way victims report domestic violence.

First, our analysis shows a decrease in the number of formal complaints of domestic violence in the months of March and April of 2020 when compared to the same period in 2018 and 2019. Heat maps show that there is a shift in the number of formal complaints of domestic violence from the modern center of Quito to the southern part, as shown in Fig. 12.4. Geary's C Index of spatial autocorrelation also shows a smaller number of clusters and a more random spatial pattern in 2020.

In terms of the demographic and socioeconomic drivers of domestic violence complaints, this analysis shows the negative effect of the lockdown on the formal reporting of domestic violence. It also shows the importance of the distance between victims and complaint units. In addition, our statistical model shows the negative effects of extreme poverty on the submission of complaints to the legal system.

This analysis is limited by several factors, which range from conceptual to methodological. Domestic violence complaints submitted to the legal system are just a part of the problem. It is accepted that only a minor portion of domestic violence cases is taken to the different components of the legal system, including the police and prosecutor. Further research will analyze calls to emergency systems. From the methodological point of view, this is a quantitative analysis based on secondary data. It would be optimal to complement this analysis with qualitative methods, such as interviews with victims of domestic violence. A specific finding is that distance to violence reporting unit is important to explain lower levels of reporting, which suggests that creating more easily accessible units will improve how victims register abuse.

Finally, there is the need to understand better the secondary-order effects of the COVID-19 pandemic, which are, like the pandemic, very dynamic and spatially explicit. Policy implications and practical solutions will depend on our level of understanding of the underlaying factors and impacts.

Acknowledgement This research was possible thanks to the Cities' COVID Mitigation Mapping (C2M2) Program, developed by the U.S. Department of State's Humanitarian Information Unit and the American Association of Geographers (AAG). We would like to thank Melinda Laituri, Laura Cline, and Candida Mannozzi for their continuous support.

References

Aldaz-Carroll, E., & Morán, R. (2001). Escaping the poverty trap in Latin America: The role of family factors. *Cuadernos de Economía, 38*(114), 155–190.

Anurudran, A., Yared, L., Comrie, C., Harrison, K., & Burke, T. (2020). Domestic violence amid COVID-19. *International Journal of Gynecology & Obstetrics, 150*(2), 255–256.

Battiston, D., Cruces, G., Lopez-Calva, L. F., Lugo, M. A., & Santos, M. E. (2013). Income and beyond: Multidimensional poverty in six Latin American countries. *Social Indicators Research, 112*(2), 291–314.

Bettinger-Lopez, C., & Bro, A. (2020). A double pandemic: Domestic violence in the age of COVID-19. *Council on Foreign Relations, 13*, 1–7.

Boserup, B., McKenney, M., & Elkbuli, A. (2020). Alarming trends in US domestic violence during the COVID-19 pandemic. *The American Journal of Emergency Medicine, 38*(12), 2753–2755.

Carrión, F., & Pinto, J. P. (2019). Production and spatial organization of old and 'new' inequalities in Quito. *Andamios, 16*(39), 101–125.

Chamorro, H. M. (2008). *Manual de protección a víctimas de violencia de género*. Editorial Club Universitario.

Comité de Operaciones de Emergencia Nacional. (2020). *Informes de situación COVID-19 Ecuador 2020–2021*. Comité de Operaciones de Emergencia Nacional. Available from https://www.gestionderiesgos.gob.ec/wp-content/uploads/2020/04/Informe-de-Situaci%C3%B3n-No030-Casos-Coronavirus-Ecuador-08042020.pdf

Defensoría del Pueblo del Ecuador. (2020). *Covid-19 y el incremento de la violencia doméstica contra las mujeres*. Defensoría del Pueblo del Ecuador.
Distrito Metropolitano de Quito. (2022). *Sistema metropolitano de infromación* 2022 [cited 2022]. Available from http://gobiernoabierto.quito.gob.ec/
Fan, C., & Myint, S. (2014). A comparison of spatial autocorrelation indices and landscape metrics in measuring urban landscape fragmentation. *Landscape and Urban Planning, 121*, 117–128.
Fiscalia General del Estado. (2022). *Violencia domestica en Quito 2019–2021* Fiscalia General del Estado. Quito.
Flake, D. F., & Forste, R. (2006). Fighting families: Family characteristics associated with domestic violence in five Latin American countries. *Journal of Family Violence, 21*(1), 19.
Gelb, J., & Palley, M. L. (2009). *Women and politics around the world: A comparative history and survey* (Vol. 1). ABC-CLIO.
Heaton, T. B., & Forste, R. (2008). Domestic violence, couple interaction and children's health in Latin America. *Journal of Family Violence, 23*(3), 183–193.
INEC. (2010). *Resultados del Censo 2010 de población y vivienda en el Ecuador*. Instituto Nacional de Estadísticas y Censos. Quito.
INEC. (2022). *Ecuador en cifras* 2022 [cited 2022]. Available from https://www.ecuadorencifras.gob.ec/
Leslie, E., & Wilson, R. (2020). Sheltering in place and domestic violence: Evidence from calls for service during COVID-19. *Journal of Public Economics, 189*, 104241.
Medina, D. C. M., Guayacundo, M. O. B., Encalada, L. M. A., & Estupiñán, A. M. M. (2020). Violencia intrafamiliar en el marco de la emergencia sanitaria por el COVID-19. *CienciAmérica: Revista de divulgación científica de la Universidad Tecnológica Indoamérica, 9*(2), 261–267.
Moran, P. A. (1948). The interpretation of statistical maps. *Journal of the Royal Statistical Society. Series B (Methodological), 10*(2), 243–251.
Olarte, S. H. (2019). Territorial exclusion as a limit to urban development: The case of Quito, Ecuador. *Theoretical and Empirical Researches in Urban Management, 14*(4), 53–69.
Ord, J. K., & Getis, A. (1995). Local spatial autocorrelation statistics: Distributional issues and an application. *Geographical Analysis, 27*(4), 286–306.
Pasquier-Doumer, L., & Brandon, F. R. (2015). Aspiration failure: A poverty trap for indigenous children in Peru? *World Development, 72*, 208–223.
Piquero, A. R., Jennings, W. G., Jemison, E., Kaukinen, C., & Knaul, F. M. (2021). Evidence from a systematic review and meta-analysis: Domestic violence during the COVID-19 pandemic. *Journal of Criminal Justice, 2021*, 101806.
Psacharopoulos, G. (1997). Child labor versus educational attainment some evidence from Latin America. *Journal of Population Economics, 10*(4), 377–386.
Sabina, C., & Figueroa, D. P. (2019). "Hay que tener suerte": Gender-based violence service provision in Quito, Ecuador. *Health and Human Rights, 21*(2), 295–307.
Sagot, M. (2005). The critical path of women affected by family violence in Latin America: Case studies from 10 countries. *Violence Against Women, 11*(10), 1292–1318.
Silverio-Murillo, A., Balmori de la Miyar, J. R., & Hoehn-Velasco, L. (2020). Families under confinement: COVID-19, domestic violence, and alcohol consumption. *SSRN Electronic Journal*. https://doi.org/10.2139/ssrn.3688384
Staggs, S. L., & Riger, S. (2005). Effects of intimate partner violence on low-income women's health and employment. *American Journal of Community Psychology, 36*(1), 133–145.
Wei, C., Barona, P. C., & Blaschke, T. (2015). Where is the poverty area? Quantifying the neighborhood effect in a deprivation index estimation: A case study in Quito, Ecuador. *GI_Forum, 2015*, 625–634.
Wilson, T. D. (2014). *Violence against women in Latin America*. SAGE Publications.

Chapter 13
Mapping COVID-19: Should It Be Based on the Incidence Rate? A Case Study in China

Meifang Li, Xuru Peng, and Xun Shi

Personal Story

During the last 2 years, I personally experienced two very different policy responses to the COVID-19 pandemic, one in China and one in the United States, which made me think quite a lot. I went back to China during the winter break of my job in an American college for a planned short family visit in January 2020. I bumped into the outbreak of the pandemic and got stuck there, because of the sudden and strict travel restriction. I managed to come back to the states after 13 months. When I was in China in 2020, because of my academic specialty in spatial epidemiology, I got invited by my hometown CDC (the CDC of Nanchang City) to help them develop a GIS-based system for dealing with the pandemic. This opportunity brought by the unfortunate delay of the trip and returning to my normal life allows me to get out of the ivory tower and gain the first-hand experience of a real-world war against a contemporary pandemic. With the implementation of extremely strict lockdown policies, China effectively abated nationwide transmission of the disease in about 2 months. Nanchang City, the capital of my home province, Jiangxi Province, has a population of 5 million people and is the one among

M. Li (✉)
Department of Geography, Dartmouth College, Hanover, NH, USA

School of Geography and Environment, Jiangxi Normal University, Nanchang, Jiangxi Province, China

Nanchang City Center for Disease Control and Prevention, Nanchang, Jiangxi Province, China
e-mail: meifang.li@dartmouth.edu

X. Peng · X. Shi
Department of Geography, Dartmouth College, Hanover, NH 03755, USA

M. Laituri et al. (eds.), *The Geographies of COVID-19*, Global Perspectives on Health Geography, https://doi.org/10.1007/978-3-031-11775-6_13

all provincial capitals in China nearest to the epicenter, Wuhan, but eventually had only 248 cases up to January 31, 2021. When I came back to the states in early February 2021, I felt I was like a soldier who just came back from a fierce battlefield suddenly plunged into a relaxing and soothing normal life. I was not required to quarantine although I just had traveled internationally. Almost all stores and restaurants were open. Except for the masks on people's faces, I could not tell much difference between what I saw and what had been before the pandemic. As a public health professional, I understood that such loose policies may lead to a higher risk of transmission, but why did not I want to refuse to enjoy this much more normal life? This question made me think about what the optimal policies and implementations would be, whether the strict policies in China are worth the cost in all aspects, and what we could do better next time (hope we do not have to answer the last question ever). I understand all these are complicated questions and I do not expect to have satisfactory answers to them shortly. This is why I am so glad to have this chance to contribute to this book by telling a little about the pandemic story in China. I see this contribution as a small part of a much bigger effort that seeks answers to those questions. *Meifang Li*

Introduction

Many spatial analyses of communicable diseases start with cluster detection, which seems to be a mindset carried over from the environmental health research of chronic diseases. Furthermore, many researchers directly applied the spatial autocorrelation tools available in commercial GIS or cluster detection methods developed for chronic diseases to communicable diseases. Ever since the outbreak of the COVID-19 pandemic, numerous studies have used that approach. For example, Shariati et al. (2020) directly used the Getis-Ord Gi* and Anselin's Local Moran's I in ArcGIS to detect hot spots of the cumulative incidence rate and cumulative mortality rate of all countries affected by the COVID-19 pandemic at the end of April 2020. Das et al. (2021) used the Getis-Ord Gi* to identify hot spots of COVID-19 incidence in the Kolkata megacity, India. Wang et al. (2021) applied global Moran's I, Anselin's Local Moran's I, Getis-Ord Gi*, and the spatiotemporal scan statistics to the city-level COVID-19 incidence in China, from January to October 2020. Xiong et al. (2020) used similar methods to explore the spatial patterns and the influencing factors of COVID-19 incidence at both prefecture and county levels in Hubei Province, China. Using daily case data at the county level, Desjardins et al. (2020) detected space-time clusters of COVID-19 in the United States with SaTScan for two overlapping periods: January 22 to March 9, 2020, and January 22 to March 27, 2020. Using the same method, Hohl et al. 2020 updated that work by extending the study period to April 27, 2020 (Hohl et al., 2020a), and then further to May 31,

2020 (Hohl et al., 2020b). Martines et al. (2021) used a similar method to detect daily COVID-19 clusters and examined the relative risk between several periods in 5570 Brazilian municipalities.

With no exception, all these works ran spatial autocorrelation or cluster detection analyses essentially based on the incidence rate, essentially a ratio between the case count and the population size, apparently a methodological setting directly taken from the analyses of chronic disease. However, the assumption beneath the analyses of chronic diseases based on the incidence rate is that there is a strong association between the case count and the population. In other words, a certain disease has a normal range or a baseline of rate in a certain population, and thus, we expect the case count in a certain population to be within a certain range. The disease cluster detection then detects if the case count within a population, location, and time period is abnormally high or low, so as to suggest local environmental or social risk factors (Mclafferty, 2015). Therefore, the influence of the population size needs to be eliminated or at least minimized, to reveal the impact of environmental risk factors (Shi, 2009). The most common way to eliminate the influence of population size is to use the rate rather than the case count. However, the assumption of a normal rate is problematic to an acute human-to-human communicable disease like COVID-19, because usually, we can hardly claim that there is a normal incidence rate for such a disease, and thus, the relationship between the case count and population size will not be straightforward. This problem has three major sources: First, the number of cases in a region is highly related to the number of primary cases, which are the cases that start the transmission in a region. For example, if there are no primary cases at all in a region, then no matter how large the population the region has, the number of cases in the region will be zero. Second, the number of cases is highly related to the chance of a patient to infect a susceptible person and in turn highly related to the population density, as a high population density increases the chance of contact between individuals. Third, the transmission process can be prevented and controlled, and thus, an epidemic may not have a natural development in the population and thus can hardly demonstrate the "real" relationship between the disease and the population. This last point is especially true in a contemporary epidemic/pandemic.

There have been studies exploring the relationships among the population size, population density, and COVID-19. Bhadra et al. (2021) found a moderate association between the population density and COVID-19 incidence rate and mortality rate at the district level of India. Kadi and Khelfaoui (2020) found a strong correlation between the population density and COVID-19 incidence in Algeria. Similar findings were also reported in the United States (Carozzi, 2020; Sy et al., 2021; Wong & Li, 2020), Turkey (Baser, 2021; Coşkun et al., 2021), Japan (Kodera et al., 2020), and Brazil (Pequeno et al., 2020). However, Sun et al. (2020) and Fang and Wahba (2020) found that the population density or urban density was not related to the COVID-19 incidence in China, and they attributed it to the strict lockdown policy implemented by the government. Barr and Tassier (2020) found that the timing of early case arrivals is much more important than that of the population density. Until now, there have been no consistent findings or conclusions for COVID-19 on the associations among the population size, population density, and case count, let alone for general communicable diseases. There is also a lack of in-depth exploration and discussion on this issue, which hampers our revealing, characterization, and understanding of the meaningful spatiotemporal variation of communicable diseases.

Here we used the COVID-19 data in China to explore its associations with population size and population density. Since China had the earliest outbreak of COVID-19 and also is a country that has implemented one of the strictest control policies in the world, the findings of this study will not only be indicative and have important academic and practical implications, but also be meaningful to the general study of contemporary communicable diseases.

Data and Methods

COVID-19 Epidemic Data

We obtained the COVID-19 data of China from an R package compiled by Wu et al. (2021), from which we extracted the daily cumulative number of confirmed cases since December 1, 2019, at the city and provincial levels. We inferred the daily new case count from the daily cumulative count. Figure 13.1 shows the daily new case count and daily cumulative case count from December 1, 2019, to December 31, 2021. As of December 31, 2021, 115,042 cases were reported in China, including 11,139 cases imported from outside China. The period from mid-January 2020 to the beginning of March 2020 was the primary peak period of the COVID-19 epidemic in China, followed by three small peaks occurring in August 2020, January 2021, and August 2021. These later peaks were mainly caused by imported cases

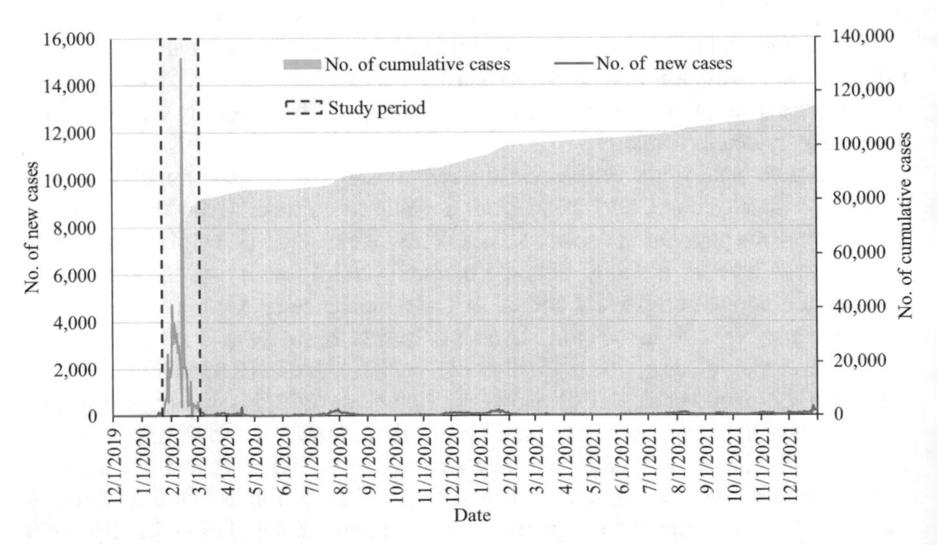

Fig. 13.1 Temporal distribution of daily numbers of COVID-19 cases in China from December 1, 2019, to December 31, 2021

and were quickly abated. In this study, we focus on the period of the primary peak, specifically, the 7 weeks from January 18, 2020, to March 6, 2020. A total of 80,751 indigenous cases were reported in China during this period. We conducted analyses at both city and provincial levels.

Population Data

To correspond to the disease data, we retrieved 2020 China's Population Count data (Bondarenko et al., 2020; WorldPop, 2021a) from the WorldPop, an open-access spatial demographic dataset (WorldPop, 2021b). The data was produced based on the 2020 population census and constrainedly disaggregated based on building areas determined by the Built-Settlement Growth Model (BSGM) (Nieves et al., 2020). It is a raster layer with about 100-m resolution, and each cell contains the estimated number of people in that small area represented by the cell.

We used the zonal statistics tool in ArcGIS to calculate the population and area of each city and each province. We then calculated the population density of each city and each province by dividing the population by the area of the city or province.

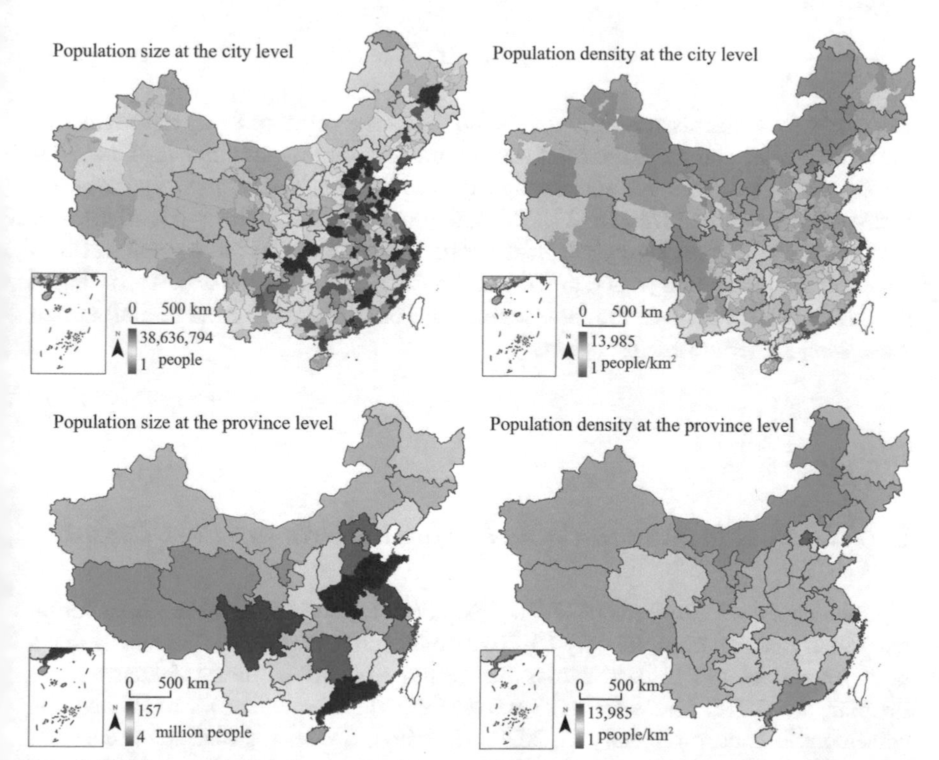

Fig. 13.2 The population size and population density at the city and provincial levels in China in 2020

Notably, the area only includes the building area that is likely to have a permanent population. Figure 13.2 shows the spatial distribution of population size and population density at the city and provincial levels.

Statistical Methods

For our seven-week study period, we ran the linear regression to explore the association between the COVID-19 epidemic and the population size or population density on a weekly basis. This analysis was applied at both city and provincial levels. We applied the regression analysis to the weekly case data of those provinces or cities with at least one case in that week, to avoid the zero-inflation issue.

Besides the case count, we also evaluated the association between the transmission ability and population size or population density. Due to the limitation of data, in this study we simplistically calculated the transmission ability by dividing the case count of the *later weeks* of the 7 weeks by the case count of the *earlier weeks*, assuming that cases in the *earlier weeks* were imported and the cases in the *later weeks* were indigenous. The calculation of transmission ability is shown in Eq. (13.1):

$$r = C_{\mathrm{l}} / C_{\mathrm{e}} \tag{13.1}$$

where r is the transmission ability, C_{l} is the total case count of a city or province in the *later weeks*, and C_{e} is the total case count of a city or province in the *earlier weeks*. We tested three specifications of the *earlier weeks* (i.e., the imported-case period): the first week (Jan 18–24, 2020), the first 2 weeks (Jan 18–31, 2020), and the first 3 weeks (Jan 18–Feb 7, 2020). The specification of the *earlier weeks* is based on the incubation period of COVID-19 (McAloon et al., 2020). Accordingly, the three specifications of the *later weeks* (i.e., the indigenous-case period) are all those weeks after the *earlier weeks*.

Results

COVID-19 Epidemics in China During the Primary Peak Period

Figure 13.3 shows the weekly COVID-19 epidemic at the city level of China during the primary peak period, January 18, 2020, to March 6, 2020. The maps show that a rapid spread of the disease across the country occurred during January 25 to February 14, 2020. Those are the 3 weeks following the lockdown of Wuhan, the epicenter of China, on January 24, 2020. After those 3 weeks, the epidemic declined sharply. Spatially, the transmissions mostly occurred in the eastern part of China, especially the cities close to Wuhan.

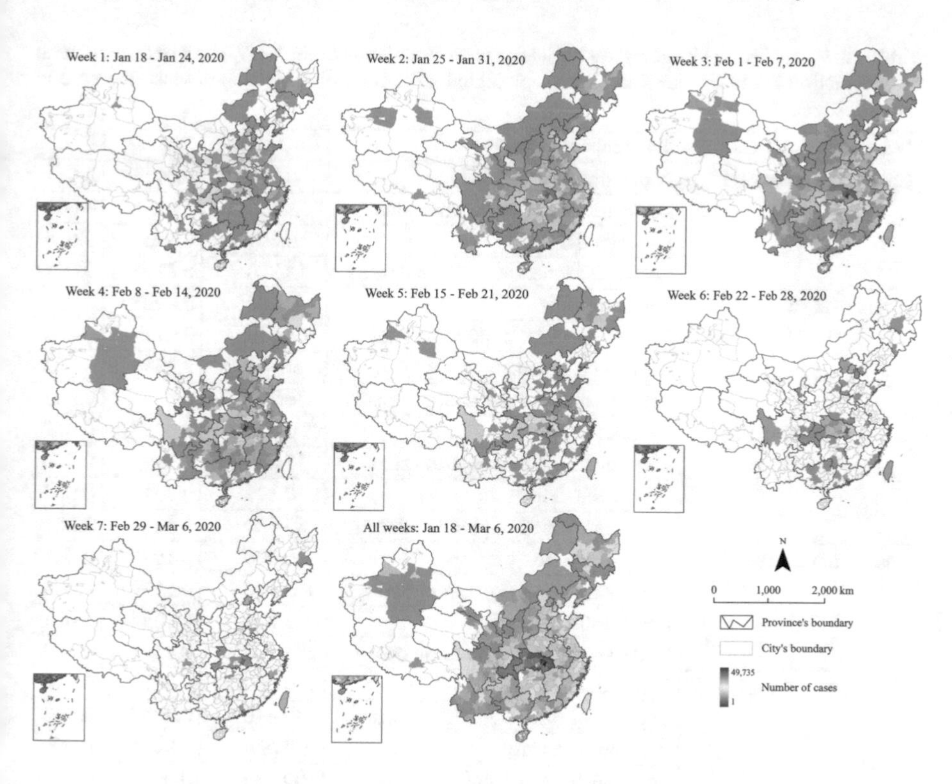

Fig. 13.3 Spatial distribution of weekly COVID-19 case counts in China during our study period, January 18 to March 6, 2020

Association of COVID-19 Epidemic with the Population in China at the City Level

Tables 13.1 and 13.2 show the city-level statistical results on the associations of the population size and the population density with the COVID-19 epidemic in China, evaluated using the linear regression analysis. To explore the potential nonlinear relationship between the population and case count/transmission ability, we also conducted regression analysis using log10-transformed population values.

Table 13.1 shows that only weeks 1, 2, and 3 (the period observing the majority of the cases) and the entire study period claim statistically significant associations between the case count and the original population size, and only weeks 1 and 2 have statistically significant associations about the log-transformed population size. All associations about population size are quite weak ($R^2 \leq 0.044$), and those about the population density are even weaker ($R^2 \leq 0.007$), and none of them are significant. The log10 transformation does not have a considerable impact on the results. Generally, the log-transformation slightly lowers R^2 and reduces the statistical significance for the population size, whereas it slightly raises R^2 and increases the

Table 13.1 Weekly results of regression analysis on association of the COVID-19 case count with the population at the city level in China, calculated with those cities having at least one case in that week

Dependent variable	Independent variable	Coefficient	*P*-value	R^2
Case count of week 1	Population size	0.000	0.007**	0.044
	Log-transformed population size	21.430	0.028*	0.030
	Population density	0.191	0.376	0.005
	Log-transformed population density	13.970	0.275	0.007
Case count of week 2	Population size	0.000	0.002**	0.034
	Log-transformed population size	68.370	0.011*	0.022
	Population density	0.704	0.357	0.003
	Log-transformed population density	36.337	0.348	0.003
Case count of week 3	Population size	0.000	0.025*	0.017
	Log-transformed population size	77.050	0.195	0.006
	Population density	0.677	0.761	0.000
	Log-transformed population density	83.480	0.521	0.001
Case count of week 4	Population size	0.000	0.103	0.011
	Log-transformed population size	412.100	0.153	0.008
	Population density	3.314	0.665	0.001
	Log-transformed population density	295.900	0.453	0.002
Case count of week 5	Population size	0.000	0.207	0.010
	Log-transformed population size	49.840	0.484	0.003
	Population density	0.686	0.803	0.000
	Log-transformed population density	100.830	0.529	0.003
Case count of week 6	Population size	0.000	0.459	0.010
	Log-transformed population size	24.750	0.599	0.005
	Population density	0.169	0.935	0.000
	Log-transformed population density	71.880	0.628	0.004
Case count of week 7	Population size	0.000	0.651	0.009
	Log-transformed population size	21.130	0.656	0.009
	Population density	−0.318	0.882	0.001
	Log-transformed population density	42.890	0.793	0.003
Case count of all weeks	Population size	0.000	0.029*	0.015
	Log-transformed population size	321.800	0.220	0.005
	Population density	3.841	0.698	0.000
	Log-transformed population density	425.400	0.446	0.002

Significant codes: 0 "***" 0.001 "**" 0.01 "*" 0.05 "" 1

statistical significance for the population density. There is also a trend that the associations became weaker as the epidemic evolved.

As for the transmission ability, Table 13.2 shows that only the setting of using the first 2 weeks as the imported-case period results in slight albeit significant negative associations with the original and log-transformed population density ($R^2 \leq 0.020$), and using the first 3 weeks as the imported-case period results in significant negative associations with the log-transformed population size and log-transformed

Table 13.2 Results of regression analysis on the association of the COVID-19 transmission ability with the population at the city level in China, calculated with those cities having at least one case in the imported-case period

Dependent variable	Independent variable	Coefficient	P-value	R^2
Infection rate calculated using week 1 as the imported-case period	Population size	−0.000	0.293	0.007
	Log-transformed population size	−47.940	0.135	0.014
	Population density	−0.538	0.447	0.004
	Log-transformed population density	−23.220	0.581	0.002
Infection rate calculated using weeks 1–2 as the imported-case period	Population size	0.000	0.991	0.000
	Log-transformed population size	0.288	0.555	0.001
	Population density	−0.030	0.030*	0.016
	Log-transformed population density	−1.666	0.016*	0.020
Infection rate calculated using weeks 1–3 as the imported-case period	Population size	−0.000	0.216	0.005
	Log-transformed population size	−0.231	0.001***	0.033
	Population density	−0.000	0.947	0.000
	Log-transformed population density	−0.379	0.012*	0.020

Significant codes: 0 "***" 0.001 "**" 0.01 "*" 0.05 "" 1

population density. All the other settings do not result in significant associations. The log-transformation generally gives somewhat stronger associations.

Association of COVID-19 Epidemic with the Population in China at the Provincial Level

Tables 13.3 and 13.4 show the province-level results. At the provincial level, the associations become even weaker than their counterparts at the city level. None of the associations at the provincial level are significant. Another notable pattern is that, at the provincial level, the association between the case count and the population density is consistently negative across different settings and for both original and log-transformed values (Table 13.3). For the transmission ability, this pattern is not consistent: the original values and transformed values of population density result in opposite signs for the association. However, since all the associations are very weak and not significant, it is hard to judge if this negative-association pattern is meaningful.

Table 13.3 Weekly results of regression analysis on the association of the COVID-19 case count with the population at the provincial level in China, calculated with those provinces having at least one case in that week

Dependent variable	Independent variable	Coefficient	*P*-value	R^2
Case count of week 1	Population size	0.000	0.363	0.028
	Log-transformed population size	8.430	0.498	0.015
	Population density	−0.282	0.681	0.006
	Log-transformed population density	−4.227	0.938	0.000
Case count of week 2	Population size	0.000	0.333	0.029
	Log-transformed population size	78.080	0.502	0.014
	Population density	−2.892	0.650	0.006
	Log-transformed population density	−93.270	0.852	0.001
Case count of week 3	Population size	0.000	0.539	0.012
	Log-transformed population size	165.400	0.616	0.008
	Population density	−8.161	0.653	0.007
	Log-transformed population density	−443.700	0.757	0.003
Case count of week 4	Population size	0.000	0.712	0.005
	Log-transformed population size	227.600	0.723	0.004
	Population density	−17.020	0.689	0.006
	Log-transformed population density	−737.300	0.787	0.003
Case count of week 5	Population size	0.000	0.699	0.005
	Log-transformed population size	69.490	0.725	0.004
	Population density	−5.305	0.684	0.006
	Log-transformed population density	−237.200	0.777	0.003
Case count of week 6	Population size	0.000	0.839	0.002
	Log-transformed population size	25.910	0.741	0.006
	Population density	−2.274	0.668	0.010
	Log-transformed population density	−117.300	0.742	0.006
Case count of week 7	Population size	0.000	0.846	0.004
	Log-transformed population size	19.680	0.686	0.015
	Population density	−1.545	0.636	0.021
	Log-transformed population density	−58.690	0.790	0.007
Case count of all weeks	Population size	0.000	0.563	0.011
	Log-transformed population size	592.000	0.631	0.007
	Population density	−28.340	0.675	0.006
	Log-transformed population density	−1498.000	0.777	0.003

Significant codes: 0 "***" 0.001 "**" 0.01 "*" 0.05 "" 1

Discussion

With the COVID-19 data of China during the first and primary peak period (January 18–March 6, 2020), this research explored how the spatial variation of COVID-19, as an acute human-to-human communicable disease, was associated with the population size and population density. This study was initiated by the consideration that

Table 13.4 Results of regression analysis on the association of the COVID-19 transmission ability with the population at the provincial level in China, calculated with those provinces having at least one case in the imported-case period

Dependent variable	Independent variable	Coefficient	*P*-value	R^2
Infection rate calculated using week 1 as the imported-case period	Population size	0.000	0.177	0.060
	Log-transformed population size	2.819	0.144	0.070
	Population density	−0.178	0.093	0.091
	Log-transformed population density	−11.329	0.177	0.060
Infection rate calculated using weeks 1–2 as the imported-case period	Population size	−0.000	0.610	0.008
	Log-transformed population size	−0.156	0.411	0.021
	Population density	0.005	0.646	0.007
	Log-transformed population density	−0.229	0.778	0.003
Infection rate calculated using weeks 1–3 as the imported-case period	Population size	−0.000	0.235	0.044
	Log-transformed population size	−0.172	0.004**	0.228
	Population density	0.005	0.117	0.075
	Log-transformed population density	−0.034	0.903	0.000

Significant codes: 0 "***" 0.001 "**" 0.01 "*" 0.05 "" 1

the case count of a human-to-human communicable disease would not have a simple relationship with the population size like its counterpart of a common chronic disease, and thus, typical spatial analytical procedures originally developed for chronic diseases in environmental health studies, e.g., cluster detection based on disease rate, may not be directly applicable to a communicable disease.

This study found that neither the population size nor the population density is strongly associated with the case count or the transmission rate of COVID-19 in China, at both city and provincial levels. These findings are consistent with two previous studies (Fang & Wahba, 2020; Sun et al., 2020) that point out that the population density is not associated with the variation of the COVID-19 epidemic in China and attribute this lack of association to the country's strict lockdown policies. "Lockdown" in China means the implementation of some very strict measures. For example, when a case was identified and located, the entire apartment building or even the entire neighborhood this person resided in would be locked: no one would be allowed to come out from the building or the neighborhood, and daily necessities would be delivered by certain community managers to the door of each family. Our elaboration on this is that such measures apparently had effectively reduced the contact frequency among people, and thereby, the transmission got rapidly stopped and rarely formed a natural transmission process in the local population. Moreover, the early lockdown of the epicenter, the Wuhan City and Hubei Province, had considerably lessened the spread of the disease to other areas. For example, Nanchang City in Jiangxi Province, the nearest provincial capital from

Wuhan, with a population of about 5 million, only saw 248 cases as of December 31, 2021. This results in the small number problem that makes the incidence or the relationship between case count and population highly statistically unstable.

The finding of this study implies that an analysis based on the incidence rate of COVID-19 in China becomes less meaningful. For example, it may be not necessary, or even misleading, to run cluster detection using the incidence rate to identify the high-risk area of COVID-19 in China. Instead, depending on the research question to answer, simple use of the case count may be more informative. Ideally, if the available data can support it, the optimal measurement should be the case-reproduction ratio (R_t) (Li et al., 2019). This notion can be extended to other areas and/or other communicable diseases with similar characteristics to the COVID-19 epidemic in China (strictly controlled situation and limited case count).

This study is a preliminary exploration with limitations, from the data to the methods. Due to the data (un)availability, we have only obtained the case count aggregated to the city and provincial levels, with which we cannot tell when and where each patient got infected. As a result, we could only calculate a very simplistic measurement of the transmission ability. Also, we do not have the demographic and socioeconomic information about those cases, e.g., age, gender, and career, which hinders us to explore the differences across them. The primary statistical analysis used in this study is the linear regression that took in all provinces or cities having cases as subjects. More sophisticated analytical techniques, e.g., nonlinear models and localized models, can be further explored.

Conclusion

The population size and population density were not closely related to the COVID-19 case count and transmission ability in China, at both city and provincial levels. This is likely due to the strict prevention measures and the limited number of cases in the administrative units. This finding brings up the notion that when conducting spatial analysis for COVID-19 and similar human-to-human communicable diseases, e.g., cluster detection or environmental-risk detection, we should be more cautious and considerate, rather than simply take methods and processes developed for chronic diseases.

Acknowledgment This research was funded by the National Natural Science Foundation of China (grant no.: 42001342), the 2020 Scholar Innovation and Advancement Awards of Dartmouth, and the Fellowship of the Geospatial Software Institute (GSI) Conceptualization Project sponsored by NSF.

References

Barr, J., & Tassier, T. (2020). *Are crowded cities the reason for the COVID-19 pandemic? Placing too much blame on urban density is a mistake*. Available at: https://blogs.scientificamerican.com/observations/are-crowded-cities-the-reason-for-the-covid-19-pandemic/ (accessed December 1, 2021).

Baser, O. (2021). Population density index and its use for distribution of COVID-19: A case study using Turkish data. *Health Policy, 125*(2), 148–154.

Bhadra, A., Mukherjee, A., & Sarkar, K. (2021). Impact of population density on COVID-19 infected and mortality rate in India. *Modeling Earth Systems and Environment, 7*(1), 623–629.

Bondarenko, M., Kerr, D., Sorichetta, A., et al. (2020). *Census/projection-disaggregated gridded population datasets, adjusted to match the corresponding UNPD 2020 estimates, for 183 countries in 2020 using Built-Settlement Growth Model (BSGM) outputs*. The University of Southampton.

Carozzi, F. (2020). *Urban density and COVID-19*. Available at SSRN 3643204.

Coşkun, H., Yıldırım, N., & Gündüz, S. (2021). The spread of COVID-19 virus through population density and wind in Turkey cities. *Science of the Total Environment, 751*, 141663.

Das, A., Ghosh, S., Das, K., et al. (2021). Living environment matters: Unravelling the spatial clustering of COVID-19 hotspots in Kolkata megacity, India. *Sustainable Cities and Society, 65*, 102577.

Desjardins, M. R., Hohl, A., & Delmelle, E. M. (2020). Rapid surveillance of COVID-19 in the United States using a prospective space-time scan statistic: Detecting and evaluating emerging clusters. *Applied Geography, 118*, 102202.

Fang, W., & Wahba, S. (2020). *Urban density is not an enemy in the coronavirus fight: Evidence from China*. Available at: https://blogs.worldbank.org/sustainablecities/urban-density-not-enemy-coronavirus-fight-evidence-china (accessed December 1, 2021).

Hohl, A., Delmelle, E., & Desjardins, M. (2020a). Rapid detection of COVID-19 clusters in the United States using a prospective space-time scan statistic: An update. *Sigspatial Special, 12*(1), 27–33.

Hohl, A., Delmelle, E. M., Desjardins, M. R., et al. (2020b). Daily surveillance of COVID-19 using the prospective space-time scan statistic in the United States. *Spatial and Spatio-temporal Epidemiology, 34*, 100354.

Kadi, N., & Khelfaoui, M. (2020). Population density, a factor in the spread of COVID-19 in Algeria: Statistic study. *Bulletin of the National Research Centre, 44*(1), 1–7.

Kodera, S., Rashed, E. A., & Hirata, A. (2020). Correlation between COVID-19 morbidity and mortality rates in Japan and local population density, temperature, and absolute humidity. *International Journal of Environmental Research and Public Health, 17*(15), 5477.

Li, M., Shi, X., Li, X., et al. (2019). Epidemic forest: A spatiotemporal model for communicable diseases. *Annals of the American Association of Geographers, 109*(3), 812–836.

Martines, M. R., Ferreira, R. V., Toppa, R. H., et al. (2021). Detecting space-time clusters of COVID-19 in Brazil: Mortality, inequality, socioeconomic vulnerability, and the relative risk of the disease in Brazilian municipalities. *Journal of Geographical Systems, 23*(1), 7–36.

McAloon, C., Collins, Á., Hunt, K., et al. (2020). Incubation period of COVID-19: A rapid systematic review and meta-analysis of observational research. *BMJ Open, 10*(8), e039652.

Mclafferty, S. (2015). Disease cluster detection methods: Recent developments and public health implications. *Annals of GIS, 21*, 127–133.

Nieves, J. J., Sorichetta, A., Linard, C., et al. (2020). Annually modelling built-settlements between remotely-sensed observations using relative changes in subnational populations and lights at night. *Computers, Environment and Urban Systems, 80*, 101444.

Pequeno, P., Mendel, B., Rosa, C., et al. (2020). Air transportation, population density, and temperature predict the spread of COVID-19 in Brazil. *PeerJ, 8*, e9322.

Shariati, M., Mesgari, T., Kasraee, M., et al. (2020). Spatiotemporal analysis and hotspots detection of COVID-19 using geographic information system (March and April 2020). *Journal of Environmental Health Science and Engineering, 18*(2), 1499–1507.

Shi, X. (2009). A geocomputational process for characterizing the spatial pattern of lung cancer incidence in New Hampshire. *Annals of the Association of American Geographers, 99*, 521–533.

Sun, Z., Zhang, H., Yang, Y., et al. (2020). Impacts of geographic factors and population density on the COVID-19 spreading under the lockdown policies of China. *Science of the Total Environment, 746*, 141347.

Sy, K. T. L., White, L. F., & Nichols, B. E. (2021). Population density and basic reproductive number of COVID-19 across United States counties. *PLoS One, 16*(4), e0249271.

Wang, Q., Dong, W., Yang, K., et al. (2021). Temporal and spatial analysis of COVID-19 transmission in China and its influencing factors. *International Journal of Infectious Diseases, 105*, 675–685.

Wong, D. W., & Li, Y. (2020). Spreading of COVID-19: Density matters. *PLoS One, 15*(12), e0242398.

WorldPop. (2021a). *The spatial distribution of population in 2020 with country total adjusted to match the corresponding UNPD estimate, China*. Available at: https://www.worldpop.org/geodata/summary?id=49919 (accessed December 1, 2021).

WorldPop. (2021b). *Open spatial demographic data and research*. Available at: https://www.worldpop.org/ (accessed December 1, 2021).

Wu, T., Hu, E., Ge, X., et al. (2021). nCov2019: An R package for studying the COVID-19 coronavirus pandemic. *PeerJ, 9*, e11421.

Xiong, Y., Wang, Y., Chen, F., et al. (2020). Spatial statistics and influencing factors of the COVID-19 epidemic at both prefecture and county levels in Hubei Province, China. *International Journal of Environmental Research and Public Health, 17*(11), 3903.

Chapter 14
Regional Patterns of the Pandemic: A View from Aotearoa New Zealand

Malcolm Campbell, Lukas Marek, Jesse Wiki, Matthew Hobbs, Lindsey Conrow, and Simon Kingham

Personal Story

At the very outset of the pandemic, the televised address from the New Zealand Prime Minister, Jacinda Ardern, signalled something serious was coming, akin to a wartime broadcast. Those distant memories of mild confusion and the first national lockdown now feel like a different era. Almost two years later, COVID-19 has fully arrived in our midst and is very much an unwelcome guest. New Zealand has felt like a different planet compared to many other places around the world. But an inability to travel, even with very ill family overseas, has been cruel, but necessary. The emotional burden of separation has become a constant background noise, being disconnected from family and friends that you would rather be beside. The personal burden also sits alongside the loss of professional networks, conferences and visiting colleagues around the world.

For me, a key point of hopefulness for the future has been the strength of the collective efforts that have been focused on COVID-19 from the research community around NZ. Seeing just how rapidly a sense of solidarity and

M. Campbell (✉) · L. Conrow · S. Kingham
GeoHealth Laboratory, University of Canterbury, Christchurch, New Zealand

School of Earth and Environment, University of Canterbury, Chrsitchurch, New Zealand
e-mail: Malcolm.campbell@canterbury.ac.nz

L. Marek
GeoHealth Laboratory, University of Canterbury, Christchurch, New Zealand

J. Wiki
GeoHealth Laboratory, University of Canterbury, Christchurch, New Zealand

School of Population Health, University of Auckland, Auckland, New Zealand

M. Hobbs
GeoHealth Laboratory, University of Canterbury, Christchurch, New Zealand

School of Health Sciences, University of Canterbury, Christchurch, New Zealand

M. Laituri et al. (eds.), *The Geographies of COVID-19*, Global Perspectives on Health Geography, https://doi.org/10.1007/978-3-031-11775-6_14

mission can take over when most needed, academics mapping and modelling disease spread in tandem with those in government. The story continues….
Malcolm Campbell

Introduction

This chapter aims to explore how the COVID-19 pandemic unfolded in Aotearoa New Zealand (NZ). It provides an overview of a range of research carried out during the COVID-19 pandemic, with a specific focus on NZ. First, we begin with a short review of the emergence of the SARS-CoV-2 virus in early 2020 and its impacts on NZ. The chapter takes a journey through the (still) unfolding pandemic from a NZ perspective, recognising some of the important and unique features of the NZ context, such as the position of Māori within NZ.

Second, we turn to explore the role of place, or a geographical perspective, a well-known determinant of disease and illness (Macintyre et al., 2002; Bowie et al., 2013; Campbell et al., 2014; Campbell & Ballas, 2016; Marek et al., 2021b), that can either ameliorate or exacerbate health and well-being outcomes. In addition, we discuss the linked existing and accelerating inequalities (Ahmed et al., 2020) and inequities generated by the pandemic in NZ, arguably giving rise to an emerging syndemic.

Third, we explore how geography has become a critical area of research in understanding how to contain the SARs-CoV-2 virus. Often, NZ has been isolated from the rest of the world using a series of border controls that only allow NZ citizens and permanent residents to enter, placing those people without the prerequisite visas in a precarious position. The international borders have also been supplemented by regional borders within NZ. Additionally, the *locations of interest*, a term used to describe potential exposure events, have become a regular feature of daily life for some, requiring individuals to get tested or to isolate from the rest of the community.

Further, we discuss some of the more emergent social issues that have been exacerbated or have accelerated, during the unfolding pandemic, before turning to focus on the subsequent waves of COVID-19 emanating from new variants of concern in NZ. Further, emerging variants of concern (VoC) have altered the government's approach to the pandemic from an elimination to a suppression strategy. The penultimate section of the chapter focuses on the vaccination experience of NZ, demonstrating how inequalities persistently arise in health care delivery.

We conclude the chapter with a summary of the most salient aspects of COVID-19 in NZ, highlighting potential policy solutions to some of the inequalities and patterns we have observed.

The Emergence of COVID-19 in Aotearoa New Zealand

The emergence of COVID-19 in early 2020 was a challenge to many governments worldwide facing an unprecedented public health challenge. In NZ, the first case of COVID-19 was reported on the 28th of February 2020. This gave rise to the use of several Nonpharmaceutical Interventions (NPIs), as a primary tool of the public health response. NPIs are used because no vaccination or medication to treat COVID-19 existed (or was approved by regulating agencies at that time). The NPIs involved measures such as social distancing and practicing good hand hygiene as an attempt to contain the rapid spread of a virus that could lead to morbidity or mortality. In the initial phase of COVID-19, NZ closed its borders (Table 14.1) and managed to eliminate the virus within its borders, reducing the need for movement restrictions in NZ for large parts of 2020 (Fig. 14.1), different from the international context.

Table 14.1 Key dates NZ alert level changes. Alert levels specified the public health and social measures to be taken, level 1 being the base level and level 4 the most stringent

Date	Nonpharmaceutical intervention/alert level change
16-Mar-20	(BR): border restrictions, compulsory self-isolation, cruise ships prohibited
20-Mar-20	(BC): border closed to all but NZ citizens and residents
22-Mar-20	(L2): level 2
24-Mar-20	(L3): level 3
26-Mar-20	(L4): level 4
28-Apr-20	(L3): level 3
14-May-20	(L2): level 2
9-Jun-20	(L1): level 1
12-Aug-20	(L2, AKL L3): Auckland level 3, rest of NZ level 2
31-Aug-20	(L2): level 2
22-Sep-20	(L1, AKL L2): Auckland level 2 (24 Sep), rest of NZ level 1 (22 Sep)
8-Oct-20	(L1): level 1
14-Feb-21	(L2, AKL L3): Auckland level 3, rest of NZ level 2
18-Feb-21	(L1, AKL L2): Auckland level 2, rest of NZ level 1
23-Feb-21	(L1): level 1
28-Feb-21	(L2, AKL L3): Auckland level 3, rest of NZ level 2
7-Mar-21	(L1, AKL L2): Auckland level 2, rest of NZ level 1
12-Mar-21	(L1): level 1
23-Jun-21	(L1, WLG L2): Wellington level 2, rest of NZ level 1
30-Jun-21	(L1): level 1
18-Aug-21	(L4): level 4
1-Sep-21	(L3, AKL NLD L4): Auckland and Northland Level 4, rest of NZ level 3
3-Sep-21	(L3, AKL L4): Auckland level 4, rest of NZ level 3
8-Sep-21	(L2, AKL L4): Auckland level 4, rest of NZ level 2
22-Sep-21	(L2, AKL L3): Auckland level 3, rest of NZ level 2
4-Oct-21	(L2, AKL HN L3): Auckland and parts of Waikato level 3, rest of NZ level 2
6-Oct-21	(L2, *AKL HN L3): Auckland (eased restrictions) and parts of Waikato level 3, rest of NZ level 2
9-Oct-21	(L2, AKL NLD HN L3): Auckland, Northland and parts of Waikato level 3, rest of NZ level 2
20-Oct-21	(L2, AKL HN L3): Auckland and parts of Waikato level 3, rest of NZ level 2

Source: New Zealand Government (2021b)

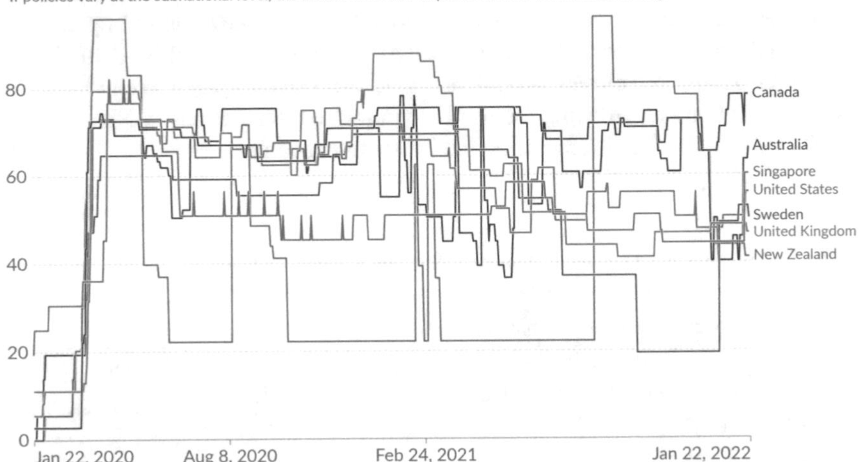

Fig. 14.1 COVID-19 stringency index, selected countries. (Source: https://ourworldindata.org/covid-stringency-index. Accessed Feb 2022)

NZ's early experience of COVID-19 is an important point of departure; the trajectory of COVID-19 was rather more impactful in many places worldwide, even amongst developed countries, such as the United Kingdom (UK), the United States of America (USA) or Sweden, opting instead for mitigation of COVID-19 (Baral et al., 2021). We can see (Fig. 14.1) that amongst similar countries, NZ has utilised the most stringent level of restrictions during its outbreaks and then removed restrictions rapidly once the outbreaks were contained. The term 'going hard and early' was coined to characterise the policy of rapid and wide-ranging restrictions such as lockdowns in the public imagination to communicate the rapid restrictions that would be used. This has given rise to a rather different experience of COVID-19, compared to other countries, with the population enjoying relative freedom and normalcy (except at the international border) for large periods of 2020 and 2021. Travellers to NZ were only allowed entry if they were NZ citizens or residents and had to undergo a period of mandatory quarantine for 14 days. The mandatory quarantine system (Managed Isolation and Quarantine [MIQ]) was effective in preventing spread into NZ from the outside, but led to a host of problems for those who needed to travel internationally for a variety of legitimate reasons. The MIQ system used hotels to house those who were returning from overseas before being released back into the community. At the time of writing the border controls were phased out at 11.59 pm on 31st July 2022, more than 2 years after introduction.

The Geography of COVID-19 in Aotearoa New Zealand

In this section, we discuss the introduction of geography as the pandemic has unfolded in NZ, specifically, the use of regional lockdowns and movement data which has highlighted inequities between people and places in NZ.

Early in the pandemic, the response was a single coherent nationwide policy that had no geographic nuance. This made the impact of restrictions on movement apply to every citizen equally across NZ. There was a simplicity to this single, national non-spatial approach, which was employed successfully to eliminate COVID-19 in the first wave of the pandemic alongside other measures such as closed borders. The impacts of nationwide restrictions on movement are evident in mobility data collected since the pandemic began, with drastic reductions in visits to many types of places during higher alert levels (Fig. 14.2). However, in subsequent outbreaks of the SARS-CoV-2 virus, a more nuanced and specific approach was required as the virus impacted some places more than others.

For instance, on 12 August 2020 (see Table 14.1), it was announced that the Auckland region would move to a more stringent series of restrictions (alert level 3), while the rest of NZ would not. This divergence and the introduction of a geographically targeted approach was new to NZ. A similar scenario emerged when on 28 February 2021, the largest city of Auckland, again experienced a higher

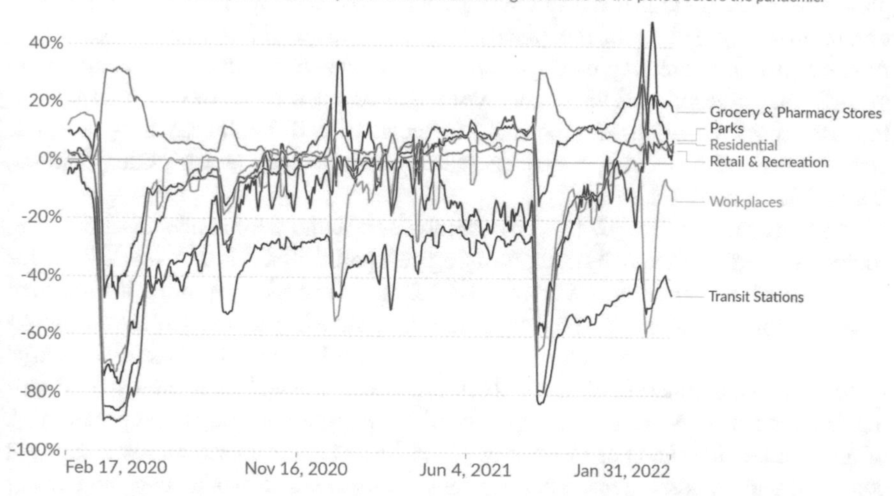

Fig. 14.2 Visitors by location type. (Note that residential is measured in duration, Google Community Mobility Trends, NZ (Ritchie et al., 2020). Accessed Feb 2022. https://ourworldindata.org/covid-google-mobility-trends)

level of movement restriction than the rest of NZ. These new geographically distinct restrictions meant that the capital city, Wellington, experienced a similar fate as Auckland on 23 June 2021, when a visiting Australian tourist with COVID-19 meant that this region experienced a differing level of restrictions than the rest of NZ (Table 14.1).

In previous research (Campbell et al., 2021), we reported on the important distinctions in movement that occurred before and during a national lockdown. The research showed that those places which were relatively more deprived (as measured by a socio-economic deprivation index) could not reduce their movements by as much as relatively less deprived places. This rendered those neighbourhoods with the most movement more susceptible to the virus in their communities. This is an important inequity related to a neighbourhood deprivation status, a proxy for socio-economic disadvantage. We have previously hypothesised that this is partially due to the types of jobs that people undertake being 'essential', for example, those working in supermarkets, who were (or are) unable to work from home. There are established patterns of health outcomes and behaviours being strongly related to the level of neighbourhood deprivation, including for COVID-19 (Beale et al., 2021).

Though disrupted movement patterns and behaviours are expected in lockdown scenarios, the NZ experience nonetheless highlights ongoing justice and inequity issues around transport and accessibility. Public transport (PT) service capacities were reduced to accommodate social distancing requirements during alert levels 3 and 4 and PT users in NZ tend to be younger and living in areas with higher deprivation (Shaw et al., 2017). Adults with disabilities have continued using public transport at significantly higher rates than those with no disability and are more likely to report difficulty getting to the places they need to go (New Zealand Transport Agency, 2021). There may be disproportionately negative impacts for people who already face transport disadvantage with reduced access to modal alternatives. Further access reductions run the risk of exacerbating difficulty reaching essential destinations and increasing social exclusion (Rojas-Rueda and Morales-Zamora, 2021; Palm et al., 2021).

Previous research in NZ has also demonstrated the confluence of underlying social and medical factors by area, utilising geospatial methods to visualise metrics of area-level vulnerability (Wiki et al., 2021b). Through mapping, it identified areas that may experience disproportionately negative impacts of disease outbreaks based on age, underlying health conditions, sociocultural factors, socio-economic factors and a combination of these. Results showed unequal social and spatial vulnerabilities across NZ, identifying areas where populations may be at greater risk of adverse health outcomes (Marek et al., 2021c). This is a particularly important consideration as such areas typically have less access to health care and fewer resources.

Ongoing Outbreaks: An (Ever) Evolving Pandemic

The arrival of the delta (Δ/δ) variant of concern (VoC) in NZ changed the approach of the main actor during the pandemic, the government, in attempting to contain and/or eliminate the virus. Delta was a more transmissible variant, spreading more easily than prior VoCs and caused people to develop more serious COVID-19 illness than prior variants with higher risk of needing hospital care (New Zealand Ministry of Health, 2021a). In addition to this, people with delta infections carried a higher viral load for longer than previous VoCs. The time from exposure to the virus until the first positive test was also shorter for the delta variant. This combination of circumstances presented NZ with a much more challenging job to eliminate the virus.

At the time of writing, it is unclear what the future trajectory of the pandemic will be in NZ, with predictions ranging from elimination to the virus becoming endemic with the arrival of the omicron (O/o) VoC. However, one key lesson is already emerging, that the strategy that served NZ well in the earlier outbreaks is not going to be as effective against a more transmissible variant of SARs-COV-2. A key aspect of this has been the confusing messages that have begun to emerge around the 'rules' of the new alert levels (discussed in an earlier section). The alert level system that worked well in previous outbreaks (Table 14.1) was altered to include new categories within level 3 (of four levels) to include a series of steps, 1, 2 and 3, progressively less restrictive as the steps increased. This would gradually increase the ability of those under restrictions to increase their movements and social contacts, from 10 to 25 to 50 contacts. The government announced the next stage of the COVID-19 response plan was to provide a pathway out of lockdown and give vaccinated citizens more freedom: the COVID-19 Protection Framework.

The COVID-19 Protection Framework introduced a new flexible three-level approach to managing COVID-19 in the community based on three levels: green, orange and red (New Zealand Government, 2021a). Green signals there are some COVID-19 cases in the community and sporadic imported cases, limited community transmission and manageable hospitalisations. At the orange level, there will be increasing community transmission that is putting pressure on the health system. The whole health system shifts focus to pandemic response and there may also be an increasing threat for at-risk people. Finally, the red level means action needs to be taken to protect both at-risk people and the health system from an unsustainable number of hospitalisations and working from home will be encouraged.

What is distinct about more recent outbreaks has been the important role of vaccination in being able to ameliorate some of the worst effects of the pandemic (see Table 14.2). The table demonstrates the strong link between not being vaccinated and hospitalisation (187/1806) = 10.4% vs (3/240 = 1.3%), as observed worldwide (Tartof et al., 2021). The observations show that there are also more cases amongst the unvaccinated.

Table 14.2 Vaccination by hospitalisation status for COVID-19 cases

Vaccination status	Total cases	Cases hospitalised	%
No doses	1806	187	10.4
Not eligible (less than 12 years)	709	12	1.7
1 dose only (less than 14 days)	269	26	9.7
1 dose only (at least 14 days)	398	27	6.8
Fully vaccinated (2nd dose less than 14 days)	88	5	5.7
Fully vaccinated (at least 14 days)	240	3	1.3
Total	*3510*	*260*	*7.4*

Source: MoH as of 1 Nov 2021

Vaccination: Vax to the Max

The vaccination programme in NZ experienced a slow start in comparison to many other Organisation for Economic Co-operation and Development (OECD) countries and was formulated in an environment where there was no widespread (or often any) community transmission within the international border (New Zealand Ministry of Health, 2021b). This situation gave rise to an approach that prioritised those at the border as the most at-risk group from potential exposure to the virus. Group 1 (border workers and their household contacts) was prioritised given this context. Subsequently, Group 2 were health workers such as emergency response services or working in a residential facility as well as defence force personnel (and their household contacts) with border roles. The final category included those aged over 65 years and with underlying health conditions in the Counties Manukau area (south of Auckland, in proximity to the largest international airport). The next group in order of priority was those at the highest risk of serious illness, including anyone aged over 65 years as well as people with underlying health conditions. Finally, vaccination would be opened in stages for descending age bands to the rest of the NZ population.

A key aim of the vaccination programme was protecting those people in the population most vulnerable and likely to be most exposed. As vaccinations gathered pace, it became clear that even given differing ethnic age profiles, inequalities in vaccination uptake were becoming apparent. For example, the NZ European population is older than the Māori or Pacific Peoples. In NZ, ethnicity is prioritised, meaning that those people who identify with more than one group are assigned an ethnic identity in an order of priority from Māori, Pacific Peoples, Asian, to European/other. We can observe inequalities in the second vaccination doses, child vaccinations and, more recently, booster uptake (see Table 14.3), which shows the national (NZ) rates. Most of NZ has fully vaccinated rates (two doses) over 90%, with Māori reaching at least 85%. Turning to geographical patterns, we use the District Health Board (DHB) areas which are regional organisations responsible for the provision of health and disability services. The DHBs around the main urban areas of Auckland (51.7%), Wellington (52.8%) and Christchurch (39.4%) are at the top of the geographic distribution for childhood vaccinations, but lower for the

Table 14.3 National vaccination rates by ethnic group (%). Fully vaccinated (two doses), childhood vaccination and booster (third doses)

Ethnicity	Second doses, 12+ (%)	First doses, children 5–11 (%)	Booster, 18+ (%)
Māori	85.33	21.24	66.08
Pacific Peoples	94.13	28.13	54.89
Asian	>99.00	54.95	53.20
European / Other	92.93	43.69	77.37
Total	94.26	38.57	70.51

Source: Ministry of Health, New Zealand (2021b)

booster vaccination rate. The South Island DHBs of South Canterbury (76.2%) and West Coast (75.0%) are first and second, respectively, in NZ in the booster uptake (New Zealand Ministry of Health, 2022).

To some extent, the differences in vaccination rates reflect the group-based vaccination programme outlined above as well as the differing ethnic and age structure of the population in each DHB. Meaning that those DHBs with the least vaccinated areas are often more deprived and contain higher proportions of Māori and Pacific Peoples (see Fig. 14.3). These differences can be seen in Fig. 14.3, showing maps of the vaccination rates for children 5–11 years and booster uptake by DHB and ethnicity. Childhood immunisation rates are always lower in Māori than in European, Asian and other ethnicities, with differences ranging from 4.23% (West Coast) to 34.05% (Counties Manukau). A similar situation exists for Pacific Peoples, with the exception of the South Canterbury DHB where childhood immunisation rates are higher. Pacific Peoples also have a considerably lower booster uptake than Māori and the combined European, Asian and other ethnicities (New Zealand Ministry of Health, 2022).

Wider Impacts: From Pandemic to Syndemic

A concerning development during the pandemic has been the colliding forces of social and spatial inequalities and the COVID-19 pandemic to produce an emergent syndemic. A syndemic is the simultaneous impact of the social determinants of health and an epidemic (Singer, 2009). Syndemic is a term that joins two words: ‘synergy’ and ‘demic.’ Synergy means two or more agents working together to create a larger compound effect than each acting alone and ‘demic’ means essentially people (Singer, 2009). The key idea is the interaction of two or more diseases or health conditions, especially because of social inequity, acting on a population to cause greater harm than otherwise would have been the case from an epidemic, for example, the compounding effects of inequality, obesity, overcrowding and COVID-19, simultaneously. A disease that began amongst international travellers but became a problem for those people and places in the most deprived parts of our country is one way of understanding the outworking of a syndemic. A recent commentary (Campbell, 2021) highlighted the importance of the digital divide as just

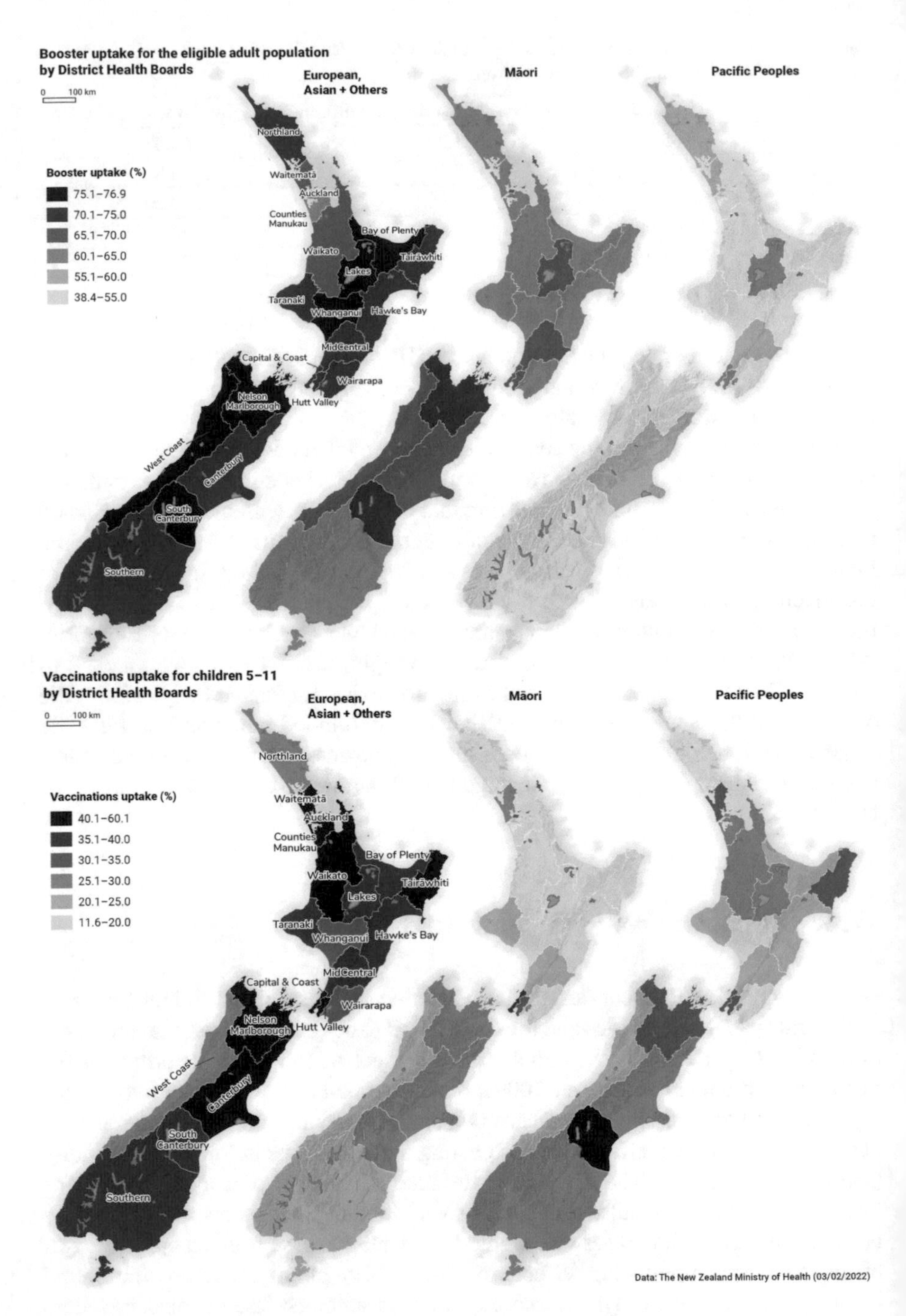

Fig. 14.3 Vaccination coverage by District Health Boards. (Source: MoH as of 3 Feb 2022)

one component of pandemic response which alienates particular sections of society from important information, adding further social inequality to a pandemic. The digital divide is especially important when a key aspect of NPIs is the use of smartphone applications for contact tracing as well as the use of online services to book vaccinations, or to find locations of interest, rendering those excluded more vulnerable to COVID-19.

In NZ and other developed countries, there are existing health inequalities stemming from prevailing socio-economic, demographic, cultural, historical or geographical conditions (Wiki et al., 2021a; Hobbs et al., 2021; Bowie et al., 2016). These inequalities are often reflected in worsened health outcomes observed in certain populations and localities (Bowie et al., 2013). The inequities that adversely impact the indigenous Māori population (Hobbs et al., 2019), approximately 16.5% of the population, as well as the Pacific Peoples, 8.1% of the population, are particularly important in the NZ context. These inequities occur alongside additional socio-economic, environmental and geographical factors. This can result in situations in which vulnerable communities are also hard to reach and/or are underserviced due to physical, financial and other barriers. An example of health service barriers and accessibility challenges are the utilisation of health services in the central North Island of NZ where people living in poor conditions and in highly deprived areas are those who experience deteriorated health conditions but also underuse health services, a situation that is exacerbated for Māori (Marek et al., 2021a).

Even though the overall population density of NZ is low, the majority of the population (88.4%) live in urban environments (Statistics New Zealand, 2021). The urban environments vary not only in size or type, but also in their function and composition of services as well as in access to natural environments. Previous research suggests that both the degree of urbanicity (Marek et al., 2020) and combined access to health-promoting and health-constraining built and natural environments affect health outcomes and physical and mental health (Marek et al., 2021b; Hobbs et al., 2021). Small towns of NZ also experience greater adverse health outcomes compared with either urban centres or rural areas. The small towns often have limited service accessibility and, simultaneously, are subject to unique population dynamics such as declining populations, ageing, missing employment opportunities and high proportions of ethnic minority groups (Marek et al., 2020), complicating the role of geographical inequalities.

As a pathway out of the pandemic depends on vaccination, we turn to look at some examples of patterns of immunisation for NZ. Childhood immunisation coverage shows spatial and spatiotemporal patterns (Marek et al., 2021c) where the most deprived areas have the lowest level of immunisation. Current inequities exist in localities with highly vulnerable populations to COVID-19 (Wiki et al., 2021b) and low spatial access to vaccination services, which implies that populations with the highest need of vaccinations also have very limited access to vaccination services (Whitehead et al., 2021).

A particularly problematic area for NZ even in the period before the pandemic was the role of the housing 'market' in creating social inequalities. As the COVID-19 pandemic has evolved alongside social and economic inequalities in NZ, we have reached a point where approximately half of households' assets are housing (Reserve Bank of New Zealand, 2021) and often framed as investment assets for renters or tourists (Campbell et al., 2019) rather than as homes. What is of particular concern is that from an already 'crisis' state in terms of housing affordability, prices have increased a further quarter to one-third, further exacerbating housing affordability in NZ from already exceptionally high levels by international comparisons with over half of all households renting privately financially strained. The wilful neglect from policymakers to address this most fundamental need, housing, will ensure that the impacts post-pandemic is felt by those 'locked out' of housing due to decreasing affordability even after 'lockdowns' end.

In tandem with increasing housing prices, the government has enacted a plethora of policy changes alongside the central bank of NZ, which has attempted to control the rapid housing price escalation. In the rental market, the introduction of a rent freeze to curtail and ameliorate the financial challenges arising from COVID-19, from 26 March to 25 September 2020, when landlords were not allowed to increase rents, may have only postponed larger rent price rises (New Zealand Government, 2020). This is a twofold disadvantage, as it further pushes back those with home ownership aspirations by blunting their ability to save whilst paying more in rent, as house prices increase. The trends in house and rental prices during the pandemic have compounded a growing social crisis for those who find themselves at the mercy of social housing provision. Unsurprisingly, the housing waiting list in NZ has also grown concurrently with house prices (and rents). This has stretched many individuals' and families' ability to pay for housing, leading to a growing stress on the public housing system. This has all lead to much more widespread social impacts, arguably moving from a pandemic to a syndemic.

Conclusions: Policy Implications and Future Work

The lesson that needs to be learned is that prevention is a key aspect of a public health response to COVID-19 and not just in NZ. By allowing social and spatial inequalities to grow in the period before, and even during, the pandemic, the pandemic has and will continue to impact on those more and less disadvantaged markedly, developing into a syndemic. The inequalities and inequities in rates of cases and severity of COVID-19 (Steyn et al., 2021) and in vaccination rates between Māori and non-Māori reflect broader inequities in the health experience of New Zealanders highlight that there is much work to be done in building a more equal society. One long-standing suggested solution is that Māori health should be managed and delivered by Māori with Green and Berentson-Shaw stating '*Over the course of the Covid-19 pandemic, we have been repeatedly shown that when Māori lead, we get things done, we keep our communities well, and we get our whānau vaccinated*' (Green & Berentson-Shaw, 2021). Partially in response to this, in April 2021 the government announced it would establish a Māori Health Authority as a Departmental Agency within the Ministry of Health.

As the pandemic has progressed there have also been calls that encouragement is given to increasing the types of environment people spend time in, specifically spending more time in outdoor environments where transmission is much lower (Wild & Woodward, 2021). This has included implications for transport where many countries of the world have seen significant investment in active transport, which can be seen as a COVID-19, physical activity and climate change solution simultaneously (Jáuregui et al., 2021). A 'natural experiment' of lockdowns can lead to fundamental changes in how we live, work and play and how we move around our environments, but perhaps only for a short period of time.

We argue that future work should explicitly attempt to quantify and/or model the role of the social determinants of health in ameliorating or accelerating the pandemic. For example, this could include modelling the social differences in the ability to restrict movement, in vaccination rates or access to health services relevant to the pandemic. A key feature of the pandemic in NZ so far has been the ongoing modelling and quantification by various research groups of the likely scenarios and forecasts of disease trajectories and dissemination of these models to the public. We propose that more detailed modelling of the underlying social and spatial inequalities is needed to better understand the prevention of a myriad of diseases and illnesses and to promote well-being. This would move beyond modelling disease (SEIR) to other tools like agent-based models or simulation of social policy (Campbell & Ballas, 2016; Campbell, 2011) to understand the potential for prevention in a pandemic environment.

There is also potential for future research into the post-acute sequelae of SARS CoV-2 infection (PASC) or 'long-COVID' to better understand the long-term impacts of this condition which is only beginning to emerge. Long-COVID may impact populations for quite some time into the future and may well prove to be vindictive of the NZ approach of an elimination strategy for as long as was feasible.

Overall, a key lesson emerging is that the consequences of neglecting social inequalities will lead to a more unequal and longer pandemic, or more accurately a syndemic. The adage is correct; prevention is better than cure. We would therefore urge policymakers to pay careful attention to the inequalities and inequities that exist and to consider investment and redistribution to those people and places that often fall behind in health and well-being, especially during a pandemic.

Acknowledgement The GeoHealth Laboratory has had input from a variety of funders, stakeholders and clinicians. We appreciate and wish to acknowledge their contributions in shaping our skills, projects and outputs.

References

Ahmed, F., Ahmed, N. E., Pissarides, C., & Stiglitz, J. (2020). Why inequality could spread covid-19. *The Lancet Public Health, 5*, E240.

Baral, S., Chandler, R., Prieto, R. G., Gupta, S., Mishra, S., & Kulldorff, M. (2021). Leveraging epidemiological principles to evaluate Sweden's covid-19 response. *Annals of Epidemiology, 54*, 21–26.

Beale, S., Braithwaite, I., Navaratnam, A. M. D., Hardelid, P., Rodger, A., Aryee, A., Byrne, T. E., Fong, E. W. L., Fragaszy, E., Geismar, C., Kovar, J., Nguyen, V., Patel, P., Shrotri, M., Aldridge, R., & Hayward, A. (2021). Deprivation and exposure to public activities during the covid-19 pandemic in England and Wales. *Journal of Epidemiology and Community Health*, Jech-2021-217076.

Bowie, C., Beere, P., Griffin, E., Campbell, M., & Kingham, S. (2013). Variation in health and social equity in the spaces where we live: A review of previous literature from the geohealth laboratory. *New Zealand Sociology Journal, 28*, 164–191.

Bowie, C., Campbell, M., Beere, P., & Kingham, S. (2016). Social and spatial inequalities in rotaviral enteritis: A case for universally funded vaccination in New Zealand. *The New Zealand Medical Journal (Online), 129*, 59–66.

Campbell, M. (2011). *Exploring the social and spatial inequalities of ill-health in Scotland: A spatial microsimulation approach.*

Campbell, M. (2021). Pandemics and emergent digital inequalities. *New Zealand Geographer, 77*, 180–184.

Campbell, M., & Ballas, D. (2016). Simalba: A spatial microsimulation approach to the analysis of health inequalities. *Frontiers in Public Health, 4*, 230.

Campbell, M., Apparicio, P., & Day, P. (2014). Geographic analysis of infant mortality in New Zealand, 1995–2008: An ethnicity perspective. *Australian and New Zealand Journal of Public Health, 38*, 221–226.

Campbell, M., Mcnair, H., Mackay, M., & Perkins, H. C. (2019). Disrupting the regional housing market: Airbnb in New Zealand. *Regional Studies, Regional Science, 6*, 139–142.

Campbell, M., Marek, L., Wiki, J., Hobbs, M., Sabel, C. E., Mccarthy, J., & Kingham, S. (2021). National movement patterns during the covid-19 pandemic in New Zealand: The unexplored role of neighbourhood deprivation. *Journal of Epidemiology and Community Health, 75*, 903–905.

Green, J., & Berentson-Shaw, J. (2021). Supporting Māori-led covid-19 vaccination conversations. *Radio New Zealand.*

Hobbs, M., Ahuriri-Driscoll, A., Marek, L., Campbell, M., Tomintz, M., & Kingham, S. (2019). Reducing health inequity for Māori people in New Zealand. *The Lancet, 394*, 1613–1614.

Hobbs, M., Kingham, S., Wiki, J., Marek, L., & Campbell, M. (2021). Unhealthy environments are associated with adverse mental health and psychological distress: Cross-sectional evidence from nationally representative data in New Zealand. *Preventive Medicine, 145*, 106416.

Jáuregui, A., Lambert, E. V., Panter, J., Moore, C., & Salvo, D. (2021). Scaling up urban infrastructure for physical activity in the covid-19 pandemic and beyond. *The Lancet, 398*, 370–372.

Macintyre, S., Ellaway, A., & Cummins, S. (2002). Place effects on health: How can we conceptualise, operationalise and measure them? *Social Science & Medicine, 55*, 125–139.

Marek, L., Wiki, J., Campbell, M., Kingham, S., Sabel, C., Tomintz, M., & Hobbs, M. (2020). Slipping under the radar: Worsened health outcomes in semi-urban areas of New Zealand. *The New Zealand Medical Journal, 133*, 121–125.

Marek, L., Greenwell, J., Hobbs, M., Mccarthy, J., Wiki, J., Campbell, M., Kingham, S., & Tomintz, M. (2021a). *Combining large linked social service microdata and geospatial data to identify vulnerable populations in New Zealand.* Edward Elgar Publishing.

Marek, L., Hobbs, M., Wiki, J., Kingham, S., & Campbell, M. (2021b). The good, the bad, and the environment: Developing an area-based measure of access to health-promoting and health-constraining environments in New Zealand. *International Journal of Health Geographics, 20*, 16.

Marek, L., Hobbs, M., Wiki, J., Mccarthy, J., Tomintz, M., Campbell, M., & Kingham, S. (2021c). Spatial-temporal patterns of childhood immunization in New Zealand (2006–2017): An improving pattern but not for all? *European Journal of Public Health, 31*, 561–566.

New Zealand Government. (2020). *Covid-19: Rent increase freeze and more protection for tenants* [Online]. Available: https://www.beehive.govt.nz/release/covid-19-rent-increase-freeze-and-more-protection-tenants. Accessed Oct 2021.

New Zealand Government. (2021a). *Covid-19 protection framework (traffic lights)* [Online]. Available: https://covid19.govt.nz/alert-levels-and-updates/covid-19-protection/. Accessed Oct 2021.

New Zealand Government. (2021b). *History of the covid-19 alert system* [Online]. Wellington. Available: https://covid19.govt.nz/alert-levels-and-updates/history-of-the-covid-19-alert-system/. Accessed Oct 2021.

New Zealand Ministry of Health. (2021a). *Covid-19: About the delta variant* [Online]. Available: https://www.health.govt.nz/our-work/diseases-and-conditions/covid-19-novel-coronavirus/covid-19-health-advice-public/about-covid-19/covid-19-about-delta-variant. Accessed Oct 2021.

New Zealand Ministry of Health. (2021b). *Covid-19: Current cases* [Online]. Wellington. Available: https://www.health.govt.nz/our-work/diseases-and-conditions/covid-19-novel-coronavirus/covid-19-data-and-statistics/covid-19-current-cases. Accessed Oct 2021.

New Zealand Ministry of Health. (2022). *Covid-19: Vaccine data* [Online]. Wellington. Available: https://www.health.govt.nz/our-work/diseases-and-conditions/covid-19-novel-coronavirus/covid-19-data-and-statistics/covid-19-vaccine-data. Accessed 03 Feb 2022.

New Zealand Transport Agency. (2021). *Covid-19 transport impact* (Fieldwork wave 26 deep dive analysis covid-19 impacts by disability).

Palm, M., Allen, J., Liu, B., Zhang, Y., Widener, M., & Farber, S. (2021). Riders who avoided public transit during covid-19. *Journal of the American Planning Association, 87*, 455–469.

Reserve Bank of New Zealand. (2021). *Household balance sheet* [Online]. Reserve Bank of New Zealand. Available: https://www.rbnz.govt.nz/statistics/c22. Accessed Oct 2021.

Ritchie, H., Mathieu, E., Rodés-Guirao, L., Appel, C., Giattino, C., Ortiz-Ospina, E., Hasell, J., Macdonald, B., Beltekian, D., & Roser, M. (2020). Coronavirus pandemic (Covid-19). *Our World in Data*.

Rojas-Rueda, D., & Morales-Zamora, E. (2021). Built environment, transport, and covid-19: A review. *Current Environmental Health Reports, 8*, 138–145.

Shaw, C., Keall, M., & Guiney, H. (2017). What modes of transport are associated with higher levels of physical activity? Cross-sectional study of New Zealand adults. *Journal of Transport & Health, 7*, 125–133.

Singer, M. (2009). *Introduction to syndemics: A critical systems approach to public and community health*. Wiley.

Statistics New Zealand. (2021). *Functional urban areas – Methodology and classification*.

Steyn, N., Binny, R. N., Hannah, K., Hendy, S., James, A., Lustig, A., Ridings, K., Plank, M. J., & Sporle, A. (2021). Māori and pacific people in New Zealand have higher risk of hospitalisation for covid-19. *Medrxiv*, 2020.12. 25.20248427.

Tartof, S. Y., Slezak, J. M., Fischer, H., Hong, V., Ackerson, B. K., Ranasinghe, O. N., Frankland, T. B., Ogun, O. A., Zamparo, J. M., Gray, S., Valluri, S. R., Pan, K., Angulo, F. J., Jodar, L., & Mclaughlin, J. M. (2021). Effectiveness of mRNA BNT162b2 covid-19 vaccine up to 6 months in a large integrated health system in the USA: A retrospective cohort study. *The Lancet, 398*, 1407–1416.

Whitehead, J., Carr, P. A., Scott, N., & Lawrenson, R. (2021). Spatial inequity in distribution of covid-19 vaccination services in aotearoa. *Medrxiv*, 2021.08.26.21262647.

Wiki, J., Kingham, S., & Campbell, M. (2021a). A geospatial analysis of type 2 diabetes mellitus and the food environment in urban New Zealand. *Social Science & Medicine, 288*, 113231.

Wiki, J., Marek, L., Hobbs, M., Kingham, S., & Campbell, M. (2021b). Understanding vulnerability to covid-19 in New Zealand: A nationwide cross-sectional study. *Journal of the Royal Society of New Zealand, 51*, S179–S196.

Wild, K., & Woodward, A. (2021). Take it outside to fight covid. *Newsroom*.

Chapter 15
Aboriginal and Torres Strait Islander Peoples and the COVID-19 Pandemic: A Spatial and Place-Based Analysis

Aryati Yashadhana, Miri Raven, Nellie Pollard-Wharton, and Brett Biles

Personal Story

Positionality of the Authors

This chapter was written from the perspectives of three Aboriginal scholars (BB, MR, NPW) who bring both lived experience and Indigenous knowledge and perspective and a non-Aboriginal scholar (AY) with expertise in Aboriginal health. To ensure transparency and validity, the authors have provided positionality statements. Aryati Yashadhana is a non-Indigenous person of Southeast Asian descent and second-generation Australian immigrant. For over a decade she has worked as an ally with Aboriginal peoples, specifically the Gamilaraay, Yuwaalaraay and Yuin peoples of New South Wales (NSW) and the Mara people of the Northern Territory in Australia in health and cultural research. Miri Raven is a Yamatji-Noongar and non-Indigenous woman from Western Australia. Miri is a geographer and researcher with two decades of experience working on Indigenous knowledge related to biological resources, biodiversity conservation and food security. Nellie-Pollard Wharton is a Kooma woman, social worker, co-founder of First Nations COVID-19 Response and public health researcher and Associate Lecturer in Aboriginal and Torres Strait Islander health. Brett Biles is a Murrawarri man now living on Wiradjuri country in Albury Wodonga, Australia. Brett's worldviews centre around being a Murrawarri man which impacts all aspects of his life. Brett has worked in academia for over 10 years.

Introduction

The region of Oceania is home to multiple, diverse Indigenous peoples, cultures and languages, which span geographic borders defined by physical landscapes, cultural distinction and law, and since invasion and colonisation, nation-states and

A. Yashadhana (✉) · M. Raven · N. Pollard-Wharton · B. Biles
University of New South Wales, Sydney, NSW, Australia
e-mail: a.yashadhana@unsw.edu.au

M. Laituri et al. (eds.), *The Geographies of COVID-19*, Global Perspectives on Health Geography, https://doi.org/10.1007/978-3-031-11775-6_15

governments. All these characteristics have spatial and geographical aspects and have impacted how Indigenous peoples across the Oceanic region have experienced and responded to the COVID-19 pandemic. In this chapter, we draw attention to Aboriginal and Torres Strait Islander (herein Aboriginal)[1] peoples in Australia, through exploring how spatiality has shaped experiences and responses to the COVID-19 pandemic. We have chosen to take this approach because that is the spatial environment in which the authors exist and in doing so aim to leverage our experiences, perspectives and knowledge of localised contexts and stories.

Aboriginal and Torres Strait Islander Population in Australia

In 2018 the Aboriginal and Torres Strait Islander population in Australia consisted of 778,064 people, accounting for 3.1% of Australia's total population (Table 15.1). NSW has the largest population of Aboriginal peoples (239,587) and the Northern Territory has the highest proportion (31%) of Aboriginal peoples among its population (Australian Bureau of Statistics, 2018). In terms of geography, 37% of Aboriginal peoples live in major cities and 19% live in remote or very remote areas (Australian Bureau of Statistics, 2018). Table 15.1 outlines Aboriginal and Torres Strait Islander population by jurisdiction in Australia in 2018.

Existing Inequities and COVID-19 Risk Among Aboriginal and Torres Strait Islander Australians

In the early stages (March 2020) of the COVID-19 pandemic, the Australian federal government closed Australia's borders to all non-citizens and non-residents, a protectionist approach reflective of pandemic responses historically. As COVID-19 spread across the globe placing pressure on health systems, governments and economies, the heightened risk and potential severity of impact among Aboriginal peoples was highlighted (Yashadhana et al., 2020). Central to this heightened risk is a continuous deficit-based narrative of comparative (to non-Aboriginal Australians) and preventable health, socioeconomic and cultural inequities that are rooted in a violent history of ongoing colonisation, dispossession, racism and forced assimilation.

[1] We use the term 'Aboriginal' to refer to the Indigenous peoples of 'mainland' Australia, 'Torres Strait Islander' to refer to the Indigenous peoples from the islands across the Torres Strait and 'Indigenous peoples' to describe First Peoples and Original Peoples from elsewhere across the globe. We understand that the use of this term is not without critique (see Smith, 1999). Tuhiwai-Smith, 2001. *Decolonizing Methodologies: Research and Indigenous Peoples.* London, Zed books). Our intention is to use this term within the spirit of collective struggle for health and cultural rights.

Table 15.1 Estimated Aboriginal and Torres Strait Islander population by location in Australia, 2018

Jurisdiction	Number of Aboriginal and Torres Strait Islander people	Proportion of total Aboriginal and Torres Strait Islander population (%)	Proportion of jurisdiction population (%)
New South Wales	239,587	31	3.0
Queensland	223,883	29	4.5
Western Australia	101,753	13	3.9
Northern Territory	76,845	9.9	31
Victoria	56,528	7.3	0.9
South Australia	43,317	5.6	2.5
Tasmania	28,328	3.6	5.4
Australian Capital Territory	7525	1.0	1.8
Australia	778,064	100	3.1

Source: Australian Bureau of Statistics (Australian Bureau of Statistics, 2018)

From a physical health perspective, inequitable rates of chronic conditions known to impact COVID-19 severity and fatality, including respiratory disease, diabetes, cardiovascular conditions, cancer or multimorbidity (more than one condition) between Aboriginal and non-Aboriginal Australians, increase risks and susceptibility (Yashadhana et al., 2020). From a sociocultural perspective, food security (Follent et al., 2021), interconnected households (Hui et al., 2021), inadequate housing conditions that accommodate for larger family groups, cultural and linguistic marginalisation from health information and the role of racism in the enforcement of public health restrictions were all highlighted as factors that may contribute to further health and social inequities in the face of COVID-19 (Yashadhana et al., 2020). From a spatial perspective, Aboriginal peoples living rurally experience increased COVID-19-related inequities, including higher chronic disease rates, incidences of inadequate housing (Yashadhana et al., 2020), higher food prices (Follent et al., 2021) and under-resourced health services that lack the labour power and infrastructure to deal with pandemic caseloads (Fitts et al., 2020). High mobility across geographic areas serve as an important mechanism of well-being through familial connection and attendance to cultural events or ceremonies, yet in the context of COVID-19 were used to justify issues of disjointed health care (Prout, 2018) and over-policing (Donohue & McDowall, 2021).

Previous pandemics such as the H1N1 outbreak in 2009, disproportionately impacted Aboriginal peoples who constituted 16% of hospitalised cases and 12% of deaths (Yashadhana et al., 2020). This was in part due to limited input from Aboriginal Community Controlled Health Organisations (ACCHOs) which hampered ‘on the ground’ responses and adequate surveillance (Griffiths et al., 2021).

Spatial aspects of Australia's 'island' geography, border closures, strict quarantine measures and effective contact tracing (due to low case numbers) meant that the initial waves of the COVID-19 pandemic were controlled among the general population. Not a single Aboriginal death was reported in the first year of the pandemic (Dudgeon et al., 2021). Rapid responses from ACCHOs, including lobbying government to restrict access to remote Aboriginal communities, increase access to personal protective equipment, testing and contact tracing (Crooks et al., 2020), and developing accessible COVID-19 health information (Finlay & Wenitong, 2020) reiterated the effectiveness and importance of Aboriginal leadership (Eades et al., 2020). Calls were also made for sustainable needs-based funding to support strengths-based and place-based approaches to tackle the encroaching mental health and well-being impact of the pandemic – labelled the second pandemic – a reality of pre-pandemic inequities reflected in mental health outcomes among Aboriginal peoples (Dudgeon et al., 2021).

Aboriginal and Torres Strait Islander COVID-19 Advisory Group and Pandemic Management

The Aboriginal and Torres Strait Islander COVID-19 Advisory Group was co-chaired by the NACCHO and Commonwealth Department of Health. This advisory group developed the *Australian Health Sector Response Plan for Novel Coronavirus* (COVID-19) Management and the Operational Plan for Aboriginal peoples (Department of Health, 2020). This plan included clinical guidelines and initiatives for remote communities to utilise in preparation for COVID-19 being active within communities (Moodie et al., 2020). As part of this management plan, spatial aspects were also considered. This is due to the different mobility patterns for Aboriginal peoples, for example, some communities have a high flow of visitors placing extra pressure on existing primary health care infrastructure (through ACCHOs), impacting workforce availability, regular provision of care and social determinants such as housing and food security (Yashadhana et al., 2020).

Aboriginal Community Controlled Health Organisations (ACCHOs) Responses to COVID-19

The goals of ACCHOs are to deliver holistic and culturally appropriate health services through enabling whole-of-community self-determination and individual spiritual, cultural, physical, social and emotional well-being, so that Aboriginal peoples can enjoy a good quality of life (National Aboriginal Community Controlled Health Organisation, 2021). ACCHOs, compared with mainstream health services, focus on tailored accessibility, culturally safe care (community-based Aboriginal staff, culturally responsive service delivery) and holistic approaches to health and well-being (Gomersall et al., 2017).

Table 15.2 Peak body organisations in the Aboriginal Community Controlled Health Sector

State/territory	Peak body organisation	COVID-19 resource
New South Wales	Aboriginal Health and Medical Research Council (AHMRC)	https://www.ahmrc.org.au/coronavirus/
Queensland	Queensland Aboriginal and Islander Health Council (QAIHC)	https://www.qaihc.com.au/resources/covid-19-coronavirus-resources
Western Australia	Aboriginal Health Council of Western Australia (AHCWA)	https://www.ahcwa.org.au/coronavirus-updates
Northern Territory	Aboriginal Medical Services Alliance Northern Territory (AMSANT)	http://www.amsant.org.au/covid-19/
Victoria	Victorian Aboriginal Community Controlled Organisation (VACCHO)	https://www.vaccho.org.au/about-us/coronavirus-information/
South Australia	Aboriginal Health Council of South Australia (AHCSA)	https://ahcsa.org.au/coronavirus-covid-19/
Tasmania	Tasmanian Aboriginal Centre Inc.	http://tacinc.com.au/stay-safe-at-home-with-palawa-kani/
Australian Capital Territory	Winnunga Nimmityjah Aboriginal Health Service	https://www.winnunga.org.au/covid-19-coronavirus-precautions-for-winnunga-clients/
Australia (National)	National Aboriginal Community Controlled Health Organisation (NACCHO)	https://www.naccho.org.au/aboriginal-health-alerts-coronavirus-covid-19

Broadly, ACCHO support has included provision of vaccinations, health advice and being advocates for their clients and communities about the impact of COVID-19 and COVID-19 responses in their communities (see Table 15.2 for state-based examples). This support has extended to raising concerns through state, territory and national media. Land Councils have assisted communities with controlling access permits to non-essential travel to remote communities and translating and developing health messages into local languages (Moodie et al., 2020). ACCHOs alongside local mayors have raised concerns around lack of support, overcrowded housing, inappropriateness of urban mainstream COVID-19 measures for remote communities and influence of social media of reduced vaccination rates (Burt, 2022; Cluff, 2022; Dick et al., 2022).

National Level

The National Aboriginal Community Controlled Health Organisation (NACCHO) is a living embodiment of the aspirations of Aboriginal communities and their struggle for self-determination. NACCHO is the national peak body representing 144 ACCHOs across the country on Aboriginal health and well-being issues. These ACCHOS are controlled by locally elected Aboriginal and Torres Strait Islander boards of management and governance. There are approximately 300 NACCHO member clinics that are providing holistic primary health to Aboriginal peoples. ACCHOs employ approximately 6000 staff and 3500 of these staff are Aboriginal peoples (National Aboriginal Community Controlled Health Organisation, 2021). Nationally, NACCHO, state, territory and regional peak organisations and member organisations participated in a national Aboriginal and Torres Strait Islander COVID-19 Advisory Group which reported directly to the Chief Medical Officer and the Australian Health Protection Principal Committee, the advice of which was given to the National Cabinet influencing COVID-19 policy at the federal level (Moodie et al., 2020).

State and Territory Levels

At the state and territory levels, there are Aboriginal and Torres Strait Islander health peak bodies which provide support for ACCHOs within their states and territories. Each peak health body has provided localised responses and resources to the COVID-19 pandemic for their member organisations to access and disseminate to each of their local ACCHOs (Table 15.2).

Aboriginal and Torres Strait Islander health professional bodies have also developed culturally safe resources for their health professionals to utilise and disseminate information (Table 15.3).

In the early stages of the pandemic, there was remarkable success in preventing COVID-19 infections in Aboriginal communities in Australia. Anecdotally, the success was largely due to ACCHOs quickly and effectively mobilising to quarantine communities and to use appropriate and targeted communications (see resources in Tables 15.2 and 15.3), including a 'care mode' which incorporated a delivery of food and essential items to households (Archibald-Binge & Geraghty, 2020).

COVID-19 Outbreaks and Aboriginal and Torres Strait Islander Peoples

In June 2021, the emergence of the COVID-19 Delta strain and a slow and politically mismanaged vaccine rollout at the federal level (De Leeuw et al., 2022) led to an outbreak across multiple Australian states and territories and for the first time in remote Aboriginal communities. Despite pandemic planning and the Australian government recommending that Aboriginal Australians were a prioritised target group for the vaccine rollout program (Carroll et al., 2021), little was done to ensure this occurred, including limited collaboration directly with Aboriginal peoples and communities.

At the time of writing, national data reporting the total numbers of COVID-19 cases and deaths among Aboriginal Australians were not available. For this reason, we turned to media articles which reported a total of 5500 COVID-19 infections

Table 15.3 Aboriginal and Torres Strait Islander health professional bodies

Aboriginal and Torres Strait Islander health professional bodies	COVID-19 resource
Indigenous Allied Health Australia	https://iaha.com.au/coronavirus-covid-19-staying-connected-stronger-together/
Congress of Aboriginal and Torres Strait Islander Nurses and Midwives	https://www.catsinam.org.au/communications/covid19
National Aboriginal and Torres Strait Islander Health Worker Association	https://natsihwa.org.au/covid-19-resources-updates-brochures/resources
Gayaa Dhuwi (Proud Spirit) Australia: National Aboriginal and Torres Strait Islander Health Leadership in Social and Emotional Wellbeing, Mental Health, and Suicide Prevention	https://www.gayaadhuwi.org.au/coronavirus

among Aboriginal peoples since the beginning of the Delta outbreak, with 550 hospitalised, 60 in intensive care, and 12 who have died (Thorpe, 2021). This counts for around 4% of all COVID-19 cases in Australia (130,000). Remote Aboriginal communities such as Wilcannia and Enngonia in the state of NSW reported between 15% and 30% of their whole population having contracted COVID-19, raising infections rates to double that of non-Aboriginal Australians (The Guardian, 2021).

In remote communities, despite repeated calls for accommodation support, residents were left to isolate in tents (The Guardian, 2021) and over-policing (excessive police presence and use of force) (Boon-Kuo et al., 2021) led to exorbitant public health fines despite justifiable reasons including sharing transport to acquire groceries (Kerin, 2021). COVID-19 has impacted remote communities with specific challenges around crowed households making quarantine difficult and those in isolation having to rely on food drops (Daly, 2022). For example, there are reports of up to as many as 20 people from three family groups in one house (Burt, 2022). Overcrowding is reported to have increased the spread of COVID-19 in communities such as Wilcannia and Enngonia in NSW and Yarrabah in far north Queensland (Burt, 2022). The availability of housing for quarantining existed for some locations, but this was sometimes government or public servant housing and it was reported that some departments were reluctant to release the housing in case they lose the accommodation (Burt, 2022).

Place-Based Examples of Localised Responses to the COVID-19 Pandemic

Early in the pandemic (February 2020), ACCHOs and Aboriginal communities mobilised rapidly to reduce the risk of local COVID-19 outbreaks; this was a key reason for the *early* success in keeping Aboriginal people safe from COVID-19 (Eades et al., 2020). Once vaccinations became available, low vaccination rates among Aboriginal peoples at the onset of the Delta outbreak were a major concern, as the inevitable easing of public health orders restricting movements and enforcing quarantine had the potential to translate into high rates of infection for unvaccinated groups (Power et al., 2020). Aboriginal organisations (NACCHO, Aboriginal Medical Services of the Northern Territory and the Central Australian Congress) called on state and territory government to delay the easing of restrictions until full vaccination rates of Aboriginal peoples aged 12 years and older reach 90–95% (Komesaroff et al., 2021). Throughout this time, state- and territory-based ACCHOs were creating vaccination campaigns and vaccination services to increase vaccination rates for local communities. Outcomes varied in different places. Below, we exemplify using localised case studies.

First Nations COVID-19 Response (NSW)

First Nations COVID-19 Response (FNR) is a small grassroots initiative run by Aboriginal community members (including NPW) and volunteers in Sydney NSW. FNR was initially established on social media in March 2020 to create an online space for mutual support, connection and care amongst Aboriginal peoples in the inner suburbs of Sydney. Running solely on donations and voluntary labour, FNR filled the gaps in place of limited government supporting Aboriginal community members with immediate crisis relief, including supply of groceries, gift cards, meals and Elder visits. The founders did not initially accept financial contributions due to administrative burdens, yet as demand increased financial donations were necessary to meet increasing demands and provide additional support to elders, individuals and families, particularly in home or hotel isolation or quarantine. Regarding partnerships with organisations such as 'Plate it Forward' and Aboriginal businesses (Tin Humpy Café), a physical space was provided for weekly community collection of meals, groceries and art packs. FNR advocates for food as a fundamental right and not a privilege and continues to provide families and communities with food relief.

VACCHO Vaccine Vans (Victoria)

In October 2021 VACCHO (see Table 15.2) launched two dedicated vaccine vans making the COVID-19 vaccine available to Aboriginal and Torres Strait Islander community members who have had difficulties accessing the vaccine. Over the span of nine weeks, the vaccination vans administered hundreds of vaccinations to members of the Aboriginal and Torres Strait Islander communities in different regional areas of Victoria, Australia. Figure 15.1 illustrates the places and distances the VACCHO vaccination vans travelled to. At the time of writing, the vaccination van

Fig. 15.1 VACCHO vaccination van sites (distances from Melbourne)

will be returning to these areas to administer first, second and third/booster doses as well as vaccines for children 5–11 years of age.

Northern Territory Response to COVID-19

Northern Territory (NT) has the highest proportion of Aboriginal peoples with approximately 30% identifying. As reported in late January 2022, nine out of ten patients with COVID-19 in NT hospitals are Aboriginal people (ABC News, 2022). The NT reopened its borders in January 2022 and Aboriginal Medical Services Alliance chief executive John Paterson has stated '*We predicted this, predicted that the virus would spread like wildfire through our remote communities and this is exactly what we're seeing now*' (ABC News, 2022).

Outbreaks increased with the Omicron variant, and problems were reported around accessing COVID-19 testing (Allam, 2022; Daly, 2022). For example, a group of people sent home from quarantine to a remote NT community, who later tested positive, raised particular concern. A locally established COVID-19 response group and community organisations and leaders expressed concern that the 'living with COVID' strategy developed for the 'urban mainstream environment' did not work for people in remote communities (Allam, 2022). A strong military presence was used, including evacuation to external facilities, with calls for 'defence force-style isolation facilities' or a tent system to be established in overcrowded communities (Dick et al., 2022). One community in Central Australia was sent into a seven-day lockdown after recording 22 coronavirus cases which coincided with possible flood warnings for key roads and threatened to block access to food and medicine during the lockdown (Dick et al., 2022). In light of environmental factors influencing food security during lockdowns, there were calls for the NT government to enact a 'standby plan' to enable food and medicine relief to communities isolated by flood waters (Dick et al., 2022).

The ongoing COVID-19 outbreak has seen an 'open letter' written to the Chief Minister of the Northern Territory from the Central Australian Aboriginal Congress (CAAC), AMSANT, the Central Land Council and the CEO of Lhere Artepe Aboriginal Corporation (CAAC, 2022) – calling for an immediate lockdown across Central Australia NT to stop movement of people and flatten the curve of new infections and hospitalisations. Prompt action was called for and recognition of coronavirus' indiscriminate nature, yet imminent inequitable outcomes where the vulnerable and the unvaccinated will be most impacted.

Vaccination Access and Uptake

In June 2020 at the onset of the Delta outbreak, only 5% of Aboriginal people in NSW (state with the largest Aboriginal population) were vaccinated (NSW Health, 2022). Initial low rates could be attributed to rapidly changing and confusing advice from the Australian government regarding eligibility (Doran, 2021), procurement and availability through ACCHOs (Naren et al., 2021) and the investment in culturally appropriate community-led immunisation services that establish vaccination as the social norm (McCalman et al., 2021). Low vaccination rates in some remote communities, regions and jurisdictions raise concerns about the impact of the Delta and Omicron variants on Aboriginal peoples. Western Australia, for instance, is reported to have the lowest Aboriginal vaccination rate in Australia with the lowest rates reported for the Pilbara and Kimberley regions (de Kruiff, 2022). While the federal government was responsible for vaccine rollout, the WA government and Rio Tinto became involved to lift vaccination rates for Indigenous people in the state (de Kruiff, 2022). The Royal Flying Doctors Service also conducted family days and outreach clinics with ACCHOs in WA with the aim to increase vaccination rates, in addition to mobile vaccination teams providing door to door services and voucher incentives (de Kruiff, 2022).

There are concerns about low vaccination rates of Aboriginal peoples experiencing homelessness, particularly in WA (where they have maintained a closed border policy), who have high mobility and social interactions and are likely to have pre-existing health conditions deeming them susceptible to spreading the virus or succumbing to it (Torre, 2022). Anecdotal evidence suggests that the low uptake of vaccines and anti-vaccination stances were the result of misinformation on social media (Cluff, 2022), fear and complacency, yet increasing outbreaks are prompting reconsideration in some communities (Archibald-Binge, 2021).

Funding for the Pandemic

In 2020, additional funding provided for the pandemic in Australia was estimated to equate to $20 for each Aboriginal and/or Torres Strait Islander person; this was compared to the Canadian response which equated to $270 per Indigenous person (Department of Health, 2020). In early 2022, the federal government also indicated that it has granted funding to protect remote communities related to essential social services ($23 million), community night patrol ($10 million), school nutrition projects (SNP) ($5 million), and four Northern Territory Land Councils ($10 million) (Australian Government, 2022). Aside from the Land Council funding, which are Aboriginal and Torres Strait Islander organisations, it is not clear what proportion of funds in the other categories are specifically earmarked for initiatives targeting Aboriginal peoples in remote communities.

Conclusion

This chapter has explored the structural and spatial contexts of the COVID-19 pandemic as related to Aboriginal peoples across Australia. In doing so we have highlighted localised responses to the COVID-19 pandemic, much of which were championed by ACCHOs and other community-led organisations. While Aboriginal peoples (alongside non-Aboriginal Australians) were protected from COVID-19 during the early waves of the pandemic, complacency on behalf of the federal government, which led to slow vaccine procurement and disorganised management of vaccination prioritisation, left Aboriginal and Torres Strait Islander communities open to Delta and Omicron strain outbreaks. Regardless, culturally tailored and swift responses from Aboriginal and Torres Strait Islander leaders, peak bodies, health services and grassroots organisations have filled the gaps, from vaccination uptake to food and medical supply relief, saving lives and supporting community. As the COVID-19 situation evolves, a focus on spatial, cultural and health equity must be upheld, to ensure Aboriginal peoples are supported.

References

ABC News. (2022). Nine out of 10 COVID patients in NT hospitals are Aboriginal people. Experts are concerned, but not surprised.

Allam, L. (2022). NT Aboriginal community demands answers after residents sent home from quarantine test positive for Covid. *The Guardian*.

Archibald-Binge, E. (2021, November 16, at 8:01 pm). Why COVID-19 is infiltrating Indigenous communities. Retrieved January 27, 2022, from https://www.abc.net.au/news/2021-11-16/why-covid-19-is-infiltrating-indigenous-communities/100625424

Archibald-Binge, E., & Geraghty, K. (2020, March 30 at 12.05 am). 'We treat them like gold': Aboriginal community rallies around elders. Retrieved January 29, 2022, from https://www.smh.com.au/national/nsw/we-treat-them-like-gold-aboriginal-community-rallies-around-elders-20200327-p54ekl.html

Australian Bureau of Statistics. (2018). Life tables for Aboriginal and Torres Strait Islander Australians 2015-2017 Canberra ACT (No. 3302.0.55.003).

Australian Government. (2022). Protecting remote communities. Retrieved January 27, 2022, from https://www.niaa.gov.au/covid-19/protecting-remote-communities

Boon-Kuo, L., Brodie, A., Keene-McCann, J., Sentas, V., & Weber, L. (2021). Policing biosecurity: Police enforcement of special measures in New South Wales and Victoria during the COVID-19 pandemic. *Current Issues in Criminal Justice, 33*(1), 76–88.

Burt, J. (2022, January 23 at 7:25 am). Indigenous mayors sound alarm over crowded housing amid COVID outbreaks. *ABC News*. Retrieved January 27, 2022, from https://www.abc.net.au/news/2022-01-23/queensland-indigenous-mayors-alarmed-by-overcrowding/100773094

CAAC, A., CLC covid in Central Aust, (2022). Open letter to the Hon Michael Gunner MLA, Chief Minister of the Northern Territory.

Carroll, S. R., Akee, R., Chung, P., Cormack, D., Kukutai, T., Lovett, R., Suina, M., & Rowe, R. K. (2021). Indigenous peoples' data during COVID-19: From external to internal. *Frontiers in Sociology, 6*, 62.

Cluff, R. (2022). Indigenous health service slams social media scam after drop in child vaccination rate. *TropicNow*.

Crooks, K., Casey, D., & Ward, J. S. (2020). First Nations people leading the way in COVID-19 pandemic planning, response and management. *Medical Journal of Australia, 213*(4), 151–152.

Daly, J. (2022). *COVID outbreak in Cherbourg's Aboriginal community defies odds as peak reached*.

de Kruiff, P. (2022). WA's Indigenous vaccination rate the worst in Australia. *WA Today*.

De Leeuw, E., McCracken, K., Harris, P., & Yashadhana, A. (2022). Under-promise and over-deliver: The failure of political rhetoric in managing COVID-19 in Australia. In R. Akhtar (Ed.), *Coronavirus (COVID-19) outbreaks, environment and human behaviour: International case studies* (Vol. 2). Springer.

Department of Health. (2020). *Australian health sector emergency response plan for novel coronavirus (COVID-19): Management plan for aboriginal and torres strait islander populations*. Opertional Plan for Aobriginal and Torres Strait Ialsnder Populations. Department of Health. Canberra, Commonwealth of Australia.

Dick, S., Abram, M., & Nowroozi, I.. (2022, January 24 at 9:43 am). Changing COVID-19 health advice, barriers to testing and isolating are disadvantaging Aboriginal Territorians, say peak bodies. Retrieved January 24, 2022, from https://www.abc.net.au/news/2022-01-24/nt-aboriginal-territorians-confusing-covid-19-health-advice/100776078

Donohue, M., & McDowall, A. (2021). A discourse analysis of the Aboriginal and Torres Strait Islander COVID-19 policy response. *Australian and New Zealand Journal of Public Health., 45*, 651–657.

Doran, M. (2021). Health officials racing to 'accelerate' COVID-19 vaccination rates in Indigenous communities as restrictions ease. *ABC News*, Australian Broadcasting Corporation.

Dudgeon, P., Alexi, J., Derry, K., Brideson, T., Calma, T., Darwin, L., Gray, P., Hirvonen, T., McPhee, R., & Milroy, H. (2021). Mental health and well-being of Aboriginal and Torres Strait Islander peoples in Australia during COVID-19. *Australian Journal of Social Issues., 56*, 485–502.

Eades, S., Eades, F., McCaullay, D., Nelson, L., Phelan, P., & Stanley, F. (2020). Australia's First Nations' response to the COVID-19 pandemic. *The Lancet, 396*(10246), 237–238.

Finlay, S., & Wenitong, M. (2020). Aboriginal Community Controlled Health Organisations are taking a leading role in COVID-19 health communication. *Australian and New Zealand Journal of Public Health., 46*, 170–172.

Fitts, M. S., Russell, D., Mathew, S., Liddle, Z., Mulholland, E., Comerford, C., & Wakerman, J. (2020). Remote health service vulnerabilities and responses to the COVID-19 pandemic. *Australian Journal of Rural Health, 28*(6), 613–617.

Follent, D., Paulson, C., Orcher, P., O'Neill, B., Lee, D., Briscoe, K., & Dimopoulos-Bick, T. L. (2021). The indirect impacts of COVID-19 on Aboriginal communities across New South Wales. *The Medical Journal of Australia, 214*(5), 199.

Gomersall, J. S., Gibson, O., Dwyer, J., O'Donnell, K., Stephenson, M., Carter, D., Canuto, K., Munn, Z., Aromataris, E., & Brown, A. (2017). What Indigenous Australian clients value about primary health care: a systematic review of qualitative evidence. *Australian and New Zealand Journal of Public Health, 41*(4), 417–423.

Griffiths, K., Ring, I., Madden, R., & Pulver, L. J. (2021). In the pursuit of equity: COVID-19, data and Aboriginal and Torres Strait Islander people in Australia. *Statistical Journal of the IAOS* (Preprint), 1–9.

Hui, B. B., Brown, D., Chisholm, R. H., Geard, N., McVernon, J., & Regan, D. G. (2021). Modelling testing and response strategies for COVID-19 outbreaks in remote Australian Aboriginal communities. *BMC Infectious Diseases, 21*(1), 1–12.

Kerin, L. (2021). 'Causing stress and anxiety': Aboriginal people in far west NSW copping hefty COVID fines. *SBS News.*

Komesaroff, P. A., Ah Chee, D., Boffa, J., Kerridge, I., & Tilton, E. (2021). COVID-19 restrictions should only be lifted when it is safe to do so for Aboriginal communities. *Internal Medicine Journal, 51*, 1806–1809.

McCalman, J., Longbottom, M., Fagan, S., Fagan, R., Andrews, S., & Miller, A. (2021). Leading with local solutions to keep Yarrabah safe: A grounded theory study of an Aboriginal community-controlled health organisation's response to COVID-19. *BMC Health Services Research, 21*(1), 732.

Moodie, N., Ward, J., Dudgeon, P., Altman, J., Casey, D., Cripps, K., Derry, K., Eades, S., Adams, K., Davis, M., Faulkner, S., Hunt, J., Klein, E., McDonnell, S., Ring, I., Sutherland, S., & Yap, M. (2020). Roadmap to recovery: Reporting on a research taskforce supporting Indigenous responses to COVID-19 in Australia. *Australian Journal of Social Issues, 56*, 1–13.

Naren, T., Burzacott, J., West, C., & Widdicombe, D. (2021). Role of Aboriginal Health Practitioners in administering and increasing COVID-19 vaccination rates in a Victorian Aboriginal Community Controlled Health Organisation. *Rural and Remote Health, 21*(4), 7043–7043.

National Aboriginal Community Controlled Health Organisation. (2021). About NACCHO, from https://www.naccho.org.au/about

NSW Health. (2022). Epidemiological week 1, ending 8 January 2022. *Covid-19 Weekly Surveillance in NSW.*

Power, T., Wilson, D., Best, O., Brockie, T., Bearskin, L. B., Millender, E., & Lowe, J. (2020). COVID-19 and indigenous peoples: An imperative for action. *Journal of Clinical Nursing., 29*, 2737–2741.

Prout, S. (2018). *The entangled relationship between Indigenous spatiality and government service delivery.* Centre for Aboriginal Economic Policy Research (CAEPR), The ....

Smith, L. T. (1999). *Decolonizing methodologies: Research and indigenous peoples.* Zed books.

The Guardian. (2021). Indigenous people infected with Covid Delta strain at twice the rate of other Australians. *The Guardian*.

Thorpe, N. (2021). COVID-19 cases are falling across NSW but new infections emerging in Indigenous communities. *ABC News*.

Torre, G. (2022). Race to protect WA's homeless against COVID. *National Indigenous Times*.

Yashadhana, A., Pollard-Wharton, N., Zwi, A. B., & Biles, B. (2020). Indigenous Australians at increased risk of COVID-19 due to existing health and socioeconomic inequities. *Lancet Regional Health-Western Pacific, 1*, 100007.

Chapter 16
Impacts of COVID-19 Pandemic Lockdown on the Livelihoods of Male Commercial Boda-Boda Motorists in Uganda

Harriet Kebirungi and Hadijah Mwenyango

Personal Story

The start of the COVID-19 pandemic found me with young twin babies and observing standard operating procedures (SOPs) seemed next to impossible. As a single mother, I had to go out of the house twice a week to buy food stuff for my babies. I developed an attitude of "what come may be." I resisted home confinement; walked a ten-kilometer deserted road to the market to buy food for my children. Only ambulances and security vehicles were allowed on the road. All roads were manned by armed security agencies and punctuated with roadblocks. The reasons for deployment were to keep law and order, implementation of SOPs and security of persons and property.

Boda-Boda motorists only transported cargo with definite destinations; those unauthorized were being shot at by security agencies leaving them dead and others injured. The horror of seeing Boda-Boda motorists tortured, their motorcycles confiscated by armed men, and other horrific scenes broadcasted on radio and television countrywide immensely haunted me. Due to social distance measures, it was difficult for me to intervene. Although I had the desire to help out, I was also struggling for my family. Boda-Boda motorists earned income through passengers and goods transportation, but due to the COVID-19 lockdown, Boda-Boda motorists lost clients. The horror of seeing Boda-Boda motorists suffering left me with an unanswered question on how the COVID-19 lockdown affected the livelihoods of male commercial Boda-Boda motorists in Uganda. It was this question that motivated me to conduct this study. *Kebirungi Harriet*

H. Kebirungi (✉)
Kyambogo University, Kampala, Uganda
email: hkebirungi@kyu.ac.ug

H. Mwenyango
Edinburgh Napier University, Scotland, United Kingdom
email: h.mwenyango@napier.ac.uk

M. Laituri et al. (eds.), *The Geographies of COVID-19*, Global Perspectives on Health Geography, https://doi.org/10.1007/978-3-031-11775-6_16

Introduction

This chapter examines the impacts of the COVID-19 lockdown on the livelihoods of Boda-Boda motorists in Kira Town Council, Wakiso District, Uganda. In Uganda, Boda-Bodas are motorcycle-based services and forms an enormous part of the public transportation network that facilitates the movement of passengers or goods from one place to another. Boda-Bodas are operated by Boda-Boda motorists and are used in the absence of efficient public transport (buses, minibus taxis and special hire taxis, etc.) especially where access is physically restricted and cannot be met by conventional public transport. After agriculture, Boda-Boda transport services is the second largest employer of youth in Uganda. It is considered one of the easiest business options for those who fail to get formal employment even after education. The Boda-Boda transport industry in Uganda is quite large. Over one million people countrywide are involved in the industry with the Kampala metropolitan area having over 300,000 people supported by the industry (Namara, 2019).

In response to strengthen security and tax base collection and decongest traffic in Kampala city center, Boda-Boda transport is regulated in the country through Boda-Boda stage registration (Kampala City Council Authority, 2020). A Boda-Boda stage is a gazetted and referenced pickup location. For example, for individual Boda-Boda motorists or companies/associations involved in Boda-Boda business, their motorcycles are expected to be registered by Kampala City Council Authority (KCCA) and operated at gazetted Boda-Boda stages of their choice. Those Boda-Bodas who do not operate within KCC or do not belong to companies/associations register at the community Boda-Boda stage of their choice through their stage leadership. Majority of Boda-Boda motorists are self-employed and earn on a day-to-day basis. This may be difficult for them to save money out of their daily earnings. Consequently, in times of social and economic shocks like pandemics including COVID-19, majority of them may have no fallback position.

In Uganda, the Boda-Boda industry was affected by the spread of COVID-19 pandemic and its associated infection and death rates. The only solution to save the population was for governments to declare lockdowns in their respective countries. Lockdowns in almost all countries were characterized by standard operating procedures like limited mobility, home confinements, social distancing, wearing face mask, shutting down all educational institutions and air transport, and introduction of curfews. Only essential workers such as healthcare personnel, a few government workers, and security organs remained in operation to contain the spread of COVID-19 pandemic.

Studies on the impact of health epidemics, pandemics, and other calamities like the case of Ebola outbreak in West Africa and other natural disasters have been associated with households becoming food insecure, low education access and attainment, domestic violence, and other deleterious effects. Similarly, the COVID-19 pandemic lockdown and associated effects in households and communities were reported. Some of the immediate aftermaths of the lockdown in many countries were spikes in gender-based violence, teenage pregnancies, hunger and starvation, student loss of learning, poor access to healthcare services, and loss of income among others (Evens & Over, 2020; Thompson et al. 2021). Long-term impacts included crippled economies with limited recovery rates, especially in the global south (Pilling, 2020), leaving some primary household breadwinners dead, while others lost employment due to the COVID-19 pandemic. These increased vulnerabilities for children of such families did not return to complete their education cycles leading to long-term earning trajectories for them and their families with overall human capital gaps (Evans & Over, 2020). Few studies have been conducted on the impacts of the COVID-19 pandemic on global, low, and middle economies (Evens & Over, 2020). However, many of these studies looked at the impact of the disease on the population in general, for example, on the health or the economy. None of the studies focused on a specific category of people and the impact the COVID-19 lockdown had on their livelihoods.

Boda-Boda motorists transport passengers and goods with a high physical interface between the motorists and their clients due to close proximity. This client-passenger interaction is believed to create a conducive environment for spreading COVID-19 pandemic in the transport sector. This industry is one of the various income-generating activities affected by COVID-19 lockdown. For example, whereas Boda-Boda motorists' businesses in east African countries (Tanzania, Rwanda, and Kenya) have access to social security services during the lockdown, there are no similar arrangements in Uganda. The absence of formal social security leaves people engaged in the informal sector, especially the Boda-Boda motorists, highly vulnerable.

To understand the impacts of the COVID-19 lockdown on the livelihoods of Boda-Boda motorists in Kira Town Council, Wakiso District, Uganda. The Kira Town Council is a bedroom community of Kampala capital city where Boda-Boda motorists provide transport services to people working and residing in Kira and those working in the capital city. Boda-Boda motorists indicated that they charge approximately 1000 Uganda shillings (equivalent to USD 0.3) for a distance covering 1 km. All Boda-Boda motorists reported that they earned a net income of less than 5000 UGX equivalent to $ 1.3.

This study had two objectives: (1) to characterize social services within the working environment of Boda-Boda motorists and (2) to explore the perceived impacts of COVID-19 lockdown on the livelihood of Boda-Boda motorists and their families. This chapter provides a brief background on Boda-Boda industry in Uganda and describes the impacts of COVID-19 lockdown. Mixed methods were employed to assess both qualitative and quantitative data and determine the results. The chapter concludes with a discussion and recommendation section.

Methodology

We utilized qualitative and quantitative methods of data collection. Four Boda-Boda stages labelled 1–4 were studied for 7 days. Quantitative data including distance to the market, health centers, police stations, and schools were generated using ArcGIS

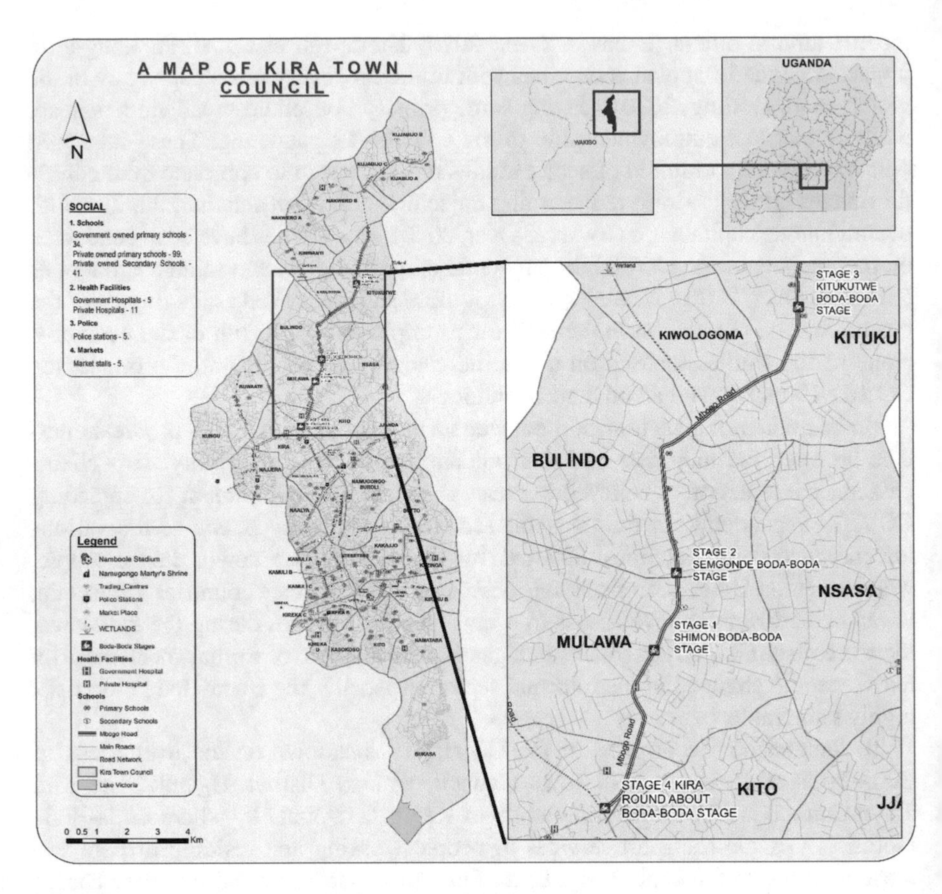

Fig. 16.1 Distribution of Boda-Boda stages and social services

10.4. These features (markets, hospital, police stations, schools, Boda-Boda stages) were georeferenced using a handheld Global Positioning System (GPS). Euclidian distance layers were generated using interpolation methods in spatial analyst tools of ArcGIS 10.4. Average distances were extracted using the extract by point function in the spatial analyst tool. Figure 16.1 shows the study area, distribution of Boda-Boda stages, and social services. The relationship of Boda-Boda motorists and the population they serve was not established due to lack of census data.

Secondary data were obtained by downloading socioeconomic services from both OpenStreetMap and ArcGIS online. Field data were processed and converted to a format usable within ArcGIS software. Secondary data were overlaid with the data collected from the field. Proximity analyses were carried out using the near tool and buffer analysis tool to determine how far the different social services were from the different Boda-Boda stages. The outputs from the proximity analysis were near table results and buffer distance results. The overall outputs were generally maps showing the locations of different services against the Boda-Boda stage locations and buffer maps showing distances within which the different services were accessible.

The study also collected data from four focus group discussions (each focus group had 8–12 participants) and four key informants. Key informant interviews were conducted to support the triangulation of the information obtained from the focus groups. Key informants included heads and deputies of the Boda-Boda stages. These were purposively selected because of their long-term experience in Boda-Boda transport, and they provided rich narratives on the effects of the COVID-19 pandemic. The choice of participants was based on accessibility, availability, and willingness to participate in the study and proximity to the researcher due to COVID-19 movement restrictions. We used qualitative content analysis to analyze data. All descriptive data were captured verbatim. We reviewed field notes and generated emerging themes, categorized data into themes, and analyzed data to interpret outcomes and reported the findings.

Results

The population characteristics of Boda-Boda motorists in our study were as follows: should be between 25 and 34 years of age, be married, and be with 2–4 years of experience in the Boda-Boda transport industry. Most of them had obtained primary school as their highest level of education. In addition, most of them bought Boda-Boda motorcycles on loan and were engaged in Boda-Boda as their sole source of income (Fig. 16.2).

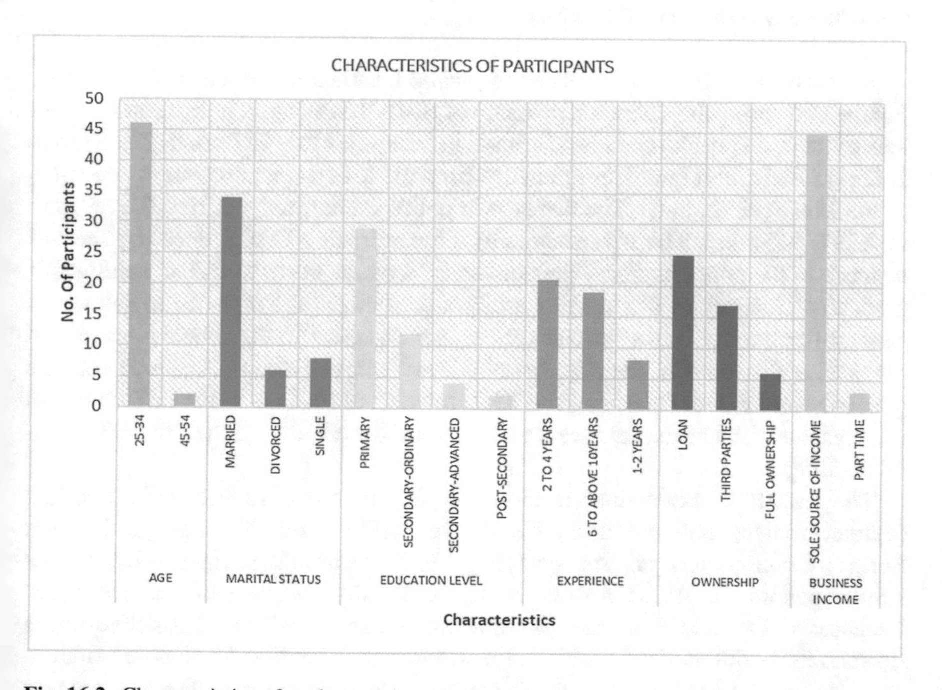

Fig. 16.2 Characteristics of study participants

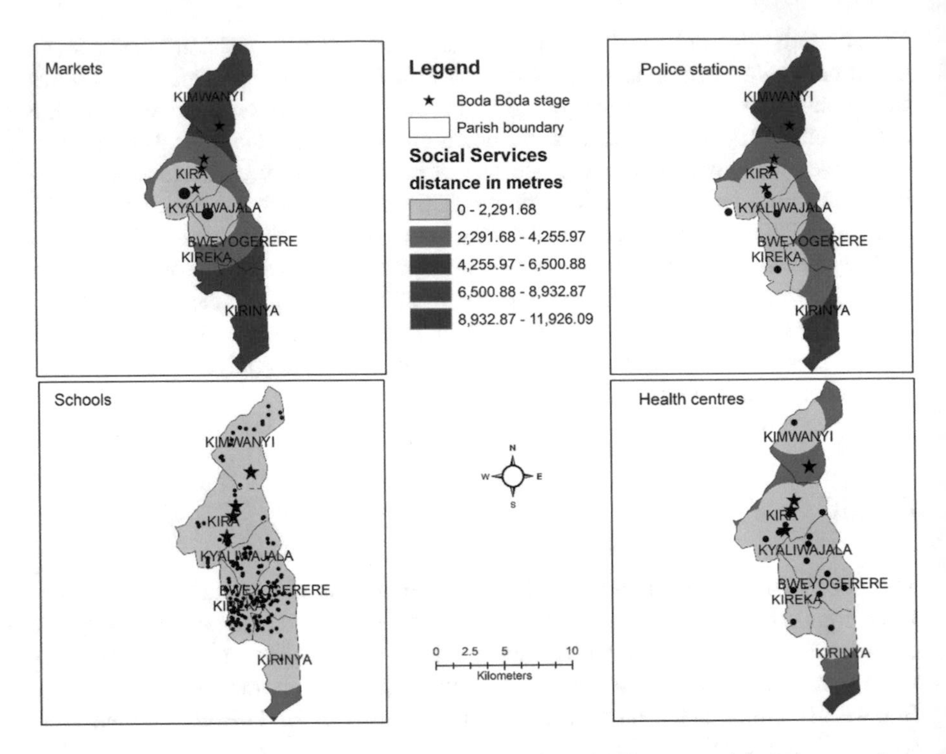

Fig. 16.3 Average distance in kilometers from the Boda-Boda stages to social services: markets, police stations, schools, and health centers

Boda-Boda stages refers to gazetted pickup locations for Boda-Boda transport. Figure 16.3 shows the Euclidian distance of Boda-Boda stages, police station, markets schools, and health centers. Distance to the market ranges from 0–2.200 km to 8.922–11.826 km. It increases gradually from the Kira stage to the north and south of the Kira Town Council. The distance to police stations ranged from 0–2.291 km to 2.291–4.255 km. The Kira stage and Kireka Trading Center have the shortest distance to the police stations. This distance increases northwards and southwards gradually. Contrary to the markets and police stations, the distance to schools is generally shorter. Areas with relatively longer distances to the schools increase in the southern portions of the city. Areas with shorter distance to health centers are generally found in the center of the Kira Town Council. Due to the limited health services, the distance to health centers increases towards the south of the city council buildings.

The Kira Boda-Boda stage is closer to all administrative and social services (market centers, police station, Namboole Stadium and Namugongo Martyrs Shrines, health centers, and schools) (Fig. 16.4). Distance to social services follow similar patterns to all Boda-Boda routes. Generally, distance to social services increased in the following order: Kira; Shimon; Semugonde and Kitikutwe stages for schools, health services, police stations, and market centers; Namboole Stadium; and Namugongo Martyrs Shrines. The shortest distance to the market centers was

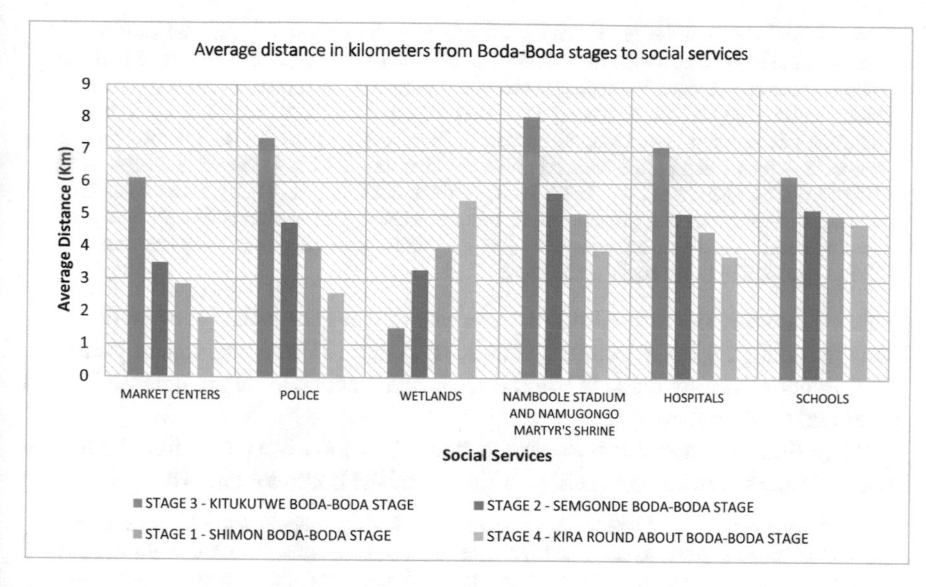

Fig. 16.4 Average distance between Boda-Boda stages and social services

approximately 1.8 km and the longest was at 6.1 km. The distance to police stations varied from 2.6 km to 7.4 km. For Namboole stadium, it varied from 3.9 km to 8 km. For health centers it ranged from 3.7 km to 7.1 km, and for schools it ranged from 4.8 km to 6.3 km. All distance to social services in the study area could not be related to the population due to lack of census data.

Perceived COVID-19 Pandemic Effects on the Livelihood of Boda-Boda Motorists

Perceived COVID-19 Pandemic on Food Security

Boda-Boda motorists who were participants in the focus groups reported that COVID-19 pandemic restrictions on mobility and confinement at home resulted in shortage of food at home. Food and other household necessities ran out and restocking became hard, the cost of food became increasingly high, and many families could only obtain cheaper food items sold on roadsides. In one of the focus groups, one participant stated:

> There is a high rise in prices for both perishable and non-perishable food commodities. At the same time, there is limited cash income. This situation has created food insecurity in households. The fact that we (Boda-Boda motorists) earn little income (hand to mouth), we find it difficult to provide enough food to our families because the Boda-Boda transport business has been locked down. In addition, our wives are not engaged in income generation activities to supplement household income and this exacerbated food insecurity in our

households. Our families are surviving on one meal a day (maize meal and beans) and because the beans take longer to cook with much-needed fuel to cook them, some families just survive on porridge. We replaced beans with mukene (silverfish) because of its nutritional values and mukene gets ready in a short time and uses less fuel compared to beans. Unfortunately, our spouses complained that mukene was not palatable with maize and that mukene is food eaten by the poor communities and animals like chickens and pigs. With this COVID-19 pandemic, we are hit hard, because our role and identity as men; heads of families the provider and protector have been collapsed. Our manhood identity has been insulted and crushed to the marrow due to our inability to provide food to our families. *Focus group, Kitukutwe Boda-Boda stage*

Demonstrating the difficulty to access markets, a geographic perspective illustrates that only the Boda-Boda stage at Kira Trading Center had a nearby market. The absence of nearby markets could have been responsible for an increase in food prices and food insecurity.

Many Boda-Boda motorists reported that when food insecurity in their homes became dire, they resorted to selling off some of the household items:

To solve the issue of food insecurity, we have sold off our household assets like televisions, beds, mattresses, music systems but the sales were not sustainable as we could not buy adequate food for our families to last a long time. We are disappointed, we have grown seeing our grandfathers and fathers provide food to their families, we have lost the provisioning role in a short time due to economic and social changes resulting from the COVID-19 pandemic lockdown. *Focus group, Shimon Boda-Boda stage*

All focus group participants reported that:

Although the Government of Uganda has been providing food to the neediest communities, it never considered us the Boda-Boda motorists as a vulnerable group for eligibility to food supplies (milled maize flour, beans, and milk). We the Boda-Boda motorists are disappointed because government authorities including our local leaders in our communities of a board have not given us an explanation why we are being excluded from benefiting from government food supplies. *Focus groups, all Boda-Boda stages*

Based on geospatial analysis used in this study area, there were no food aid program services which could have been alternative food access options for Boda-Boda motorists since they did not have any income to buy foodstuffs from the markets.

Perceived Income During the COVID-19 Pandemic

Focus group participants reported that COVID-19 lockdown measures affected their working routines in terms of working hours and created uncertainty in their daily lives because many had to abandon the Boda-Boda transport business as stated:

COVID-19 lockdown measures (Standard Operating Procedures – (SOP) contributed to major financial setbacks for us. We were restricted to only transport cargo within the restricted time of 7:00 am-2:00 pm. Even when cargo was available, it was not a guarantee to have it transported beyond curfew time. Although our work schedule was later increased in June 2020 from 7:00 am -5:00 pm, we were used to earning more money by working late evening and night hours. Night businesses were affected by police deployment and operations that covered entire Kira Town Council making it difficult for Boda-Boda movements beyond curfew time. We no longer see some of our colleagues and not sure if they quit the business or will be returning when the situation normalizes. This decreased income caused animosity in providing food to our families. On the issue of COVID-19 health impacts on our families, we survived narrowly. We depended on herbal medicine[1] which we collected from the nearby bushes. Herbal medicine worked for us well the same way it relieved most Ugandan families. *Focus group, Semugonde Boda-Boda stage*

It was further reported that due to the economic conditions, Boda-Boda motorists were brutalized as stated:

In a bid to make money, we were arrested for non-adherence to SOPs. We were tortured while in police custody, fined, while our colleagues' motorcycles were confiscated. Those colleagues that were unable to pay heavy fines and bribes imposed by the police, committed suicide. Those who bought Boda-Boda motorcycles on loans or higher purchases, such arrangements were cancelled, and vendors recalled their motorcycles. This situation has pushed most of our colleagues out of business and has created mass unemployment without any compensations. *Focus group, Kira Boda-Boda stage*

A geospatial analysis of the Kira Town Council does not provide existence of shelter homes. In addition, with increasing number of gender-based violence (GBV), the victims could not report the cases to police due to COVID movement restrictions. Boda-Boda motorists who were arrested and those whose motorcycles were confiscated found it hard to get bail from police since they could not be accessed by their families due to restricted movements.

A key informant recollected the status of the Boda-Boda business before and during the COVID-19 pandemic lockdown.

During the month of May-June of every year, we used to transport pilgrimages to the Basilica at Namugongo Martyrs Shirines. We also made money during soccer season by transporting fans to Mandela National Stadium. People prefer to use Boda-Boda transport as a fast means of transport and Boda-Bodas also navigate and pass through shortcuts through overcrowded destinations. Soccer fans pay up to 35,000 Ugs equivalent to $10 for transport to be dropped at the stadium and about 15,000 Ugs equivalent to $5 to the Basilica. The money I make in these two months (May-June) helps me to pay school fees for my children for a year. What an exciting experience we had before the COVID-19 pandemic lockdown. But all these money merry-making seasons were disrupted by COVID-19 pandemic. The Basilica was closed, and the stadium was turned into a COVID-19 treatment center. *Key informants, Kitukutwe Boda-Boda stage*

The information revealed by the key informant is consistent with the locations of the Namugongo Martyrs Shrine and Namboole National Stadium. For example, the Namboole National Stadium is the furthest administrative service from the Kitukutwe Boda-Boda stage (Fig. 16.1) suggesting that the further you go with Boda-Boda transport coupled with the prevailing business activities, the higher transport cost you pay.

[1] Boda-Boda motorists referring to herbal medicine means the use of medicinal plants (e.g., boiling and drinking concoctions of mango, lemon, guavas, eucalyptus leafs, and lemon grass) used to treat COVID-19-like symptoms including cough and cold.

Besides other challenges, participants were also worried about the new proposed Boda-Boda routes:

> As Boda-Boda motorists, we are expecting more frustrations. Kampala Metropolitan has been redesigned which has locked out Boda-Boda transport services in the city center. The implementation of this new change was planned to commence in November 2020. Only those registered associations like safe Boda-Bodas will be allowed to operate in the city area where high income is generated. Most of us who are not registered with Boda-Boda associations have counted ourselves out of business. This is a new development. Our business dwindled but authorities are demanding registration fees. *Key informants, Kira Boda-Boda stage*

This means that Boda-Boda motorists will be providing transport services to limited locations and this will increase their financial vulnerability.

Reduced income among Boda-Boda motorists due to COVID-19 pandemic restrictions also affected gender and family relations. Participants reported an upsurge of domestic violence from their female spouses and children as stated by a key informant:

> As men and heads of households, we feel our status and identity insulted by our wives and its worse when it is done before our children because we cannot provide for them. We are vulnerable, dying in silence in our own homes. We have been denied food and sex. Some women have poured boiling porridge to their husbands, this has caused death of many men. We have been accused and insulted of infidelity, witnessed our wives elope with other well-to-do men. Decreased household income due to COVID-19 lockdown has triggered emotional and physical violence. Our homes have become haven sites for terrorism. We fear to report cases of violence to authorities or seek help due to the stigma associated with being coerced by a wife. Our country, Uganda has no laws on violence against men. We are dying in silence. We have been trapped at home with our wives (abusers) the situation is more stressful, fatiguing and depressing, we need help. *FGD, Kira Boda-Boda stage*

By the time this study was conducted, all health systems and protective agents had focused on women and girl child issues and ignoring male and boys' issues. The Minister of Gender Labour and Social Development in Uganda stated that on March 30 to April 28, 2020, a total of 3280 police cases were recorded, and out of the recorded cases, gender-based violence cases totaled to 283 victims (children and women (Hauxia, 2020)). The figures were not gender-disaggregated; hence, the concerns of Boda-Boda men on violence against males should be examined.

Discussion

This study demonstrated several impacts of COVID-19 lockdown on the lives of Boda-Boda motorists in Uganda. The study revealed the existence of administrative and social services and the Boda-Boda transportation network (markets, hospital, police stations, schools, Boda-Boda stages). However, during COVID-19 lockdown, most of these services were locked down except for markets, hospitals, and police stations. Boda-Boda motorists whose livelihoods depended on these social services through transportation of passengers and goods were left jobless without any alternative sources of income. The fact that Boda-Boda motorists who earned a net income of less than 5000 Uganda shillings (equivalent to USD 1.3) had no savings to survive on during the COVID-19 pandemic.

The study further revealed that most Boda-Boda motorists became vulnerable because they lost their breadwinning roles and could hardly provide food to their families. Pereira and Oliveira (2020) affirms that males in patriarchal societies who fail to meet their ascribed roles experience increased social, psychological, and mental stress including stigma and isolation.

Further, inadequate access to healthy foods among Boda-Boda motorist's families was a risk factor that affected their physical, social, and emotional health. Food insecurity increases mortality especially those who depend on the informal sector and lack alternative means of survival. Previous studies (Breakfast & Nomarwayi, 2021) reported difficulties in obtaining food during lockdown due to intra-household inequalities. Consequently, adoption of negative coping strategies, such as sale of household assets, became the only option, which further plunges Boda-Boda motorists into poverty.

Participants also revealed that food insecurity and lack of income coupled with prolonged home confinement led to increases in domestic conflicts and partner violence in Boda-Boda motorists' homes. Similarly, Taub (2020) demonstrated that domestic violence is common due to longer stays with family members especially on vacations, in times of pandemics, or among vulnerable communities. Similarly, Van Dijk and Nkwana (2021) indicate that food insecurity has political, economic, and social dimensions rooted in power inequality and patriarchy. In most vulnerable communities, food security has been regarded as a stabilizer. Therefore, when men fail to provide for their families, they lose power and control over their families, and this may increase sexual and physical conflicts and denial of conjugal rights. Participants reported that government interventions on gender-based violence during the COVID-19 pandemic focused on women and children and failed to come up with strategies to address conditions of men who suffer violence from their spouses. The lack of programs on violence against men could be due to limited research on violence against men in Africa. There were no protective systems to address domestic violence against men in the study site. Most importantly, men face the challenge of reporting violence inflicted on them due to gender discriminatory norms that perceive them as the perpetrators (Komazec & Farmer, 2021). This implies that they do not receive the required services which culminate in poor health outcomes in the face of the pandemic (Usher et al., 2020). The findings reveal the panorama of structural oppression against people in the informal sector. Participants in this study reported police brutality in the cover of enforcing movement restrictions. Katana et al. (2021) allude that during the COVID-19 lockdown, cases of violence and discrimination intensified in both control and prevention measures. Besides being a violation of their rights, these acts indicate insensitivity of authorities to the welfare of the low-income and average earners who live hand-to-mouth.

The COVID-19 pandemic has not only affected Boda-Boda motorists but has generally affected Ugandan workers in the informal sector. Other countries like India (Estupinan & Sharma, 2020) and Indonesia (Joko, 2020) reported continued reduced wages or loss of employment especially among informal workers. All these fears combined with structural changes such as the introduction of new policies and fees on Boda-Boda motorists are manifestations of structural oppression. This situation may create social effects like disintegration of families and social networks (Gonalons-Pons & Gangl 2021), and economic costs may include funeral costs, increased hospital bills as a result of increased physical and emotional violence. It may also lead to reduced human development opportunities including schooling, teenage pregnancy, child labor, and prostitution, all of which increase poverty and vulnerabilities and oppression.

Conclusions and Recommendations

The study shows that within the Kira Town Council, Boda-Boda motorists averagely operated in a distance of between 0–2.291 km and 8.932–11.926 km. However, due to COVID-19 lockdown movement restrictions, Boda-Boda motorists lost their jobs and could not provide basic needs to their families. The impact of the COVID-19 lockdown also contributed to food insecurity and domestic violence due to lost daily income. One of the techniques to address food insecurity and income problems among vulnerable populations like Boda-Boda motorists is to create awareness about sustainable local farming, urban and rooftop gardening, and wetland resources where fresh produce is grown by community residents and may be an alternative to food access. This would promote food security with communities accessing food produce at affordable cost, close to home, without requiring them to travel to distant markets. The surplus could be sold to nearby market centers for increased household income and investment in other household developments that Boda-Boda motorists serve. There is need also to plan for shelter homes and the development of protective systems like policy reforms and awareness creation against all forms of violence against men and women. Policy reforms and awareness are needed to understand domestic violence against men as demonstrated by this study on Boda-Boda motorists.

Declaration Statement of Interest The authors of this work declare no competing interests.

References

Breakfast, N. B., Nomarwayi, T. & Titi, N. (2021). Food insecurity among African women residing in rural and informal urban areas during lockdown in South Africa. *Gender & Behaviour, 19*(1), 18331–18336.

Estupinan, X., & Sharma, M. (2020). *Job and wage losses in informal sector due to the COVID-19 lockdown measures in India.* Available at SSRN 3680379.

Evans, D., & Over, M.(2020). *The economic impact of COVID-19 in low-and middle-income countries*. Center for Global Development, Centre for Global Development, Ideas to action: Independent research for global prosperity.

Gonalons-Pons, P., & Gangl, M. (2021). Marriage and masculinity: Male-breadwinner culture, unemployment, and separation risk in 29 countries. *American Sociological Review, 86*, 465–502.

Hauxia. (2020). *Domestic violence against women increases in Uganda amid COVID-19: The Minister of Gender Labour and Social Development*. Xinttaunet. http://www.xinhuanet.com/english/2020-04/28/c_139015384.htm. Accessed 07 Oct 2020.

Joko, A. (2020). The impacts of COVID-19 pandemic to informal economic sector in Indonesia: Theoretical and empirical comparison. In *E3S web of conferences* (Vol. 200, p. 03014).

Kampala City Council Authority. (2020). Press statement on the public transport reforms to streamline and regulate the Boba Boda industry as approved by cabinet. https://www.kcca.go.ug/news/413/press-statement-on-the-public-transport-reforms-to-streamline-and-regulate-the-boba-boda-industry-as-approved-by-cabinet. Accessed 16 Mar 2020.

Katana, E., Amodan, B. O., Bulage, L., Ario, A. R., Fodjo, J. N. S., Colebunders, R., & Wanyenze, R. (2021). Factors associated with access to food and essential medicines among Ugandans during the COVID-19 lockdown: A cross-sectional study. *Journal of Interventional Epidemiology and Public Health, 4*(4), 4.

Komazec, S., & Farmer, C. (2021). Family violence homicide in Australia: The effect of victim/offender gender on social media commentary. *Journal of Gender Studies, 30*(1), 4–17.

Namara, A. (2019). Boda drivers say they'd rather flee accidents than get insurance, *Global Press Journal, 15*. Accessed at https://globalpressjournal.com/africa/uganda/boda-boda-drivers-uganda-say-theyd-rather-flee-accidents-get-insurance/

Pereira, M., & Oliveira, A. M. (2020). Poverty and food insecurity may increase as the threat of COVID-19spreads. *Public Health Nutrition, 23*(17), 3236–3240.

Pilling, D. (2020, March 12). Coronavirus and the collapse of global public health. *Financial Times*. https://www.ft.com/content/be024190-62b5-11ea-b3f3-fe4680ea68b55

Taub, A. (2020). A new COVID-19 crisis: Domestic abuse rises worldwide. *The New York Times, 6*. Available from https://chescocf.org/wp-content/uploads/2020/04/Domestic-Abuse-Rises-Worldwide-New-York-Times.pdf. Retrieved September, 2021.

Thompson, J., Ndung'u, N., Albacete, M., Suleri, A. Q., Zahid, J., & Aftab, R. (2021). *The impact of COVID-19 on livelihoods and food security* (COVID-19 responses for equity (CORE) research for policy and practice report. Institute of Development Studies. https://doi.org/10.19088/CORE.2021.001

Usher, K., Bhullar, N., Durkin, J., Gyamfi, N., & Jackson, D. (2020). Family violence and COVID-19: Increased vulnerability and reduced options for support. *International Journal of Mental Health Nursing, 29*, 549–552.

Van Dijk, H. G., & Nkwana, H. M. (2021). The face of food insecurity is female: A post-colonial feminist argument for rural women. *African Journal of Gender, Society and Development (formerly Journal of Gender, Information and Development in Africa), 10*(1), 99–116.

Part III
Lessons Learned and New Horizons (Overview)

Robert B. Richardson

Part III of this volume explores how the lessons learned from the COVID-19 pandemic set the stage for next steps as the world prepares to adapt to a post-pandemic reality. This part includes examples of mitigation strategies, adaptation pathways, and ongoing challenges.

In Chap. 17, Dr. Alexsandar Valjarević describes how different types of transportation uses have had varying impacts in terms of accelerating the transmission of the virus. In a comparison of transportation by air, rail, vehicle, and ship, the author uses geographical information systems along with other geospatial approaches to demonstrate how analysis of traffic patterns can be used to depict the spatial trajectories of the spread of COVID-19. In Chap. 18, Dr. Valentina Albanese and Dr. Giorgio S. Senesi provide an overview of digital surveillance technologies in light of the concept of surveillance capitalism, with a focus on the technological solutions deployed to contain the spread of COVID-19 in Italy. In Chap. 19, Professor Francis Koti highlights both the importance of the contribution of the urban informal sector to national economies and local livelihoods in sub-Saharan Africa and the vulnerability of that sector to political, economic, social, and environmental uncertainties such as those presented by the COVID-19 pandemic. The chapter examines the relative resilience of the urban informal sector in the face of COVID-19-imposed lockdowns and demonstrates how unique characteristics of this sector have helped facilitate reentry into the sector as economies in the region recover from the economic shocks of pandemic-related restrictions.

In Chap. 20, Professor Jason Knouft highlights the complex and often sensitive relationships between human activities and freshwater resources during the COVID-19 pandemic, providing unique insights into how humans interact with freshwater resources. The chapter describes how the combination of COVID-19 restrictions, regional climatic variability, and varying levels of social and economic

R. B. Richardson
Michigan State University, East Lansing, MI, USA

stability have implications for preemptive policies that ensure water availability and access to freshwater resources in all areas of the world. In Chap. 21, Ms. Maya Mishra argues that Earth observations will play a critical role in predicting and preventing disease incidence in the future. Using a One Health perspective, the chapter describes how Earth observations data provide an avenue for preventing the next disease pandemic before it starts. Finally, in Chap. 22, Dr. Ashley Pierce and Dr. Amanda Shores offer lessons learned from the US National Science Foundation during the pandemic and its efforts to enable the acceleration of research in times of need. The authors describe how the research community led the charge in defining the research agenda to address the pandemic. Their observations highlight the implications of inter- and transdisciplinary perspectives emphasized across COVID-19-related research projects. The response by the National Science Foundation to COVID-19 reinforces the need for interdisciplinary research and integration across the diversity of researchers to solve interwoven challenges during times of rapid change and urgent need.

In summary, the chapters in part III highlight how geospatial approaches to understanding the COVID-19 pandemic have revealed lessons that can be used to inform decisions and policies in response to future pandemics. The findings of these chapters also have important implications for adapting to the realities of a dynamic shock to public health and for mitigating its ongoing and evolving impacts at various scales ranging from local to national and global.

Chapter 17
Geospatial Techniques for Mapping the Spatial Trajectories of COVID-19

Aleksandar Valjarević

Personal Story

During the COVID-19 pandemic, there have been two complete lockdowns in Serbia. As a country located in the Balkan Peninsula, in South-East Europe, we received a smaller quantity of vaccines, but the amount increased in the spring of 2021. Although most of my friends and family members were infected, I was not. I received three doses of the vaccine, including the booster dose. So far, 2.7 million people in Serbia have received two doses of the vaccine. I have travelled multiple times this year, and as a geographer who enjoys seeing new locations and meeting new people, I managed to regain the feeling of life before COVID-19 restrictions. I sincerely hope that the pandemic will end in 2022, so that the world will be able to go back to normal.

Introduction

Studying the spatial distribution and spread of diseases is of enormous importance for human security and well-being. In the nineteenth and at the beginning of the twentieth centuries, global diseases seriously threatened humanity. The plague has given rise to at least three major pandemics. The first spread around the Mediterranean Sea in the sixth century AD, the second started in Europe in the fourteenth century and recurred intermittently for more than 300 years, and the third started in China during the middle of the nineteenth century and spread throughout the world (Stenseth et al., 2008). Today, in the twenty-first century, humanity is generally well connected and concentrated in populated, urban areas. Traffic networks are now connected better and nodes of traffic interconnected better too. More than 70% of humanity will live in urban areas by the end of the twenty-first century (Bolund & Hunhammar, 1999). The most densely populated areas will be distributed in the coastal zones of North and South America, western and central Europe, Asia, central Africa and coastal zones of Australia (Bostrom, 2009). Previous studies have

A. Valjarević (✉)
University of Belgrade, Belgrade, Serbia
e-mail: aleksandar.valjarevic@gef.bg.ac.rs

M. Laituri et al. (eds.), *The Geographies of COVID-19*, Global Perspectives on Health Geography, https://doi.org/10.1007/978-3-031-11775-6_17

shown that population growth, urbanization, global mobility and environmental change are contributing to the increased probability for pandemics to occur (Zhou et al., 2020). The COVID-19 virus originated in Wuhan City in Hubei Province of China in December 2019 (Singhal, 2020). Geographic Information Systems (GIS) and numerical methods are an essential tool for estimating pandemic characteristics through geographic space which has been employed at the local, regional and global scales (Imeri et al., 2021; Valjarević et al., 2020). There are numerous methods, including GIS methods, that can be used in the analysis of spatial distribution of COVID-19. The algorithms in GIS software applicable to the mapping of the spatial trajectory of COVID-19 are Kriging, Semi-Kriging, Modified Kriging, Kernel density, buffer and zonal statistics (Antrop, 2004). The spatial and numerical methods supported by GIS could find a prominent place in public health and epidemiology research as shown by the contemporary studies conducted worldwide (Guo, 2007; Christaki, 2015; Centers for Disease Control and Prevention, 2019; Nykiforuk & Flaman, 2011; Zhang et al., 2020; Schröder, 2006). The risk of the pandemic in modern society depends on the volume of transmission and human or environmental factors (Voigtländer & Joachim Voth, 2013). The spatial density and distribution of populations have a significant impact on the potential of viruses spreading among the public. Two hundred years ago, the world population was less than 1 billion; by 1850, it was 1262 billion; and today, it is 7.9 billion. A population growth rate of 2.2% in the last 50 years combined with projections for expanding traffic infrastructure can provide spatial trajectory for increasing viral transmission (Viviroli et al., 2007; Traphagan, 2019). Scientists predict future pandemics to spread at 1000 times faster than in the past 200 years due to expanded travelways and increasing nodal points of connection. Not only do the properties of network influence the spread of virus, but also the density of population, migration, demography, and many other factors. Today, we have frequent airlines across the world. There were two main epicentres of the pandemic in 2020, China and Italy. Today, there are 224 countries and territories with citizens infected by COVID-19 (Zajenkowski et al., 2020).

Materials and Methods

The data for this research were collected from the open-source database, Worldometer (https://www.worldometers.info/). The data are in CSV format and they represent population distribution at the global scale. Airplane trajectory data were downloaded from the International Air Transport Association (IATA) at a 10-m spatial resolution. Railroad and road data were used from the free and open-source database, Natural Earth, with a 30-m spatial resolution. Population data were collected from the United Nations at 10-m spatial resolution and included cities with more than 100,000 citizens. After initial screening for population density distribution, it was concluded that the most commonly used kinds of local transportations in urban areas are trains and buses. The high number of lines and nodes belong to air traffic, followed by road traffic, railroad traffic and then marine traffic. The densest places in the world are in Central America, east coast of North America, Central East Africa, western Europe, Southeast China, Southeast Asia, central part of Japan and east coast of Australia (see Fig. 17.1).

The high level of connection between airports occurs primarily in western, central and south Europe, the east and west coast of the USA, Southeast and East Asia, southeast Australia and urban zones in South America (Liu et al., 2020; Weigang et al., 1997). The most densely populated areas with 500 individuals/km^2 have the most connections with airports The estimated number of airplane lines was 4.1×10^{15} km in 2019. In 2020, this number was reduced to 2.4×10^{14}. The decrease is approximately 15% (Fig. 17.2).

These nodes (airports) are directly connected to road and rail networks. They have a large population density, with more than 500 inhabitants per km^2. Due to all the elements listed above, the COVID-19 virus is well dispersed on all sides of the world, excluding Antarctica. In this study, two mathematical models were used and integrated in the QGIS 3.12 software to calculate potential virus trajectories. The first one presents the fastest transmission rate of viral spread (see Eq. 17.1).

$$P(X = x) = \frac{e^{x} \cdot e^{-u}}{x!} \tag{17.1}$$

This model presents the modified exponential distribution or Poisson distribution. *X* is the Poisson distribution with mean *u*. This distribution is better than the classic exponential distribution because the parameter of measure *m* depends on the average time between occurrences. This exponential distribution more accurately depicts growth rates within natural and spatial sciences and medicine (Louzada-Neto et al., 2011; Holgate, 1964). For the analysis of transport problems in roads

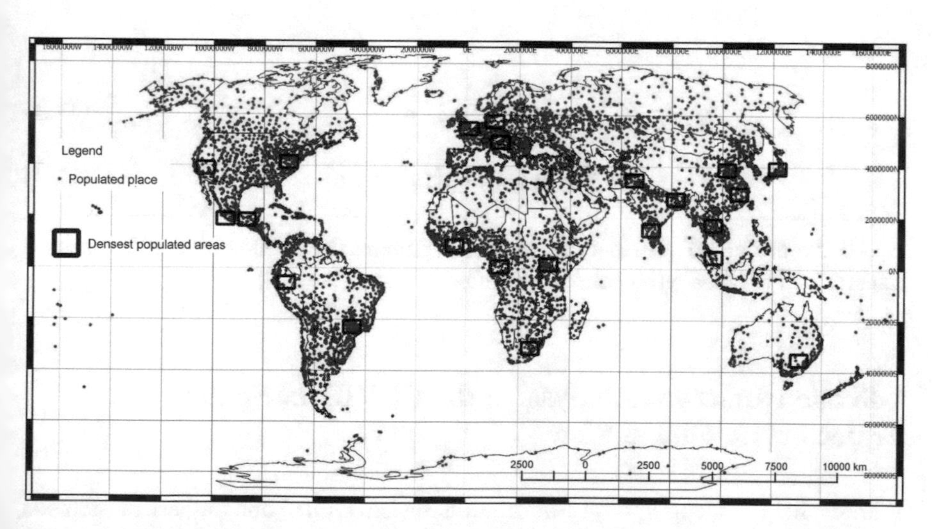

Fig. 17.1 Populated places in the world with marked density. The data used were from the open database Natural Earth. The geographical projection of this map is the Robinson Projection

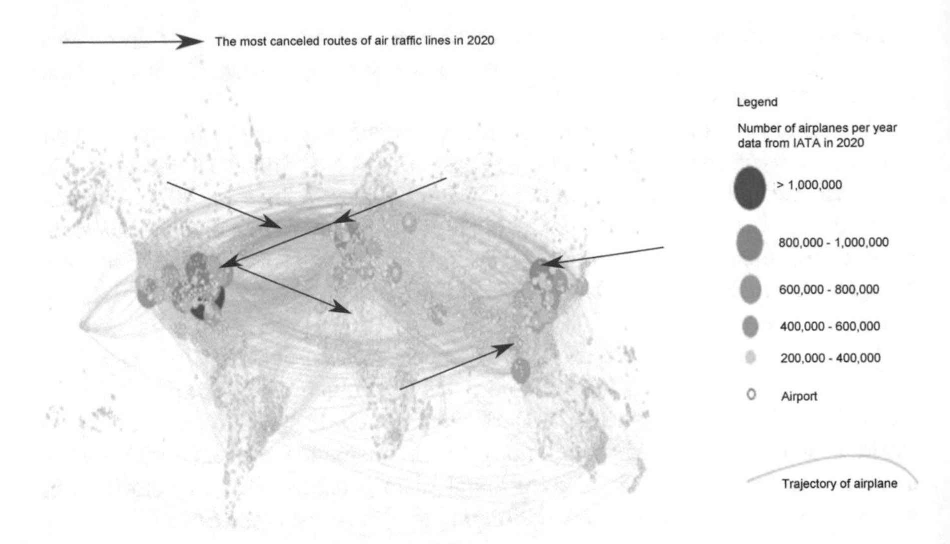

Fig. 17.2 Trajectories of airplane lines in 2019. During this time the main corridor for airline travel was between Mainland China, western Europe, and the USA. This map was generated by means of open-source GIS software 3.16.5 QGIS and Gephi 0.9.1, while data were taken from the International Air Transport Association

and railroads, we used the modified geometric distribution. The geometric distribution has the following form (Eqs. 17.2–17.4):

$$P(X = x) = q^{(x-1)} p \tag{17.2}$$

where

$$E(x) = 1 / p \tag{17.3}$$

$$\mathrm{Var}(x) = q / p^2, q = 1 - p \tag{17.4}$$

All the presented distributions were integrated into open-source software and transformed in Geo-Python and C++ codes.

GIS and Numerical Analysis of the COVID-19 Spatial Trajectory: A Global View

With the help of Geographical Information Systems (GIS) and numerical methods and proposed algorithms, spatial models of virus spatial trajectories were analysed. The distribution of the COVID-19 spatial trajectory was used in connection with the density of traffic network and the number of nodes. By using spatial methods such as Kernel and buffer, spatial properties at the time of the beginning of the viral transmission were reconstructed. The GIS software QGIS (Quantum Geographical Information System) and SAGA (System for Automated Geoscientific Analyses), along with the tools for geospatial calculations were used for the visualization of the dispersion of infected cases across the world. For spatial analysis of traffic connection (node connections), the ordinary and Semi-Kriging methods were used through QGIS and SAGA (GIS) of the Spatial Analyst. Other methods used are spatial methods conducted in GIS and include zonal algorithms. For all traffic networks at global scale and their analysis, a modified AHP (analytic hierarchy process) method was applied to provide the numerical value of the nodes' properties. The AHP or complex decision algorithms in this research had a threshold between 0 and 1. 0 which represents low connectivity, availability, direct lines, average clustering and centrality, while 1 represents high connectivity. The minimum value is 0.0 and the maximum 1.0 (Sullivan & Artino, 2013; Wright, 1955). This number represents weekly connected traffic lines on a global scale. The properties of the network were analysed by a numerical algorithm which was implemented through QGIS 3.16.5, MMGIS and graphical modeller functions (Valjarević et al., 2018). Within these functions, network properties for roads, airplanes and railroads were analysed. The nodes of airports and marine ports were analysed by Kernel distribution, buffers and zonal statistics. In that way it is possible to combine two types of data, lines and points. Buffer data used in this research were divided into two types. The first type of buffer was rounded with a radius of 20 km, 50 km and 100 km. The second type was a circular buffer with the same radius (Dong et al., 2003). The median radius of the buffer was 57.5 km and the arithmetic mean was 85 km. The generalized formula for the distribution of nodes across time and space is shown in Eq. (17.5):

$$r = \frac{\sum_{i=1}^{n}(x_i - \bar{x})(y_i - \bar{y})}{\sqrt{\sum_{i=1}^{n}(x_i - \bar{x})^2 \times \sum_{i=1}^{n}(y_i - \bar{y})^2 \times K_d}} \tag{17.5}$$

where the value r is a summary measure relating to an entire set of paired observations. In this study the r varied between 0.1, 0.4 and 0.6 and maximum 1.0 (Altman & Léger, 1995; Polansky & Baker, 2000). Finally, zonal statistics were used for the estimation of traffic properties and COVID-19 virus spatial trajectory at global scale. Four kinds of traffic were analysed: air, road, railroad and marine.

Results

During 2020, airplane traffic constantly decreased. Overall, the total spatial trajectory length of flights in 2020 was 3.9×10^{14} km compared with flights in 2019. The greatest number of flights or (33%) was between America and Europe, followed by

Europe and Asia (29%), North America and South America (9%), Asia and East Asia (11%), Europe and Africa (6%) and others (12%). The total number of nodes is 5623, and the total number of nodes is 58,560. The results of the modified AHP method revealed the following results: the connectivity, 0.4; mobility of air traffic, 1.0; availability, 0.6; density of the graph which represents direct lines, 0.001; modularity, 0.0; average clustering coefficient, 0.170; and centrality, 0.02. The number of weekly connected lines in 2020 was 1806. In 2020 in the USA, the most frequented node was Atlanta's airport. In Europe, the most frequented nodes were identified in the UK (Heathrow), Germany (Frankfurt Airport), the Netherlands (Schiphol), France (Charles De Gaulle) and Turkey (New International Airport AtaTurk). In Central Asia, the most frequently visited airports are Sheremetyevo in Russia, Beijing Capital International Airport and Shanghai Pudong International Airport in China, and the International Sydney Airport in Australia.

The results of the modified AHP method analysing road traffic in 2020 gave scores for mobility (0.7), connectivity (0.9) and availability (0.6). The densest road network was located in western, south and central Europe, eastern parts of the USA and the east coast of mainland China. The number of potential connected road lines per month was 1.3×10^9. The highest number of possible connected road lines is 4.9×10^9 and this network is located in Europe. Europe has the densest and most connected road network. South America has 2.3×10^9 road lines per month. This network is the densest in the central part of South America and in densely populated areas near coast lines. Australia is a continent with less road networks. Australia has the smallest number of connected lines in comparison with other continents with 0.3×10^9 monthly line connections (see Fig. 17.3).

In Europe and North America, the network is well connected. In South America and Asia, the network is randomly distributed and convex.

The railroad network is less dense than the road network. The results of the railroad network analysis were as follows: mobility, 0.8; connectivity, 0.2; and

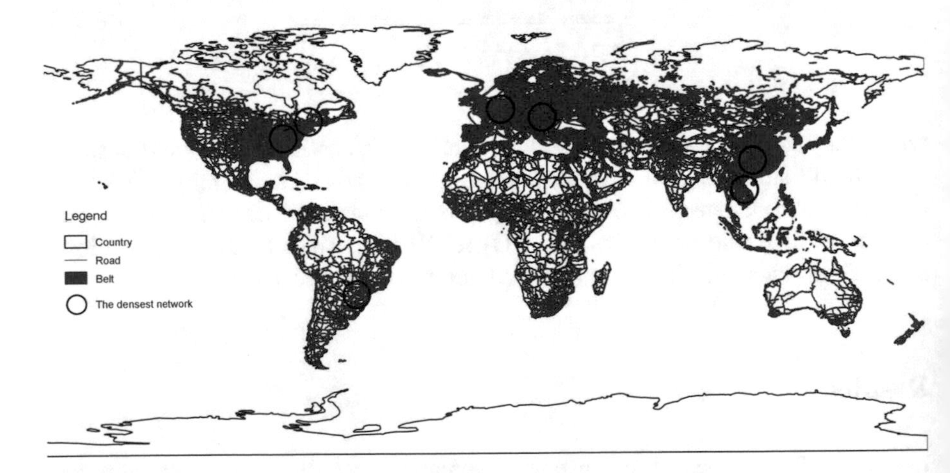

Fig. 17.3 Road network and buffer of 20 km with places marked by black circles with the densest network

availability, 0.8. The railroad traffic in central and south Europe and Southeast Asia is the densest, followed by that in North America. The number of possible connected lines per month in North America is 1.3×10^3. The general results for railroad networks are as follows: the modularity is 0.3, the coefficient of average clustering is 0.31, and the centrality is 0.04 (see Fig. 17.4). The risk of the pandemic in modern society depends on the volume of transmission and human or environmental factors (Voigtländer & Joachim Voth, 2013).

The results of the marine traffic network at global scale are as follows: the mobility is 0.3, the connectivity is 0.2, and the availability is 0.3. The highly distributed lines are in the western Europe and the east coast of the USA (see Fig. 17.5).

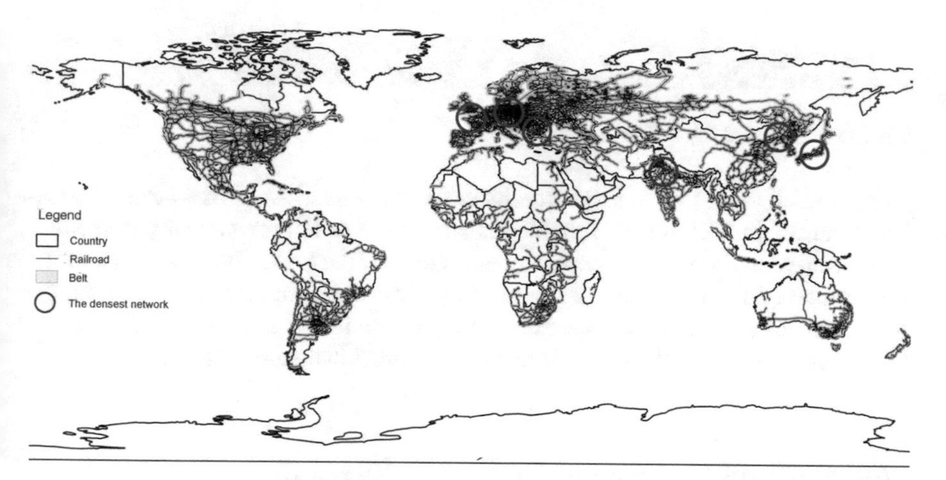

Fig. 17.4 Railroad network and buffer of 50 km with places marked by red circles with the densest network

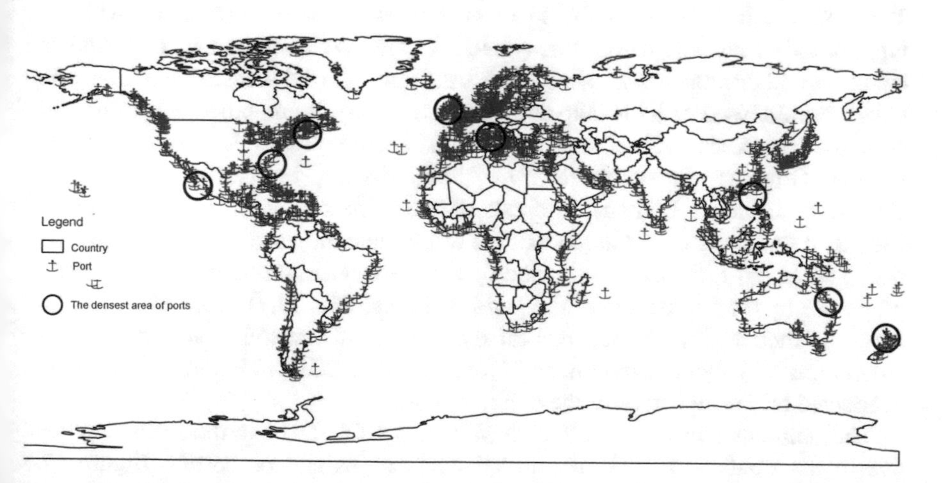

Fig. 17.5 Port networks with places marked by black circles with the densest network

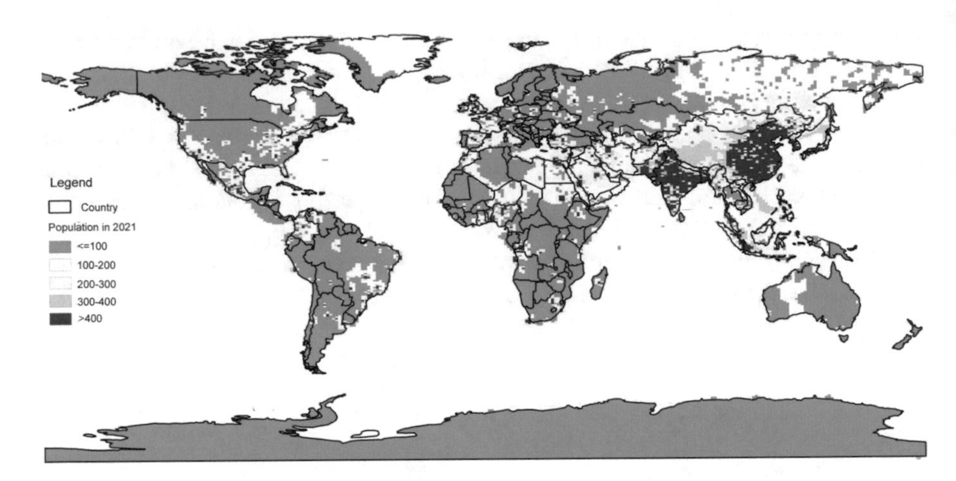

Fig. 17.6 The world population density in 2021

According to the GIS analysis conducted within this research, the average population density in 2020 was 54 inhabitants per 1 km^2. The highest density is in North India, central and western Europe, the east coast of the USA, the west coast of the USA around San Francisco, the east coast of China, the central part of China, the southeast part of Australia, urban areas in Japan and the biggest cities in Africa and South America, where there are more than 400 inhabitants/km^2 (see Fig. 17.6).

Modeling and Mapping of COVID-19 Spreading Spatial Trajectories

This research included the COVID-19 spatial trajectory in comparison with four types of traffic and population density according to the situation in 2020. A total 213 countries and territories were analysed for the purpose of this research. Distribution belts were analysed with the help of precise GIS analysis and procedures. The areas with the lowest number of registered cases include Tanzania, Comoros, Faroe Islands, Gibraltar, Saint Martin, Mauritius, Eritrea, Mongolia, North Korea, Cambodia, Monaco, Bhutan, Barbados, Brunei, Liechtenstein and small isolated islands. If humanity does not slow down or mitigate the pandemic spatial trajectories, 62% of territories can be affected by the coronavirus through the road network properties by the finalization of the pandemics (see Fig. 17.7). The risk of the pandemic in modern society depends on the volume of transmission and human or environmental factors (Voigtländer & Joachim Voth, 2013). This investigation was conducted on the trajectory of the dengue virus.

The railroads will be less affected (49%) by COVID-19 in the following year, due to the smaller network of railroad connections and properties. Figure 17.8 depicts the maximum of COVID-19 spatial trajectory and covered territories in comparison to railroad traffic.

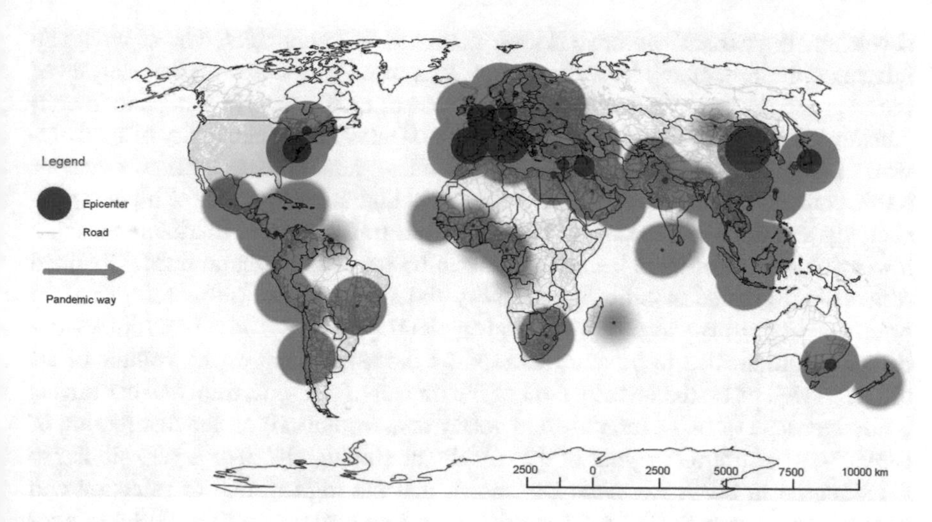

Fig. 17.7 The maximum COVID-19 spatial trajectory and covered territories compared to road traffic with a medium rate of virus transmission

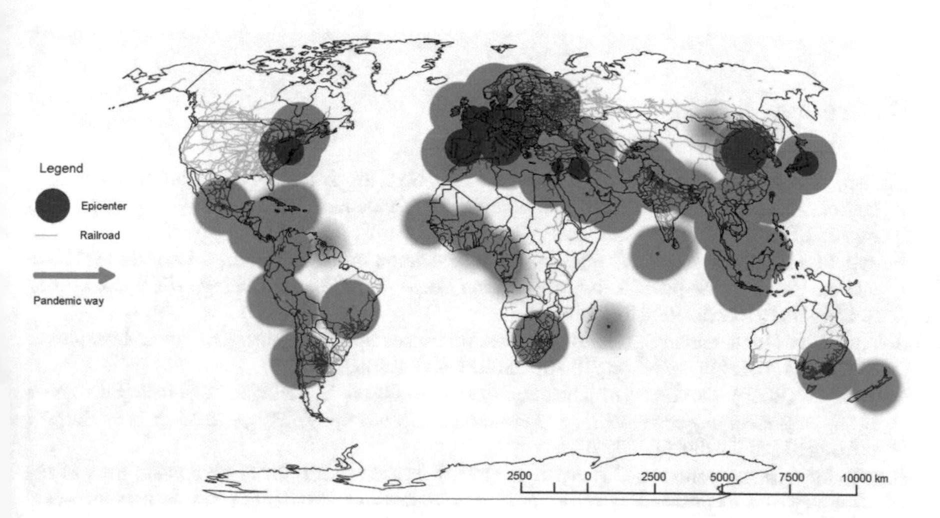

Fig. 17.8 The maximum of COVID-19 spatial trajectory and covered territories in comparison to railroad traffic

Conclusion

This research has not included the mobility of pedestrians. Another limiting factor is that not all data on all connections by means of all types of traffic have been downloaded. The research has shown that densely populated areas in urban zones, with developed traffic network, have a higher rate of COVID-19 spread. The higher the number of nodes, the higher the transmission rate. If the traffic network is less developed, the transmission rate is lower. Isolated and unpopulated areas would be infected only in interactions with visitors. This research focuses on the analysis of traffic properties at the global scale. Four types of traffic were investigated in conjunction with the pandemic dispersion of COVID-19. The connectivity of road networks is more extensive than the railroad traffic system, but less extensive than air traffic connections. Road traffic has a relatively high mobility and the highest connectivity and reliable availability. Other types of traffic such as marine represent a lower risk of transmission because ports can be very easily quarantined. Obtained results highlight that mobility, connectivity and availability of traffic resources can be crucial in the analysis of virus spread spatial trajectories. Therefore, future studies are recommended to be more focused on the assessment of the volume of air travel, as well as on the investigation of the impact of sea, bus, train and car travels (with the usage of daily, monthly and yearly data resolution) on the distribution of COVID-19 in different regions of the world. The situation in 2020 is very similar to the situation in 2019. Research has shown that the highest risk of infection and transmission rate come from air, road, railroad and marine traffic. GIS and geographical analyses may contribute to better understanding and prevention of COVID-19.

References

Altman, N., & Léger, C. (1995). Bandwidth selection for kernel distribution function estimation. *Journal of Statistical Planning and Inference, 46*(2), 195–214. https://doi.org/10.1016/0378-3758(94)00102-2

Antrop, M. (2004). Landscape change and the urbanization process in Europe. Landscape change and the urbanization process in Europe. *Land Scape and Urban Planning., 67*, 9–26. https://doi.org/10.1016/S0169-2046(03)00026-4

Bolund, P., & Hunhammar, S. (1999). Ecosystem services in urban areas. *Ecological Economics, 29*(2), 293–301. https://doi.org/10.1016/S0921-8009(99)00013-0

Bostrom, N. (2009). The future of humanity. In J. K. B. Olsen, E. Selinger, & S. Riis (Eds.), *New waves in philosophy of technology* (New waves in philosophy). Palgrave Macmillan. https://doi.org/10.1057/9780230227279_10

Centers for Disease Control and Prevention. (2019). *The deadliest flu: The complete story of the discovery and reconstruction of the 1918 pandemic virus*. Available from: https://www.cdc.gov/flu/pandemic-resources/reconstruction-1918-virus.html

Christaki, E. (2015). New technologies in predicting, preventing and controlling emerging infectious diseases. *Virulence, 6*(6), 558–565. https://doi.org/10.1080/21505594.2015.1040975

Dong, P., Yang, C., Rui, X., Zhang, L., & Cheng, Q. (2003). An effective buffer generation method in GIS, IGARSS 2003. In *2003 IEEE International Geoscience and Remote Sensing Symposium. Proceedings, 6*, 3706–3708. https://doi.org/10.1109/IGARSS.2003.1295244.

Guo, D. (2007). Visual analytics of spatial interaction patterns for pandemic decision support. *International Journal of Geographical Information Science, 21*(8), 859–877. https://doi.org/10.1080/13658810701349037

Holgate, P. (1964). Estimation for the bivariate Poisson distribution. *Biometrika, 51*(1–2), 241–287. https://doi.org/10.1093/biomet/51.1-2.241

Imeri, H., Jadhav, S., Barnard, M., & Rosenthal, M. (2021). Mapping the impact of the COVID-19 pandemic on pharmacy graduate students' wellness. *Research in Social and Administrative Pharmacy, 17*(11), 1962–1967. https://doi.org/10.1016/j.sapharm.2021.02.016

Liu, Y., Gayle, A. A., Wilder-Smith, A., & Rocklöv, J. (2020). The reproductive number of COVID-19 is higher compared to SARS coronavirus. *Journal of Travel Medicine, 27*, 1. https://doi.org/10.1093/jtm/taaa021

Louzada-Neto, F., Cancho, V. G., & Barriga, G. D. (2011). The Poisson–exponential distribution: A Bayesian approach. *Journal of Applied Statistics, 38*(6), 1239–1248. https://doi.org/10.1080/02664763.2010.491862

Nykiforuk, C. I. J., & Flaman, L. M. (2011). Geographic Information Systems (GIS) for health promotion and public health: A review. *Health Promotion Practice, 12*(1), 63–73. https://doi.org/10.1177/1524839909334624

Polansky, A. M., & Baker, E. R. (2000). Multistage plug—In bandwidth selection for kernel distribution function estimates. *Journal of Statistical Computation and Simulation, 65*(1–4), 63–80. https://doi.org/10.1080/00949650008811990

Schröder, W. (2006). GIS, geostatistics, metadata banking, and tree-based models for data analysis and mapping in environmental monitoring and epidemiology. *International Journal of Medical Microbiology, 296*(S1), 23–36.

Singhal, T. A. (2020). Review of coronavirus disease-2019 (COVID-19). *The Indian Journal of Pediatrics, 87*, 281–286. https://doi.org/10.1007/s12098-020-03263-6

Stenseth, N. C., Atshabar, B. B., Begon, M., Belmain, S. R., Bertherat, E., Carniel, E., Gage, K. L., Leirs, H., & Rahalison, L. (2008). Plague: Past, present, and future. *PLoS Medicine, 5*(1), e3. https://doi.org/10.1371/journal.pmed.0050003

Sullivan, G. M., & Artino, A. R., Jr. (2013). Analyzing and interpreting data from Likert-type scales. *Journal of Graduate Medical Education, 5*, 541–542.

Traphagan, J. W. (2019). Which humanity would space colonization save? *Futures, 110*, 47–49. https://doi.org/10.1016/j.futures.2019.02.016

Valjarević, A., Djekić, T., Stevanović, V., Ivanović, R., & Jandziković, B. (2018). GIS numerical and remote sensing analyses of forest changes in the Toplica region for the period of 1953–2013. *Applied Geography, 92*, 131–139. https://doi.org/10.1016/j.apgeog.2018.01.016

Valjarević, A., Milić, M., Valjarević, D., Stanojević-Ristić, Z., Petrović, L., Milanović, M., Filipović, D., Ristanović, B., Basarin, B., & Lukić, T. (2020). Modelling and mapping of the COVID-19 trajectory and pandemic paths at global scale: A geographer's perspective. *Open Geosciences, 12*(1), 1603–1616. https://doi.org/10.1515/geo-2020-0156

Viviroli, D., Dürr, H. H., Messerli, B., Meybeck, M., & Weingartner, R. (2007). Mountains of the world, water towers for humanity: Typology, mapping, and global significance. *Water Resources Research, 43*, W07447. https://doi.org/10.1029/2006WR005653

Voigtländer, N., & Joachim Voth, H. (2013). The three horsemen of riches: Plague, war, and urbanization in early modern Europe. *The Review of Economic Studies, 80*(2), 774–811. https://doi.org/10.1093/restud/rds034

Weigang, L., Jorge Pinto Alves, C., & Omar, N. (1997). An expert system for air traffic flow management. *Journal of Advanced Transportation, 31*, 343–361. https://doi.org/10.1002/atr.5670310308

Wright, J. K. (1955). Crossbreeding geographical quantiles. *Geographical Review, 45*, 52–65.

Zajenkowski, M., Jonason, P. K., Leniarska, M., & Kozakiewicz, Z. (2020). Who complies with the restrictions to reduce the spread of COVID-19?: Personality and perceptions of the COVID-19 situation. *Personality and Individual Differences, 166*, 110199. https://doi.org/10.1016/j.paid.2020.110199

Zhang, L., Yang, H., Wang, K., Zhan, Y., & Bian, L. (2020). Measuring imported case risk of COVID-19 from inbound international flights - A case study on China. *Journal of Air Transport Management, 89*, 101918. https://doi.org/10.1016/j.jairtraman.2020.101918

Zhou, C., Su, F., Pei, T., Zhang, A., Du, Y., Luo, B., Cao, Z., Wang, J., Yuan, W., Zhu, Y., Song, C., Chen, J., Xu, J., Li, F., Ma, T., Jiang, L., Yan, F., Yi, J., Hu, Y., et al. (2020). COVID-19: Challenges to GIS with big data. *Geography and Sustainability, 1*, 77–87. https://doi.org/10.1016/j.geosus.2020.03.005

Chapter 18
Digital Geographies and Digital Surveillance Technologies: Power and Space in the Italian Society Under Control for Public Health

Valentina Albanese and Giorgio S. Senesi

A Pandemic Story from Italy

Waking up in the morning represented a gamble every day: remembering where the day's flight would take me and where I would sleep the next night. An academic work conducted with a suitcase in hand. Home, planes, trains, hotels, trains, planes, home. Pandemic stopped all this. The morning alarm has been turned off for months. I prepared breakfast for my family and myself every day; I recovered the hugs, the smiles, the calm, the regular breathing. I rediscovered the beauty of the roofs of southern Italy, observed from the balcony; discovered sunsets from a window; and learned to know my refuge, my home, inch by inch. I re-meant home spaces that have gradually become a total space. In a short circuit between the principle of the perennial movement imposed by the pre-pandemic everyday life and the sedentary lifestyle required by the state of emergency, the world seemed to return to a common human dimension. The biosphere made everyone immerse in the same air, share the same voids between bodies. Stasis suddenly won over an impossible frenzy that, who knows, maybe we have learned to keep under control.

New Spatiality During the Pandemic

The COVID-19 pandemic can be described as the watershed between the organization of the world based on twentieth-century concepts and those generated by the digital revolution. We are in the middle of a digital turn era. Ash et al. (2018) advance "the notion of the 'digital turn' to capture the ways in which there has been

V. Albanese (✉)
Università degli Studi dell'Insubria, Como, Italy
e-mail: ve.albanese@uninsubria.it

G. S. Senesi
CNR, Istituto per la Scienza e Tecnologia dei Plasmi, Bari, Italy

M. Laituri et al. (eds.), *The Geographies of COVID-19*, Global Perspectives on Health Geography, https://doi.org/10.1007/978-3-031-11775-6_18

a demonstrably marked turn to the digital as both object and subject of geographical inquiry, and to signal the ways in which the digital has pervasively inflected geographic thought, scholarship, and practice" (p. 25). The digital turn had already been present for some time and, as well, technological innovations, i.e., digital technology and technological infrastructures, were already present, but they had never been used in such a massive way, as happened in these first 2 years of the pandemic. Therefore, it is reasonable to assess that the digital turn has found its momentum in the pandemic period to become a digital practice turn, that is, a turning point in practices, all practices, which have found their own digital alter ego to be maintained even during the periods of lockdown. Digital geographies are certainly protagonists of territorial reconfigurations. Geography becomes essential to understand which relationships between the individual and the physical environment, i.e., the exterior, and the individual and social environment, i.e., the otherness, have emerged also including the dynamics of local resistance.[1] The new spatialities generated by the pandemic and by the repeated lockdowns can be evaluated at different scales. The confinement of the bodies inside private homes led us to reflect on the new spatiality of COVID-19, starting from the minimum scale, that of the body, and then considering that of the house, and finally the urban one.

Bodies and Spaces During the Pandemic: The Italian Case

To understand what impact the COVID-19 pandemic has had on body practices, two essential variables must be considered, i.e., the forced confinement in space and the transfer of personal data, from body temperature to tastes and emotions. Thus, surveillance capitalism[2] has enriched its archives with an abnormal amount of data. Aiming to perform a geographical analysis, the two aspects of confinement and data transfer can be considered both expressions of forms of the power of those who generate, assemble, and organize data. For this reason, our reflection moves from the forms of power that produce and act on data, a power that Isin and Ruppert (2020) call *sensory power*.

According to Foucault (1975) since the seventeenth century, the relationship of power with life has changed sharply and can be summarized in the dichotomy of

[1] For example, "carbonare" lessons were organized in public gardens in Milan, every Sunday afternoon in parks, in green spaces of neighborhoods, until the schools of all types and levels are back in presence and safety. They consist in itinerant lessons to guarantee the right to study for those who, due to the digital divide, could not deal with distance learning. The purpose of *carbonare lessons* was also to protest against the closure of schools during the peak of the pandemic (https://video.repubblica.it/edizione/milano/lezioni-carbonare-a-milano-gli-studenti-and-8220disposti-a-tutto-pur-di-tornare-in-classe-and-8221/371606/372211?ref=vd-auto&cnt=1&fbclid=IwAR0UbVaRDTPbG9bKBkZcDw6ooco24PxxKzf5wbkcP2L3spQW5JkPFSLdy3s)

[2] Internet of things and digital technologies, recently, have been turned more and more into tools for behavioral manipulation and exploitation. This is the definition of Zuboff (2019) on the surveillance capitalism.

passing from the power of life or death (typical of feudal societies) to the power of allowing life or pushing to death. In the second case, the logic of sovereignty is modified, and power becomes *biopower*, thus fulfilling the productive task of managing life. These are the means by which modern biopower recovers the ancient sovereign right of life or death. Biopower, expressed as the set of power-knowledge devices controlling the management of lives, has historically developed in two ways: disciplinary and biopolitical. In the contemporary world, discipline and biopolitics are supported by new technologies and, during the pandemic, they have entered by right as life-saving tools in a state of exception (Agamben, 2003), so generating the legitimate doubt if the civil liberties must be limited to safeguard public health or if this is a mere biopolitical power means (Kitchin, 2020) implemented with new and even more effective tools.

In Italy, we have been witnesses of two main ways of exercising biopolitical power: on the one hand, the attitudes and utilities of individual bodies have been strengthened by regulation and governmentality of daily practices and, on the other hand, biopower has manifested itself as a population biopolitics. In Italy, which has been hit by the virus immediately after China (February 21, 2020), the measures to be taken to contain and contrast the spread of COVID-19 were formally adopted with a series of Ministerial Decrees (DPCM). Italy was colored red, orange, and yellow based on the number of sick people, in proportion to the number of places in intensive care units in hospitals. The rules for the confinement of bodies in private homes were initially referred only to the "red areas" and were then extended to the entire national territory (DPCM, 2020).

The first phase of rigid lockdown ended on May 3, 2020, and, with the transition to the so-called Phase 2, the productive, cultural, and social activities gradually began to resume, even if subject to compliance with some precautions, including social distancing and use of a mask indoors. The schools remained closed until the end of the school year 2019/2020. The spread of COVID-19 therefore resulted in a sudden and drastic reduction in social life in Italy, an extended transfer of professional activities to home private space, and an immediate suspension of educational, working (with some exceptions), sport, social, and cultural activities, some of which were allowed only at a distance. However, in all marginal areas where the digital divide is still present, these activities have been completely canceled from everyday life. Thus, at varying degrees, people experienced a condition of social isolation and limitation of freedom of movement outside home. In this scenario, the relationship between people and space has also inevitably changed, especially between bodies confined to the totalizing home space. Thus, the bodies had to adapt their rhythms of life to allow for the forced intersection of formal and factual territoriality. Formal territorialities are well represented by the place of residence, while the factual ones are expressed in the places where various kinds of activities, e.g., work, study, and sports, take place. Formal territoriality generally does not coincide with the factual territoriality. Although for a few people formal and factual territoriality overlap, for most men and women of any age the two territorialities not only do not coincide but in certain cases are even mutually exclusive.

The practices of daily life, completely digitized during the lockdowns, meant that private and public data were all fed into the web, into the databases that are the basis of surveillance capitalism (Zuboff, 2019). Reflections related to surveillance capitalism and sensory power add to the above considerations. Confinement, as mentioned above, marked the final entry into a digital age, through the dissemination of practices on a global scale and in every area of life. In Italy, the digital practice turn has highlighted issues on specific forms of exploitation, control, and surveillance favored by *digitalness* in an insufficient welfare framework characterized by a weakness in public services, including hospitals and local health assistance, school, and university.

Furthermore, home has also become the workplace, a space no longer separated between production and reproduction, but a new battleground between productive and unproductive bodies. In this framework, digital media have represented the control center through which data related to (un)productivity accumulate to be then transferred, analyzed, stored, and reused at will. The massive use of digitized daily practices has led to an unprecedented increase in what Klauser (2017) identified as automated, software-driven data analytics.

Digital technologies have been inextricably intertwined with daily life, becoming mediators of identity and belonging (Albanese & Graziano, 2020) and redefining the relationships of territoriality not only between people in the home, but also between people and suspended space, i.e., public space.

Private and Public Spaces During the Pandemic: The Italian Case

The social distancing and the interruption of the physical relationship with the city that is part of our daily life have led to reconsidering the relations with the territory in a markedly reduced perspective, that of the body, and in a very large perspective, the transcalar one. The dystopian relationship between the private space and the regulated, denied, public space is of particular interest between the two opposite poles of the small scale of the body and the large global scale. The relationship of people with the external space, the public space, has been limited in time (only a few hours) and in space (only some areas). This space has become less and less intelligible. Since the access to the public space has been denied, it has taken the appearance of an indeterminate, evanescent space, while home became a total space. The lockdown periods have transformed living spaces into many heterotypical micro-spaces, islands in the home island. In these micro-spaces factual territorialities, e.g., shopping, social events, conferences, smart working, and e-learning, and formal territorialities have overlapped. In the cases where these territorialities were incompatible with each other, e.g., working vs. keeping the TV on, Foucaultian heterotopias[3] arise. Very quickly, Italians had to reconstruct the relationship with the

[3] Foucaultian heterotopias are spaces that have the particular characteristic of being connected to all other spaces, but in such a way as to suspend, neutralize, or invert the set of relations that they themselves designate. Classic examples of heterotopies are prisons and cemeteries.

space both from an individual point of view and from a social, work point of view, etc. For this reason, it is interesting to consider the total space of home in which formal and factual territorialities have been forced to coincide during lockdown periods. The bridge with the usual, assimilated proxemics of everyday life was digital and the ability to keep alive activities once outside home depended entirely on *digitalness*, i.e., both the ability to use digital tools and as a digital infrastructure of the territory and access to information and communication technologies, in summary the presence or absence of a digital divide.

During the lockdown periods, also including eventual quarantines, an even more intense dichotomy occurred between private space and public space because we were obliged to live in a total, factual, formal space and use the external, public one as an exceptional one. The suspended space, the space waiting to be experienced, was also characterized by the destabilizing uncertainty of the continuous change of rules established by the Ministry of Health that have implied important consequences in terms of personal and social instability. The space enlarged and restricted by various ordinances has regulated the spaces of mobility, not only reduced but also variable, and those of consumption. The territoriality that has been expressed in the governed territory has highlighted very clearly an elected public space that was placed in the middle between the heterotopy of the private space and the utopia of the denied public space. The various Ministerial Decrees have exercised the political control of the territory by regulating the access to the public spaces and keeping alive some *exceptional* spaces (Agamben, 2003). The state of risk[4] measured on a weekly basis has determined the attribution of a color to each region: yellow, orange, or red, from a minimum to a maximum risk of contagion. The color of the regions determined the restrictions to be followed until the reduction of the risk indicator and consequent change of color.

This fragmentation of the national space into a colored space has generated new dichotomies on other scales, i.e., the regional ones, and has further prevented the reorganization of the practices of/in/with the public space during the pandemic crisis. Thus, not being able to manage everyday micro-practices as usual, the society has lost those elements of stability and stabilization of social models (Ehn & Löfgren, 2010) that are also indispensable for the decoding of everyday life.

[4]The document from which the basic choices underlying the DPCM and the "colors" ordinance derive was drawn up by a working group including the Istituto Superiore di Sanità, INAIL, the Spallanzani Institute and the Conference of Regions. The data based on the surveys were uploaded every week by the regions to the database of the Istituto Superiore di Sanità; the sources of the data, therefore, were the regions (https://documenti.camera.it/leg18/resoconti/assemblea/html/sed0424/stenografico.pdf)

Digital Surveillance Technologies: The Italian Case

During the first wave of the COVID-19 pandemic (from January 30 to May 4, 2020), individual lives intersected more and more closely with new technologies. The quality of survival during the periods of forced confinement depended on the technological access. In those periods, years-old issues related to the access to new technologies and digital illiteracy i.e., issues relating to the digital divide, became very evident. Italy is at the bottom of the indicators on digitization in Europe, and if we agree with Sassen (2007) that cities are the docking and boarding points of world economy, a high digital divide is a fuel that feeds even wider gaps between territories. In the pandemic, the gaps have moved on a much smaller scale, at the urban scale, of course, but also at the private scale of the home and down to the scale of the body, of individuals, who could or could not continue to work, study, communicate, and have a social life, although mediated by a technological device. The main causes of digital divide in Italy include the following: (a) the absence or inadequacy of technologies, e.g., devices and infrastructures, offered by the public sphere; (b) the absence of personal funds to purchase an Internet connection and/or electrical connections; (c) the lack of training, i.e., digital illiteracy; and (d) obsolescence by private technology producers. During the pandemic, the effects of the digital divide affected all areas of human existence, accentuating the inequalities between people with and without access to information and communication technologies, as in that historical moment the connection of homes with the remaining world made the difference between those who have maintained a relationship with the outside and those who have not.

Governmentality and Digital Surveillance Technology

Given the limiting context of the digital divide, the power structures of digital citizenship have limited individuals and communities by determining who was included and who was excluded from physical, digital, and social spaces. More than 40 years ago, Foucault (1988) considered that, starting from modernity, the biological life became the most important issue of politics, and it is more correct to talk of biopolitics and not of politics, and the task of government art, which he calls "governmentality," actually is the intervention in the biological lives of individuals. In addition to biopolitics and biopolitical power mentioned above, it is necessary to specify the term "biocapitalism," which consists in the possibility of obtaining profit from living bodies, i.e., life itself, and which is one of the most important forms of contemporary capitalism. Biocapitalism can be schematically described in two ways: the first one is that of industries linked to biology, such as pharmaceutical industries and vaccines, and the second more insidious one concerns the exploitation of emotions, values, and cultural models by which these companies sell their products. These concepts lead directly to the great theme of surveillance capitalism and to the "Big

Other" of which Zuboff (2019) deals in opposition to the "Big Brother" who guided our consumption before Internet was inextricably intertwined with our (on) life (Floridi, 2015). According to Zuboff, the "Big Others" are hired ad hoc to guide the preferences of specific groups of people by exercising a coercive power that does not require violence.

The "Big Others" generate new tastes, stimulate desires, and guide choices and, to do so, they rely on the large amount of data collected (more or less legally) in huge databases that are often at the center of global scandals (e.g., Cambridge Analytica data scandal in 2016). The network and the technologies that make it work have assumed a central role for any form of political hegemony because the control of information, the control of personal data, which means controlling people, feeds on digital knowledge (Deibert, 2015; Winseck, 2017). Therefore, the exceptional moment of the pandemic was also experienced as an exceptional moment for the collection of personal data, justified by the reason of safeguarding world health.

In this scenario, digital surveillance technologies have made their way. In recent years, numerous critical studies of digital capitalism have been conducted, which have addressed the problem of medical data collection previously to the COVID-19 pandemic. In particular, some authors (Fuchs, 2013; Terranova, 2000) have questioned who could benefit from health data. The dichotomy between the common good and private benefit has been studied in order to understand the practices of data generation and sharing which are difficult to control by individual citizens. Thus, the possibility of using digital systems to track people infected with the virus was immediately considered to counteract its spread. With the aim of protecting national and global public health, the media debate has exposed different digital solutions to monitor people infected with COVID-19 and those who have come into contact with them. The solutions proposed were different and with significantly variable impacts with respect to data justice, concerning the data-driven surveillance (Dencik et al., 2016). It was taken into consideration the digital controlling of gatherings using aggregated data and also geo-localizing and tracking individual movements, i.e., contact tracing. These hypotheses fall within a consolidated habitus that countries have historically developed, i.e., surveillance. Historically, surveillance has been of targeted type when it was necessary to react to an imminent threat, while it was of the mass type when it aimed to keeping generic state defense strategies in place. However, a relevant variable must be considered in these strategies. The frame of application of these surveillance techniques, as digital capitalism teaches, is no longer attributable to territorial borders but overcomes every demarcation of political space and moves in the cyberspace (Bethlehem, 2014). Therefore, the issue of contact tracing apps arises as a global as well as biopolitical issue, i.e., regulate the life of individuals by mass monitoring on a cyber-global scale.

The body is immersed in a political field (Foucault, 1975) and power is *a political technology of the body* (id.); thus, biopolitics is the relationship between power and the body or, more precisely, the body is seen as a target of power. The surveillance system appears as a space in which the bodies of individuals are subjected to a continuous control and end up by adhering to the imposed standards, in order to avoid normalizing sanctions.

The main strategy adopted by Italy to contain the virus was the blockade of the entire nation and the obligation of quarantine for all citizens except those involved in work activities deemed essential to guarantee the basic necessities to the population. In addition, contact tracing apps have been created, which for privacy reasons in Western democratic states, included soft control strategies, related to the informal aspects of control on their citizens. For example, cameras with facial recognition used in other countries were excluded. The Italian approach was thus distinguished by a greater softness by allowing the collection of anonymous data exclusively to analyze the progress of the virus.

As implemented in other Western democratic countries, the approach used to fight the virus has been *human driven* (Parola, 2020), i.e., investing in the human capital of citizens has meant in Italy the use of technology for most communication purposes. For example, the legislative choices of the Ministerial Decrees which, at various intervals, i.e., weekly, biweekly, or monthly, decreed each time, new rules of immobility for Italians, occurred through the massive use of traditional and novel media. The substantial mass-media communication was useful not only to allow the legislative measures be known, but also to leverage on the social sensitivity and civil responsibility of citizens. Obviously, this social communication style has also increased the normalization of digital surveillance operations by using the tool of digital communication as a consensus manufactory, the ability of democracies to assure that people's choices and orientations are structured in a way that people always do what they are told, despite the fact that, formally, they have the opportunity to participate in choice and decision-making processes (Chomsky & Herman, 1995). To all of this, the "Immuni" contact tracing app has been added.

The Italian Immuni App

The "Immuni" app (https://www.immuni.italia.it) was developed on the bases of the numerous issues related to data justice (e.g., health data repository), and no citizen has been forced to download it. It could be downloaded on a voluntary basis and no movement restrictions were imposed on those who did not want or could use it. These citizens who decided to download the app did not disclose information on their exact location as Immuni was not developed with GPS technology but with Bluetooth technology, which is unable to release information regarding the precise location of the user.

Another choice made by the developers of the Italian app to control issues related to data justice was the decentralized design model (Fig. 18.1).

Thus, another extremely important aspect was that there was not a single place for data storage, which was very relevant in addressing numerous data justice issues (Taylor, 2019).

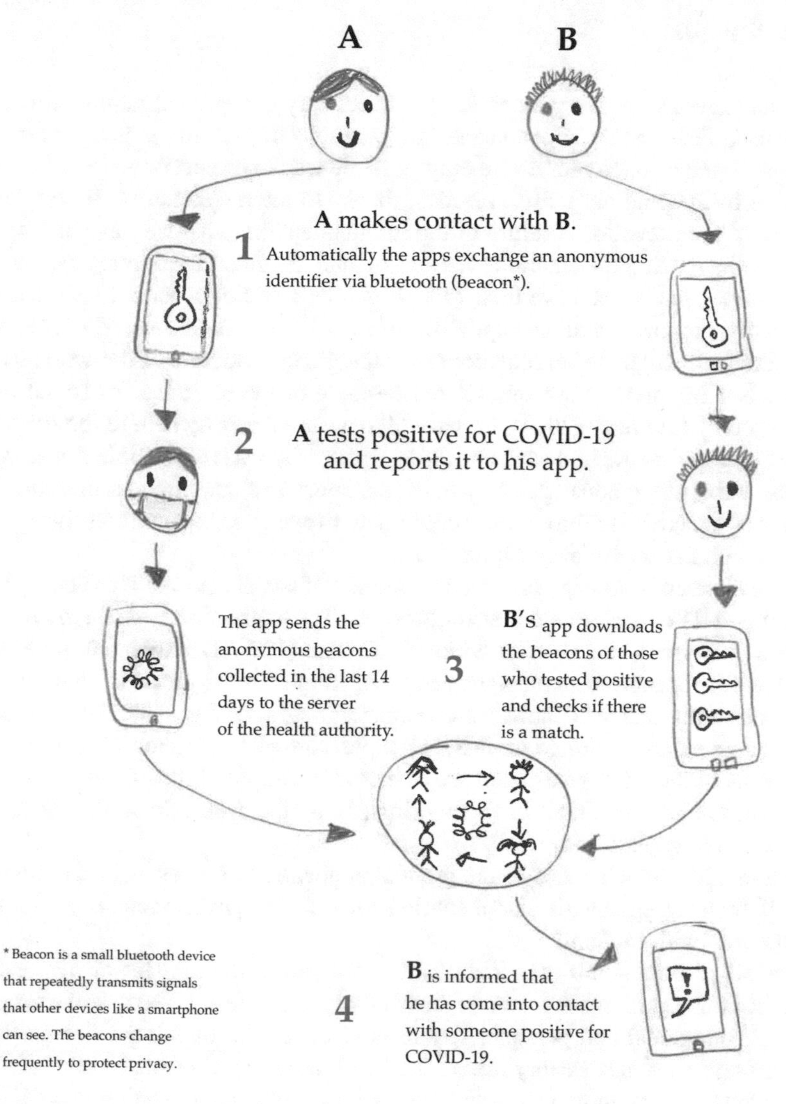

Fig. 18.1 The decentralized model of the Immuni app

The Italian protocol of a decentralized type is thus certainly not a very invasive model but is solid in terms of user privacy as the central authorities cannot identify positive citizens, their contacts, and the duration of their meeting and geo-localize their movements, which are data considered sensitive and thus protected. Differently, if these data were provided to the central server, as it occurs for apps that use a centralized model, it would be possible to trace the social networks in which individuals enter. Certainly, Immuni is not a perfect model, but is apparently among the least invasive.

Conclusions

The management of the pandemic, due to its very exceptional nature, was not an easy task. Fear and threat, connected to situations of proximity, have pushed us to accept digital technologies as the only possible way to protect ourselves, but also to survive by keeping our daily activities, thanks to the mediation of digital devices. Aware of the exceptional nature of the initial situation, what we have attempted to discuss is the implementation of control and confinement strategies, certainly exceptional, but which have their roots in a long-standing tension towards governmentality, biopower, and the capitalism of surveillance. As discussed above in ethical terms, although the Immuni contact tracing app is not one of the worst cases, it still raises important questions related to some of the ten rules for social control proposed by Chomsky (2014). In particular, we are referring here to the rules about the strategy of distraction such as talking about who was responsible for the spread of the virus, the gradualness to which lockdown and surveillance measures have been implemented, and to know people better than they know themselves, strictly linked with the surveillance capitalism.

The distraction strategy is the main element of social control. This concept is not new, but it dates back to the first century AD. The expression "*panem et circenses*" (bread and circus games) with which Giovenale (127 AD) indicated the elements necessary to sedate and control the people of Rome. This expression continues to be used currently and still indicates a demagogic attitude by which good food and entertainment are exploited to distract people from real problems. Generalist social networks of the latest generation are a relevant example of this. In social media, in fact, communications tend to focus primarily on the spectacle of the facts, rather than on their explanation.

Graduality, which refers to the gradual application of measures that, otherwise, would be unacceptable, is also a subtle instrument of governmentality. Would this be the case with Immuni?

Finally, from Chomsky to Zuboff, the problem of the transfer of personal data and the consequent control over bodies in their sociopolitical sense and over people in their emotional and personal sphere is at the center of a huge debate. Digital technologies are inextricably intertwined with our daily life and, in an urgent way, are increasingly assuming the role of mediators of identity and belonging (Albanese & Graziano, 2020) in replacing social, economic, cultural, and playful spaces. The loss of privacy and mass surveillance are the most important issues on which reflection is urgently needed. Starting from the academy, these issues invest as many areas as possible in search of a problematization and, subsequently, of a solution. In the case of contact tracing apps, all of them, in every nation, should have the sole purpose of monitoring the population for the sole purpose of reducing contagion and intercepting dangerous situations from a virological point of view in the shortest time. All new technologies, however, from contact tracing apps to any digital tool that retrieves personal data, involve several doubts and issues related to the control, the power, and the slow but inexorable release of personal information in the boundless databases. Whose database? Where from? For what use?

References

Agamben, G. (2003). *Stato di eccezione*. Bollati Boringhieri.

Albanese, V., & Graziano, T. (2020). *Place, cyberplace e le nuove geografie della comunicazione*. BUP.

Ash, J., Kitchin, R., & Leszczynski, A. (2018). Digital turn, digital geographies. *Progress in Human Geography, 42*(1), 25–43.

Bethlehem, D. (2014). The end of geography: The changing nature of the international system and the challenge to international law. *European Journal of International Law, 25*(1), 9–24.

Chomsky, N. (2014). *Media e potere*. Bepress.

Chomsky, N., & Herman, E. S. (1995). *Manufacturing consent*. Vintage Books.

Deibert, R. (2015). The geopolitics of cyberspace after Snowden. *Current History, 114*(768), 9–15. http://www.currenthistory.com/Article.php?ID=1210

Dencik, L., Hintz, A., & Cable, J. (2016). Towards data justice? The ambiguity of anti-surveillance resistance in political activism. *Big Data & Society, 3*(2). https://doi.org/10.1177/2053951716679678

DPCM. (2020). https://www.gazzettaufficiale.it/eli/id/2021/03/10/21A01552/sg

Ehn, B., & Löfgren, O. (2010). *The secret world of doing nothing*. University of California Press.

Floridi, L. (2015). *The onlife manifesto: Being human in a hyperconnected era*. Springer Nature.

Foucault, M. (1975). *Surveiller et punir*. Naissance de la prison, Éditions Gallimard.

Foucault, M. (1988). *La volontà di sapere*. Feltrinelli.

Fuchs, C. (2013). Theorising and analysing digital labour. *The Political Economy of Communication, 1*(2), 3–27.

Giovenale, S. G. (127 d.C.), *Satire*, X.

Isin, E., & Ruppert, E. (2020). The birth of sensory power: How a pandemic made it visible? *Big Data & Society, 7*(2), 2053951720969208.

Kitchin, R. (2020). Civil liberties or public health, or civil liberties and public health? Using surveillance technologies to tackle the spread of COVID-19. *Space and Polity, 24*(3), 362–381. https://doi.org/10.1080/13562576.2020.1770587

Klauser, F. (2017). *Surveillance & Space*. Sage.

Parola, J. (2020). Controllo e regolazione sui "corpi". Un confronto tra sistemi utilizzati durante l'emergenza sanitaria COVID-19. *Cambio. Rivista sulle trasformazioni sociali*, OpenLab on Covid-19. https://doi.org/10.13128/cambio-9967

Sassen, S. (2007). *Deciphering the global: Its scales, spaces and subjects*. Routledge.

Taylor, L. (2019). Ethics. In J. Ash, R. Kitchin, & A. Leszczynski (Eds.), *Digital geographies*. Sage.

Terranova, T. (2000). Free labor. *Social Text, 18*(2), 33–58.

Winseck, D. (2017). The geopolitical economy of the global internet infrastructure source. *Journal of Information Policy, 7*, 228–267. http://www.jstor.org/stable/10.5325/jinfopoli.7.2017.0228

Zuboff, S. (2019). *The age of surveillance capitalism*. Profile Books.

Chapter 19
Resilience Amid Uncertainty: COVID-19 Pandemic, the Urban Informal Sector, and Livelihoods in Sub-Saharan Africa

Francis T. Koti

A Personal Pandemic Story from Murfreesboro, TN

When you are born and raised on the equator (like I am), the annual trappings of the winter weather of North America always presents an unpleasant experience. But the isolation prompted by COVID-19 pandemic lockdowns was quite different. In fall 2019, I had just accepted a new faculty position at Middle Tennessee State University and moved to the City of Murfreesboro, Tennessee (USA), with my wife Margaret, our 12-year old daughter Arianna, and a slew of pets. My son Milton had just graduated from college and taken a job in the nearby city of Nashville, TN. Barely 4 months into the new environment, COVID-19 introduced us to a new normal of self-isolation. Admittedly however, working remotely has given me a necessary break to appreciate family and home space. Between teaching and doing research remotely, Ping-Pong games with Arianna in the garage, regular walks with Margaret in a nearby greenway, and back again to my computer, I got to appreciate and reflect on important things in life. I have, however, used the convenience of working from home to attend virtual professional development opportunities and present papers at virtual conferences. And yes, I picked up golf!

Introduction

The contribution of the informal sector to national economies and local livelihoods in sub-Saharan African countries cannot be overstated. Recent studies have estimated that the informal sector comprises well over 89.2% of the total labor force in sub-Saharan Africa, with some citing higher percentages for women and young workers compared to those of adult men (CGAP, 2020; Elgin et al., 2021; ILO, 2021; UNDP, 2021; Zeufack et al., 2020). Despite its relative importance to national

F. T. Koti (✉)
Middle Tennessee State University, Murfreesboro, TN, USA
e-mail: francis.koti@mtsu.edu

M. Laituri et al. (eds.), *The Geographies of COVID-19*, Global Perspectives on Health Geography, https://doi.org/10.1007/978-3-031-11775-6_19

economies and local livelihoods, the informal sector in sub-Saharan Africa is vulnerable to economic, social, political, and environmental uncertainties such as those presented by the COVID-19 pandemic. Two years into the global pandemic, there has been a growing interest among the academic community and development practitioners in various aspects of the informal sector in sub-Saharan Africa. While this list is not exhaustive, the literature shows a focus on various aspects such as vulnerability and precariousness (Hamilton, 2020; ILO, 2020a, b, c; Heitzig et al., 2021; Meagher, 2021), employment and livelihoods (Kinyanjui, 2020; Nguimkeu & Okou, 2020; Ohnsorge & Yu, 2021; Webb & McQuaid, 2020), containment measures and mobility patterns (Daniel & Sacchetto, 2020; Pinchoff et al., 2021), gender implications (Moussié & Staab, 2020; Mukhtarova, 2020), and microenterprises, relief for workers, and social protections (Adams et al., 2021; CGAP, 2020; Lakuma & Sunday, 2020; Shurkin et al., 2021; UNDP, 2021).

While the literature is replete with examples from across sub-Saharan Africa, there has been limited focus on the relative resilience of the urban informal sector in the first 2 years of the global pandemic despite the limited support from the respective national and local governments. This study examines the relative resilience of the urban informal sector (UIS) in sub-Saharan Africa in the first 2 years of the COVID-19 pandemic. The main goal of the study is to understand the unswerving resilience of the UIS in sub-Saharan African cities in the face of COVID-19-imposed lockdowns and closures. A survey of literature is followed by analysis of published material, media content, internet sources, and other official and nonofficial government sources.

We set the stage by establishing the definition and conceptualization of the UIS, making a key distinction between the nature of informality as a form of employment, on the one hand, and as an economic sector, on the other. After giving an account of the advent of the COVID-19 pandemic within the sub-Saharan Africa region, the focus shifts to country examples where Kenya and other countries in sub-Saharan Africa are used for illustration. These accounts are followed by explanations for the relative resilience of the urban informal sector in the face of COVID-19 lockdowns despite limited support from the national and subnational levels of the government. We argue that the lack of social protections for the informal sector is no longer an option for sub-Saharan African economies where over 89% of the labor force is in the informal sector (ILO, 2020a).

The Urban Informal Sector in Context

In their assessment of the implications of the COVID-19 pandemic on employment in the informal economy, Webb and McQuaid (2020) summarized the conceptualization of the informal sector under three competing frameworks: (1) modernization theory framework that focuses on the lack of development in the respective contexts, (2) neoliberal frameworks that interpret informal economies as a response to state regulation, and (3) political economy approaches that see informal economies as an outcome of limited state intervention and a lack of social protection for workers (Meagher, 1995; Webb & McQuaid, 2020). In this conceptualization, two types of informal workers are identified: (1) those self-employed in informal enterprises and (2) the wage employed in informal jobs. As the literature indicates however, there seems to be no standard definition of informality and/or the informal sector (Meagher, 1995; Potts, 2008; Webb & McQuaid, 2020; Zeufack et al., 2020). The International Labor Organization (ILO), for example, defines the informal sector as "all types of nonformal employment involving workers who are not protected by national labor laws" (Rigon et al., 2020).

The stated ILO definition of the informal sector seems to imply the diversified set of economic activities, enterprises, jobs, and workers that are not regulated or protected by labor laws. Despite its flaws however, this definition has been adopted by both the academic community and development practitioners and has been the basis for the characterization of the informal sector in sub-Saharan Africa. Because of the focus on employment rather than the sector itself, a wide range of misleading terms are used to refer to the sector including but not limited to the following: *hidden, shadow, precarious, contraband, irregular, invisible, unregistered, underground, illegal or illegitimate, unregulated*, and so on (Webb & McQuaid, 2020).

Whether the informal sector is defined from an employment standpoint or as an economic sector in and of itself may not be as important as the fact that the UIS is a key source of livelihood for millions of people and a backbone to the economies of many countries in sub-Saharan Africa (CGAP, 2020; Heitzig et al., 2021; Ng'weno & Porteous, 2018; Ohnsorge & Yu, 2021). In sub-Saharan Africa, the UIS comprises workers involved in a variety of activities including but not limited to street vendors, artisans, waste pickers, construction workers, home-based workers, domestic jobs, transportation workers, construction workers, and other short-term contractors. As workers in this sector lack social protections and government support, the UIS is often sensitive to political, environmental, economic, and social uncertainties (CGAP, 2020; Hamilton, 2020; Heitzig et al., 2021; UNDP, 2021). The precarious nature of the informal sector has been substantively covered by Meagher (2021) but well captured in their assessment when Zeufack et al. (2020) noted that:

> Informal workers lack benefits such as health insurance, unemployment insurance, and paid leave. Most informal workers, particularly the self-employed, need to work every day to earn their living and pay for their basic household necessities. A prolonged lockdown will put at risk the subsistence of their households. Additionally, the majority of workers hired are in a precarious situation, and most of these jobs are temporary and with low remuneration, do not offer social security, and put workers at a greater risk of injury and ill health. (Zeufack et al., 2020:74)

Regardless of the conceptual lens through which we view the UIS, it still remains a key employer and a major source of livelihood for millions of people in sub-Saharan Africa.

COVID-19 Pandemic in Sub-Saharan Africa

The first confirmed case of COVID-19 on the African continent was reported in Egypt in mid-February in 2020 and was linked to travel from China (Adams et al., 2021). In sub-Saharan Africa however, the first confirmed case was reported in Nigeria towards the end of February 2020, and by the end of the month of May the same year, the virus had been reported in the rest of the countries in the region (Adams et al., 2021; Nguimkeu & Okou, 2020; Shurkin et al., 2021). Because of poor healthcare infrastructure, high poverty rates, and high unemployment in sub-Saharan African countries, among other factors, it was feared that the COVID-19 pandemic would not only spread rapidly, but would also have a devastating effect on national economies and local livelihoods (CGAP, 2020; ILO, 2020a, b, c, 2021; Shurkin et al., 2021; Zeufack et al., 2020). However, the number of COVID-19 cases, hospitalizations, and deaths remained relatively low compared to other regions of the world (Adams et al., 2021; Heitzig et al., 2021; Moussié & Staab, 2020; Nguimkeu & Okou, 2020). While the reasons behind low COVID-19 cases in sub-Saharan Africa has spurred an interesting debate among the academic community and development practitioners (Adams et al., 2021), there seems to be considerable agreement that African governments were effective in preparing for the pandemic and implementing mitigation strategies to curb its initial spread. Adams et al. (2021), for example, highlighted the remarkable swiftness with which African governments and health organizations moved in to respond to the initial threat of COVID-19 – citing examples from Nigeria, Kenya, South Africa, Ghana, Rwanda, and Côte d'Ivoire. As Adams et al. (2021) note, "By March, (2020) almost all African nations had suspended flights from China… After March 2020, most cases imported to Africa originated from Europe, as the epicenter of the disease had shifted there. By May 2020, more than 40 African nations had closed their borders to all but cargo" (Adams et al., 2021:438).

As of the writing of this chapter (December 2021), cumulative COVID-19 cases, new cases, deaths, and hospitalizations still remain relatively low compared to other regions of the world (see Table 19.1).

As Table 19.1 shows, COVID-19 infections in the continent of Africa have not been as severe as initially expected. What is more notable is that the stringent control measures and mitigation strategies employed to curb the spread of the pandemic in the region directly threatened the operation of activities of the UIS (Shurkin et al., 2021; Zeufack et al., 2020). A few examples are discussed below.

COVID-19 Pandemic Mitigation Strategies, Government Responses, and the Exclusion of the Informal Sector: Examples from Kenya

Based on written accounts in daily newspapers, peer-reviewed publications, Internet sources, and government reporting, the first confirmed case of COVID-19 in Kenya

Table 19.1 Regional incidence of COVID-19 cases on December 3, 2021

	Cases–cumulative total	Cases–newly reported in the last 7 days	Deaths–cumulative total	Deaths–newly reported in the last 7 days	Persons fully vaccinated per 100 population
Global	**269,468,311**	**4,075,903**	**5,304,248**	**48,191**	**44.5**
Americas	98,690,724	932,459	2,373,674	12,060	58.25
Europe	91,903,122	2,547,150	1,601,656	28,084	54.14
Southeast Asia	44,748,558	95,493	714,546	2609	35.01
Eastern Mediterranean	16,947,823	88,906	312,514	1564	27.55
Western Pacific	10,612,501	214,891	148,037	3362	70.88
Africa	6,564,819	197,004	153,808	512	5.87

Source: WHO Dashboard, December 13, 2021

was reported on March 13, 2020, and 2 days later, the President of Kenya Uhuru Kenyatta issued COVID-19 virus control measures and guidelines (Omolo, 2020; GoK, 2021; Pinchoff et al., 2021). These immediate measures included but were not limited to closure of schools, working from home, cashless transactions, banned gatherings, and strict social distancing guidelines among others. On April 6, 2020, President Kenyatta further announced a cessation of movement in and out of the Nairobi Metropolitan Area followed by a nationwide curfew (Fig. 19.1). On June 6, 2020, the cessation of movement was extended and expanded to include other large cities and the curfew extended from dusk to dawn. Other more specific restrictions included closing schools for 1 year, restricting the number of passengers in public service vehicles to at most 60% of the vehicle capacity, banning of international passenger flights, imposing quarantines, and emphasizing cleanliness and hygiene guidelines among others.

As these restrictive measures were expected to affect work, school, and play and by extension hurt families financially, the Government of Kenya introduced seven measures aimed at buffering Kenyans against financial hardships created by government-imposed COVID-19 control measures. As documented by Omolo (2020), the seven measures included:

1. 100% tax relief to Kenyans earning KSh 24,000 (US$228) and below
2. Pay as you earn (PAYE) reduction from a maximum of 30% to 25%
3. Reduction of turnover tax rate from 3% to 1% for all micro-, small, and medium enterprises
4. Reduction of resident income tax to 25%
5. Making available KSh 10 billion (US$95 million) to vulnerable groups including the elderly and orphans, among others
6. Temporary suspension of the listing of loan defaulters for any person, micro-, small, and medium enterprise and corporate entities whose loan account is in arrears effective April 1, 2020
7. Reduction of value-added tax (VAT) from 16% to 14% effective April 1, 2020

Fig. 19.1 Media headlines on the COVID-19 pandemic and the informal sector in sub-Saharan Africa

The intent of these measures to cushion Kenyan citizens was welcome nationally, especially considering that they included a provision for orphans and the elderly. However, something about the government-instituted relief program for workers was notable – the measures only applied to Kenyans in formal employment and almost exclusively omitted the informal sector of the economy which employs well over 83% of the Kenyan population (Donovan & Zhu, 2020; Kinyanjui, 2020). Other than where vaguely implied such as measure #3, #5, and #6, the informal sector was otherwise omitted from government relief measures. It came as no surprise that by May 20, 2020, over 8000 people were reported as dislodged from informal settlements.

What the COVID-19 Control Measures and Exclusion from Kenyan Government Relief Measures Meant for the Urban Informal Sector

In addition to loss of life, it was expected that the immediate measures used to control the spread of COVID-19 in Kenya would have a devastating effect on the UIS (Fig. 19.1). First, the face-to-face and physical nature of activities in the UIS does not lend itself to remote operations. Second, as a key source of business for the informal sector, massive layoff of employees in the formal sector was a bad thing and so was the economic slowdown at the national level. Third, the cessation of movement initially within the Nairobi Metropolitan Area and subsequently in other cities around the country meant cutting off supplies from the rural areas which many workers in the informal sector depend on. Add to that the disruption of local, regional, and global supply chains which are the backbone of the UIS. Because the

UIS thrives when other sectors of the economy thrive, the poor performance of the UIS led some to declare that "Kenya's labor market was not made for a pandemic" (Donovan & Zhu, 2020). The exclusion of the UIS from government-instituted COVID-19 relief measures was thought to be the proverbial "last nail in the coffin" for the UIS (Guyen & Karlen, 2020).

Other Examples from Sub-Saharan Africa

Elsewhere in sub-Saharan Africa, similar accounts and experiences of the UIS have been reported in online newspapers, academic work, and policy briefs among others (Zeufack et al., 2020). In April 2020, for instance, the World Bank dedicated Volume 21 of its "*Africa's Purse*" to the assessment of the economic impact of COVID-19 in sub-Saharan Africa. In this study, Zeufack et al. (2020) predicted that economic growth in sub-Saharan Africa would decline from 2.4% in 2019 to −2.1% to −5.1% in 2020, citing that this would be the "first recession in the region in 25 years" with an estimated cost "between US$37 billion and US$79 billion in terms of output losses for 2020" (Zeufack et al., 2020). Although the focus of the study was overall national economies, it recommended implementing social protection programs to support workers, especially those in the informal sector.

While the country-specific examples highlighted are by no means exhaustive, they give a snapshot into similar experiences in other countries in sub-Saharan Africa (CGAP, 2020). As early as April 2020 and due to high COVID-19 infection rates, the Federal Government of Nigeria had imposed lockdowns in Lagos and Ogun states as well as Abuja. These were followed by lockdowns at other subnational levels of government. In their assessment of the impact of COVID-19 control measures and restrictions in Nigeria, Onyekwena and Ekeruche (2020) predicted declines in household income, consumption, investment, and exports. More specifically, they identified a looming devastating effect that these lockdowns would have on the informal sector, which contributes more than 65% of the country's economic output. As an immediate relief measure, the Central Bank of Nigeria (CBN) arranged a fiscal stimulus package, designed to cushion workers in the formal sector (Onyekwena & Ekeruche, 2020; Zeufack et al., 2020). As evident in these efforts, there was no direct mention of the urban informal sector as a recipient of the government relief measures despite its contribution to the national economy.

In South Africa where lockdown measures were in place as early as January 2020, they implemented targeted social protection measures to cushion workers. In these programs, the social security agency extended social grants to older people and those with disabilities while the government offered to pay sick leave to those workers affected by the initial lockdown or getting sick as a result of the COVID-19 outbreak. According to De Groot and Lemanski (2020), the President of South Africa Matamela Cyril Ramaphosa announced a specific socioeconomic response plan in April 2020 aimed at supporting the economy during the lockdown. Despite the nationally published relief measures and the establishment of a national body to oversee payment of unemployment insurance benefits to workers in South Africa, the pandemic benefits still remained confined to those in formal employment and did not reference the UIS (Zeufack et al., 2020).

Similar social protections were also announced in neighboring Namibia as early as March 2020. Here, workers losing jobs as a result of initial lockdowns and who were not receiving any other grants from the government would receive a one-off payment of N$750, taxpayers could borrow money against their tax payment for the previous year, and most importantly, the government would keep water supply points open. Namibia was one of the few countries in sub-Saharan Africa where the initial relief measures expressly mentioned informal workers (Zeufack et al., 2020).

A survey conducted in Uganda in the early stages of the pandemic indicated that 75% of the surveyed businesses had laid off employees due to the risks presented by COVID-19. The results of the same survey seemed to suggest that the initial lockdown measures had reduced business activity by more than 50% with most of the decline occurring in micro and small businesses (Lakuma & Sunday, 2020; Zeufack et al., 2020). Following these early signs, the National Social Security Fund (NSSF) of Uganda announced provisions that would allow distressed businesses and employers to reschedule NSSF contributions without penalties. Still, pandemic relief measures focused primarily on employers and registered businesses but not the informal sector.

What we can learn from these few examples is that despite the intent by national and subnational government levels of government to cushion workers affected by COVID-19 control measures, workers in the urban informal sector remained largely excluded from most relief programs across the continent. What is fascinating, however, is that when lockdowns were lifted and economies opened back up in the latter part of 2020, the UIS in most sub-Saharan African countries gradually picked up despite the lack of government support. In the next section we explain the unlikely resilience of the UIS amid the uncertain social, economic, and political climate.

What May Explain the Relative Resilience of the Urban Informal Sector in Sub-Saharan Africa in the Face of Uncertainties Prompted by COVID-19-Related Lockdowns and Closures?

First, the very nature and character of the UIS may partly explain its relative resilience to the economic shocks of COVID-19-related closures and lockdowns. Take, for example, the flexible work arrangements and the fact that much of the informal activity is conducted outdoors. This made it possible to observe the required social distancing guidelines and still conduct business, but most importantly to avert the potential for community spread of the virus. This position, however, contradicts an argument by Elgin et al. (2021) who have argued that extreme poverty and the crowded nature of the UIS setting aggravate the spread of COVID-19. Their argument is, however, not entirely incorrect as it references post-lockdown UIS when social distancing restrictions had been lifted.

Second, the UIS in sub-Saharan Africa is community oriented, serves local needs, and comprises social networks that go far beyond mere casual interactions at the market. As such, even under government-instituted lockdowns and closures, activities and provision of services still continued at personal level and outside business settings. In Kenya, for example, advancements in e-commerce through the M-Pesa money transfer system made remote business transactions possible for urban informal workers during lockdowns. Third, and even more importantly, majority of businesses in the UIS in sub-Saharan Africa are family operated, do not involve large investments, deal with relatively inexpensive goods, thrive on family or casual labor, and hardly ever involve debt spending. As rightly noted by Ng'weno and Porteous (2018), the capital required to (re)enter an informal business is almost negligible (as low as US$7 in the Kenyan context) and so are the losses on exit. This implies that losses incurred during COVID-19-related closures were relatively manageable for owners of businesses and small microenterprises.

The UIS in sub-Saharan African countries is also characterized by low-skill as well as low-technology jobs. From tailors of school uniforms, drivers of taxis, street vendors, and artisans to mechanics, these forms of employment have low entry skills. This very feature made it possible for workers in the sector to reenter the economy when lockdowns were lifted towards the end of 2020. Likewise, it made it relatively easy for others laid off from formal employment to enter the sector.

The impact of the intense pressure mounted by the nongovernmental sector, international organizations, development practitioners, academic community, local leaders, the media, etc. cannot be understated. This pressure played a key role in the sudden shift in attitudes towards support for the informal sector by national and subnational levels of government. Whether it was in the context of policy briefs, academic journals, blogs, conferences, or other forms of media outlets, there were sustained campaigns that called upon governments in the region to recognize and include the informal sector in their COVID-19 relief measures (Fig. 19.2) (CGAP, 2020; Daniel & Sacchetto, 2020; Dube & Katende, 2020; Guyen & Karlen, 2020; ILO, 2020a, b, c, 2021; Kinyanjui, 2020). The World Bank, for example, dedicated Volume 21 of *Africa's Purse* publication to examining the impact of the COVID-19 pandemic specifically on the urban informal sector (Zeufack et al., 2020). The fourth issue of Volume 54 of the *Development and Change* journal was also dedicated to impacts of COVID-19 on the informal sector in which Meagher (2021) and others called for the economic inclusion of informal workers observing that "the will to include (the informal sector/workers) has become paramount, and the recent global crisis of the COVID-19 pandemic has only made matters of economic inclusion more critical than ever."

As part of the intense campaign for the inclusion of the informal sector in social protections against COVID-19-related hardships, the United Nations Development Program (UNDP) Regional Bureau for Africa in collaboration with the International Labor Organization (ILO) Regional Office for Africa dedicated an entire publication into assessing (and making recommendations for) social protection systems in

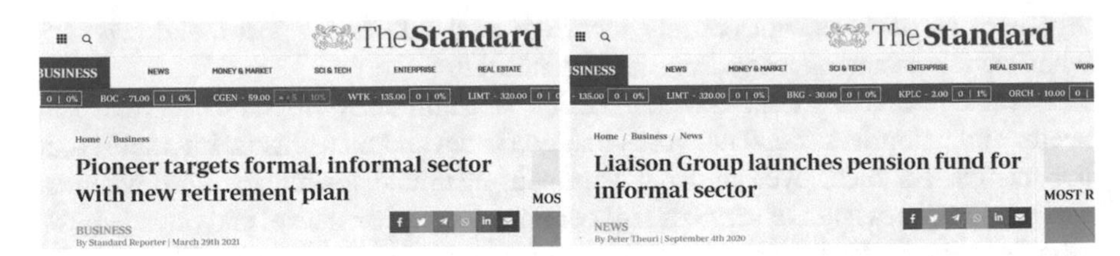
The Standard
BUSINESS
Home / Business
Pioneer targets formal, informal sector with new retirement plan
BUSINESS
By Standard Reporter | March 29th 2021

The Standard
BUSINESS
Home / Business / News
Liaison Group launches pension fund for informal sector
NEWS
By Peter Theuri | September 4th 2020

Fig. 19.2 Reports from a local Kenyan newspaper show increasing recognition of the informal sector

selected countries in sub-Saharan Africa (UNDP, 2021). In their edited volume, "*The Long Shadow of Informality: Challenges and Policies*," Franziska Ohnsorge and Shu Yu call for inclusive COVID-19 relief approaches that recognize the informal sector. While this list of advocates for the informal sector in sub-Saharan Africa is not exhaustive, it attracted the attention of national and subnational levels of government prompting some to designate financial support specifically for the informal sector. As shown in Fig. 19.2, the *Standard Newspapers* twice reported instances where local financial institutions in Kenya had initiated programs to shore up informal sector enterprises.

Conclusion and Recommendations for Further Work

This study set out to understand the relative resilience by the urban informal sector in the face of COVID-19 pandemic-related closures in sub-Saharan Africa. First, the study laid out the conceptual framework under which the UIS is discussed distinguishing between informality as a form of employment on the one hand, and as an economic sector on the other. The ensuing section gave an account of the advent of the COVID-19 pandemic in sub-Saharan Africa highlighting the major challenges associated with the initial measures to curb the spread of the pandemic. Using examples from Kenya and elsewhere in sub-Saharan Africa, the study also highlighted the exclusion of the UIS in government-sponsored measures to cushion workers. This exclusion was best captured by Meagher (2021) in this remark,

> What the COVID-19 pandemic has exposed so starkly is not the economic exclusion of informal populations, but their precarious terms of inclusion in essential production and service activities, at the heart of contemporary market societies, where they labor as casualized workers, last-mile distributors, gig workers, ersatz entrepreneurs, while remaining cut off from basic economic and social rights. (Meagher, 2021: 731)

Despite the exclusion from government-sponsored relief measures however, the UIS exhibited unswerving resilience by emerging from pandemic closures (albeit gradually) when the respective governments lifted stay-at-home measures. In the study, we offer several explanations for this unwavering resilience of the UIS, some of which include the very nature and character of operation of the sector itself. Most importantly, however, and the key lesson from the relative resilience of the UIS from the rubble of COVID-19 lockdowns is that structural and meaningful economic inclusion of the informal sector is no longer an option for sub-Saharan African countries. Because of its sheer size and contribution to local and national economies, the performance of the informal sector may well be the key to sustainable economic development in sub-Saharan Africa because when the informal sector thrives, so does other sectors of the economy.

As this is a preliminary study, the work would be incomplete without identifying further areas of inquiry. First, the study recognizes that the impacts of the COVID-19 pandemic on the UIS are far from uniform across social and spatial scales. A study is needed to further investigate how the second-order impacts of the pandemic will differ across social and spatial scales and also examine the necessary adaptations needed in order to respond to secondary impacts of the COVID-19 pandemic. Second, due to massive layoffs in the formal sector, this study predicts that the UIS is likely to experience an entry of more skilled actors as well as feature new and more sophisticated activities previously unseen in the informal sector. It will be interesting to see what the new urban informal landscape will look like. Third, most studies assessing the economic and social vulnerability of the UIS have been conducted using models developed specifically for social contexts outside sub-Saharan Africa. It would be interesting to see what a social and spatial vulnerability index for the sub-Saharan African context would look like. A further study is needed that would utilize the power of geospatial technologies to develop an Africa-specific index for assessing socio-spatial vulnerabilities prompted by downstream impacts of the COVID-19 pandemic.

Acknowledgement I wish to thank the Department of Global Studies and Human Geography at Middle Tennessee State University for the time release to conduct research and develop the manuscript and the resources that made the work possible. I am deeply indebted to my wife Margaret and daughter Arianna for putting up with me during the research and writing phase. I owe gratitude to the organizers of Cities' COVID Mitigation Mapping (C2M2) Symposium, Harvard University, in summer 2021 where this idea was first presented.

References

Adams, J., MacKenzie, M. J., Amegah, A. K., Ezeh, A., Gadanya, M. A., Omigbodun, A., Sarki, A. M., Thistle, P., Ziraba, A. K., Stranges, S., & Silvermana, M. (2021). The conundrum of low COVID-19 mortality burden in sub-Saharan Africa: Myth or reality? *Global Health Science Practice Journal, 9*(3), 433–443. https://doi.org/10.9745/GHSP-D-21-00172

Consultative Group to Assist the Poor (CGAP). (2020, August). Relief for informal workers: Falling through the cracks in the COVID-19 crisis. *COVID-19 Briefing*. https://www.cgap.org/research/COVID-19-briefing/relief-informal-workers-falling-through-cracks-covid-19. Accessed 19 Nov 2021.

Daniel, E., & Sacchetto, C. (2020, May 20). *COVID-19 and informality in sub-Saharan Africa: Containing an economic crisis*. International Growth Center (IGC) Blog Post. https://www.theigc.org/blog/covid-19-and-informality-in-sub-saharan-africa-containing-an-economic-crisis/. Accessed 21 June 2021.

De Groot, J., & Lemanski, C. (2020). COVID-19 responses: Infrastructure inequality and privileged capacity to transform everyday life in South Africa. *Environment and Urbanization, 33*(1), 255–272. https://doi.org/10.1177/0956247820970094

Donovan, L., & Zhu, A. (2020, April 10). Kenya's labor market wasn't made for a pandemic. *Foreign Policy*. https://foreignpolicy.com/2020/04/10/kenya-labor-coronavirus-pandemic-informal-workers-economic-crisis/. Accessed 29 Jan 2021.

Dube, K., & Katende, C. N. (2020, May 29). An inclusive response to COVID-19 for Africa's informal workers. *World Bank Blog*. https://blogs.worldbank.org/africacan/inclusive-response-covid-19-africas-informal-workers. Accessed 22 Nov 2021.

Elgin, C., Kose, M. A., Ohnsorge, F., & Yu, S. (2021). Chapter 2: Understanding the informal economy: Concepts and trends. In F. Ohnsorge & S. Yu (Eds.), *The long shadow of informality: Challenges and policies*. Advance edition. License: Creative Commons Attribution CC BY 3.0 IGO. The World Bank Group, Washington, D.C.

Government of Kenya (GoK) Ministry of Health. (2021). https://www.health.go.ke/ Accessed 21 June 2021.

Guyen, M., & Karlen, R. (2020, December 3). *Supporting Africa's urban informal sector: Coordinated policies with social protection at the core*. https://blogs.worldbank.org/africacan/supporting-africas-urban-informal-sector-coordinated-policies-social-protection-core. Accessed 13 Nov 2021.

Hamilton, B. (2020, April 14). Informal sector in dire situation, yet contributes billions to economy. *Low Velder*. Opinion. https://lowvelder.co.za/619561/informal-sector-in-dire-situation-yet-contributes-billions-to-economy/. Accessed 19 Nov 2021.

Heitzig, C., Uche Ordu, A., & Senbet, L. (2021, October). Sub-Saharan Africa's debt problem Mapping the pandemic's effect and the way forward. *Brookings: Africa Growth Initiative*. https://www.brookings.edu/research/sub-saharan-africas-debt-problem-mapping-the-pandemics-effect-and-the-way-forward/. Accessed 19 Nov 2021.

International Labor Organization (ILO). (2020a, April 14). *The impact of the COVID-19 on the informal economy in Africa and the related policy responses*. ILO Brief. https://www.ilo.org/wcmsp5/groups/public/%2D%2D-africa/%2D%2D-ro-abidjan/documents/briefingnote/wcms_741864.pdf. 19 Nov 2021.

International Labor Organization (ILO). (2020b, April 29). *As job losses escalate, nearly half of global workforce at risk of losing livelihood*. ILO Press Release. https://www.ilo.org/global/about-the-ilo/newsroom/news/WCMS_743036/lang%2D%2Den/index.htm. Accessed 19 Nov 2021.

International Labor Organization (ILO). (2020c, May 2020). *COVID-19 crisis and the informal economy: Immediate responses and policy challenges*. ILO Brief. https://www.ilo.org/wcmsp5/groups/public/@ed_protect/@protrav/@travail/documents/briefingnote/wcms_743623.pdf. 19 Nov 2021.

International Labor Organization (ILO). (2021, February 10). *Amidst the COVID-19 pandemic, a new UNDP-ILO report calls for social protection for workers in the informal economy*. Press Release. https://www.ilo.org/africa/media-centre/pr/WCMS_770207/lang%2D%2Den/index.htm. Accessed 22 Nov 2021.

Kinyanjui, N. (2020, March 22). How the COVID-19 pandemic will affect informal workers. Insights from Kenya. *The Conversation*. https://theconversation.com/how-the-covid-19-pandemic-will-affect-informal-workers-insights-from-kenya-134151. Accessed 29 Jan 2021.

Lakuma, C. P., & Sunday, N. (2020, May 19). Impact of COVID-19 on micro, small, and medium businesses in Uganda. *Brookings: African in Focus*. https://www.brookings.edu/blog/africa-in-focus/2020/05/19/impact-of-covid-19-on-micro-small-and-medium-businesses-in-uganda/. Accessed 13 Nov 2021.

Meagher, K. (1995). Crisis, informalization and the urban informal sector in sub-Saharan Africa. *Development and Change, 26*, 259–284.

Meagher, K. (2021). Informality and the infrastructures of inclusion: An introduction. *Development and Change, 52*(4), 729–755.

Moussié, R., & Staab, S. (2020, May 18). Three ways to contain COVID-19's impact on informal women workers. *United Nations Women Gender Data Story*. https://data.unwomen.org/features/three-ways-contain-covid-19s-impact-informal-women-workers. Accessed 19 Nov 2021.

Mukhtarova, T. (2020, July). COVID-19 and the informal sector: What it means for women now and in the future. *GIWPS Policy Brief*. https://giwps.georgetown.edu/resource/covid-19-and-the-informal-sector/. Accessed 29 Jan 2021.

Ng'weno, A, & Porteous, D. (2018, October). *Let's be real: The informal sector and the gig economy are the future, and the present, of work in Africa*. Center for Global Development (CGD) Note. https://www.cgdev.org/publication/lets-be-real-informal-sector-and-gig-economy-are-future-and-present-work-africa. Accessed 23 Nov 2021.

Nguimkeu, P., & Okou, C. (2020). *A tale of Africa today: Balancing the lives and livelihoods of informal workers during the COVID-19 pandemic*. Africa Knowledge in Time Policy Brief; Issue 1; No. 3, Washington, D.C.: World Bank Group. https://openknowledge.worldbank.org/bitstream/handle/10986/34582/A-Tale-of-Africa-Today-Balancing-the-Lives-and-Livelihoods-of-Informal-Workers-During-the-COVID-19-Pandemic.pdf?sequence=6. Accessed 22 Nov 2021.

Ohnsorge, F., & Yu, S. (eds.). (2021). *The long shadow of informality: Challenges and policies*. Advance Edition. License: Creative Commons Attribution CC BY 3.0 IGO.

Omolo, J. (2020, April 20). How the pandemic is affecting the informal sector in Kenya. *FES Blog*. https://www.fes.de/referat-afrika/neugikeiten-referat-afrika/how-the-pandemic-is-affecting-the-informal-sector-in-kenya. Accessed 29 Jan 2021.

Onyekwena, C., & Ekeruche, M.A. (2020, April 8). Understanding the impact of the COVID-19 outbreak on the Nigerian economy. *Brookings: Africa in Focus*. https://www.brookings.edu/blog/africa-in-focus/2020/04/08/understanding-the-impact-of-the-COVID-19-outbreak-on-the-nigerian-economy/. Accessed 11/13/2021.

Pinchoff, J., Kraus-Perrotta, C., Austrian, K., Tidwell, J. B., Abuya, T., Mwanga, D., Kangwana, B., Ochako, R., Muluve, E., Mbushi, F., Nzioki, M., & Ngo, T. D. (2021). Mobility patterns during COVID-19 travel restrictions in Nairobi urban informal settlements: Who is leaving home and why? *Journal of Urban Health, 98*, 211–221.

Potts, D. (2008). The urban informal sector in sub-Saharan Africa: From bad to good (and back again?). *Development Southern Africa, 52*(2), 151–167.

Rigon, A., Walker, J., & Koroma, B. (2020). Beyond formal and informal: Understanding urban informalities from Freetown. *Cities, 105*, 1–7.

Shurkin, M., Noyes, A., & Adgie, M. K. (2021). *The COVID-19 pandemic in sub-Saharan Africa: An opportunity to rethink strategic competition on the continent*. RAND Corporation. https://www.rand.org/pubs/perspectives/PEA1055-1.html

United Nations Development Fund (UNDP). (2021). *Informality and social protection in African countries: A forward-looking assessment*.

Webb, A., & McQuaid, R. (2020). Employment in the informal economy: Implications of the COVID-19 pandemic. *International Journal of Sociology and Social Policy, 40*(9/10), 1005–1019.

World Health Organization (WHO) COVID-19 Dashboard. (2021). https://covid19.who.int/. Accessed 13 Dec 2021.

Zeufack, A. G., Calderon, C., Kambou, G., Djiofack, C. Z., Kubota, M., Korman, V., & Cantu, C. C. (2020). Assessing the economic impact of COVID-19 and policy responses in sub-Saharan Africa. *Africa's Purse, 21*(April), 1–136. World Bank, Washington, DC. https://doi.org/10.1596/978-1-4648-1568-3. License: Creative Commons Attribution CC BY 3.0 IGO.

Chapter 20
Freshwater Resources and COVID-19

Jason H. Knouft

Personal Story
The COVID-19 pandemic has been, in many regards, a painful challenge. In addition to the direct health concerns associated with contracting COVID-19, everyday life has been altered by lockdowns, limited interactions with friends, and increased care for at-home children and older family members, among other issues. Nevertheless, there have been positive moments. My wife, children, and I were all at home working or attending school for at least 1 year. While this was a stressful time, I acknowledge that I am extremely fortunate to have had stable and consistent employment and easy access to healthcare, water, and food during the pandemic. Moreover, working from home provided me time with my children that I would have never had outside of the pandemic. It was wonderful to watch them grow. We had lunch together, talked during the day, played outside during lunch, and took our dog for a walk most days in the park while developing our birdwatching skills. This was a gift that I expect I will not fully appreciate until after life returns to a more normal situation, but I am still truly thankful for it now. *Jason Knouft*

Introduction

Water is fundamentally important to natural and social systems, economic activity, and human health, and it is one of the most valuable resources on the planet. However, water and human relationships are reciprocal, with human activities having significant impacts on water quantity and quality in almost all areas of the planet. Considering these reciprocal relationships and that water is inextricably linked to almost all services humans use every day (Neal, 2020), altered human activities and limited access to water can have significant direct and indirect effects on human health as well as social and natural systems. These effects have been highlighted during the COVID-19 pandemic, with access to clean and abundant

J. H. Knouft (✉)
Saint Louis University, St. Louis, MO, USA

National Great Rivers Research and Education Center, East Alton, IL, USA
e-mail: jason.knouft@slu.edu

M. Laituri et al. (eds.), *The Geographies of COVID-19*, Global Perspectives on Health Geography, https://doi.org/10.1007/978-3-031-11775-6_20

water a critical component influencing direct (e.g., disease transmission, mortality) and indirect (e.g., social, economic) impacts of COVID-19. (See Chap. 2 for a related discussion of first- and second-order impacts of the pandemic.)

Water availability is ultimately dictated by the hydrologic cycle, which represents the continuous circulation of water in the Earth system. The primary components of this circulation include evaporation, transpiration, condensation, precipitation, and runoff from the landscape, with human access to freshwater resources largely restricted to surface waters (i.e., rivers, lakes, wetlands) and groundwater. Within the hydrologic cycle, runoff of water from the landscape to rivers and lakes is a primary source of nutrients, sediment, and contaminants to surface waters. Human activities consistently elevate the occurrence of these constituents, often with detrimental impacts on water quality. This anthropogenic degradation of freshwater systems is an important example of the reciprocal nature of humans' relationship with water, as decreases in water quality driven by humans then impact the quality of the resources required by society. Understanding processes at this point of interaction between humans and the hydrologic cycle, which have likely been impacted in unusual ways over the past 2 years, is critical to realizing the impacts of the COVID-19 pandemic on society.

The local and global impacts of COVID-19, while unusual on their own, have occurred on an even more unusual warming planet, likely compounding and complicating society's response to the pandemic. The hydrologic cycle is primarily powered by thermal energy from the sun, with increases in temperature potentially serving to alter and accelerate movement of water through the Earth system. Ongoing changes in climate are having well-documented effects on water resources and freshwater ecosystems through increases in air temperature and changes in seasonal and spatial patterns of precipitation (Ficklin et al., 2016, 2018; Knouft & Ficklin, 2017). These alterations are, in many cases, contributing to changing patterns of water storage and stress around the world (Fig. 20.1, Huggins et al., 2022). Moreover, weather variability is increasing, and the nonstationarity of the Earth's climate creates greater uncertainty and limits society's ability to adequately manage water resources and predict and plan for extreme anomalies (e.g., floods and drought). These changing climate systems are responsible for the uncertainty of water availability around the world in recent years, which adds to the challenges associated with mitigating the impacts of COVID-19.

The predictable availability of clean water is largely dependent on human use and climate-driven precipitation and air temperature. The capacity of social systems to adapt to water stresses associated with climatic variability can have cascading effects on all parts of local society. Moreover, regions exhibiting greater water system resilience have a more robust ability to tolerate acute water stresses. When combined, measures of water availability and social adaptability to water resource stress levels create a mosaic across the globe (Fig. 20.2, Huggins et al., 2022), with some geopolitical regions having a much greater capacity to respond to water stress. This mosaic is the template on which the COVID-19 pandemic has occurred. Considering that ongoing changes in climate contribute, in part, to water-stressed

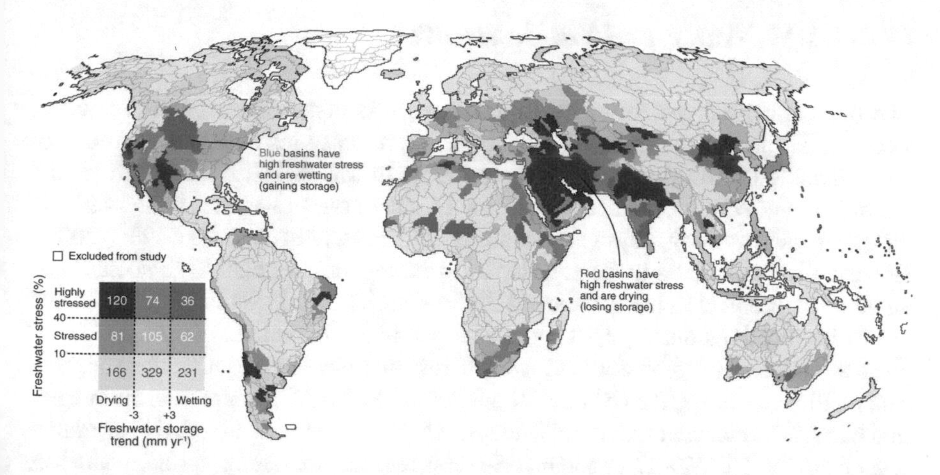

Fig. 20.1 Estimates of river basin water conditions based on trends characterizing freshwater stress and storage. Legend indicates the number of basins in each "freshwater stress × freshwater storage" category. (Figure is reproduced from Huggins et al. (2022) under the Creative Commons Attribution 4.0 International License)

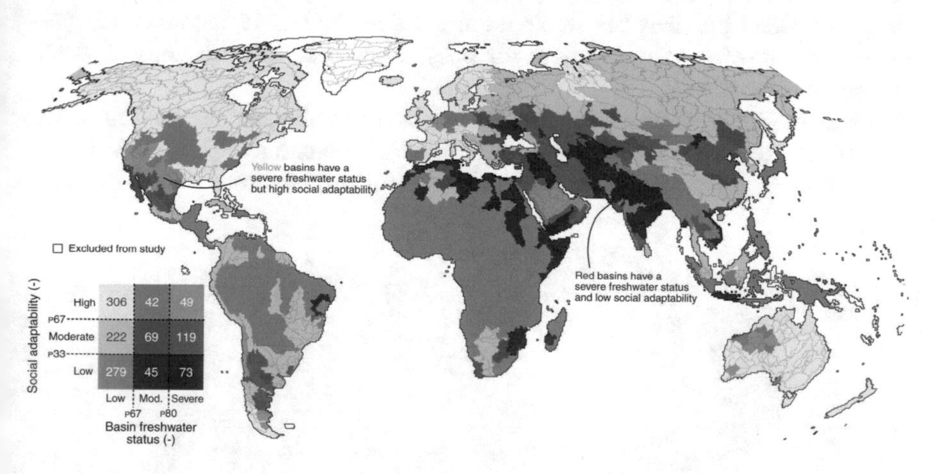

Fig. 20.2 Representation of the combined influence of freshwater status and social adaptability used to characterize the resilience of a region to water stress. Legend indicates the number of basins in each "freshwater status × social adaptability" category. (Figure is reproduced from Huggins et al. (2022) under the Creative Commons Attribution 4.0 International License)

regions, the pandemic has likely made evident how climate change can serve as a threat multiplier to human health and well-being. In this context, the challenges faced by local communities can vary among regions and include limited access to clean water for hygiene, impacts to indigenous communities, and alterations in water demand due to restrictions associated with lockdowns.

COVID-19, Water, and Public Health

The prevention and control of infectious disease is dependent on access to clean water, with disease transmission enhanced when water resources are stressed or unpredictable (WHO, 2019; Sivakumar, 2021). In March 2020, the World Health Organization (WHO) provided guidelines to aid in the prevention of COVID-19, including handwashing, physical distancing, and household cleaning (WHO, 2020). However, limited access to clean water for drinking and personal hygiene has been an ongoing problem in less-developed regions in the world, resulting in significant health implications during the COVID-19 pandemic (Donde et al., 2021). These areas are relatively less resilient to water stress and have faced particularly significant challenges during the COVID-19 pandemic. Figure 20.3 represents an integration of a freshwater status and social adaptability analysis with cumulative mortality rates from the COVID-19 pandemic. Several regions, including northern Mexico, eastern and southern South America, northern and southern Africa, and eastern Europe contain geopolitical areas that have experienced high rates of COVID-19 mortality and have been identified by Huggins et al. (2022) as water-stressed regions with limited social adaptability (Fig. 20.3). While identifying these regions does not suggest a direct link between water resources and COVID-19 mortality, the pandemic does provide an opportunity for possible detailed investigations of water-pandemic interactions.

As a baseline from which to view the implications of water stresses during the COVID-19 pandemic, Stoler et al. (2021) provides a broad pre-COVID-19 assessment of water insecurity among low- to middle-income countries. Data from 2017

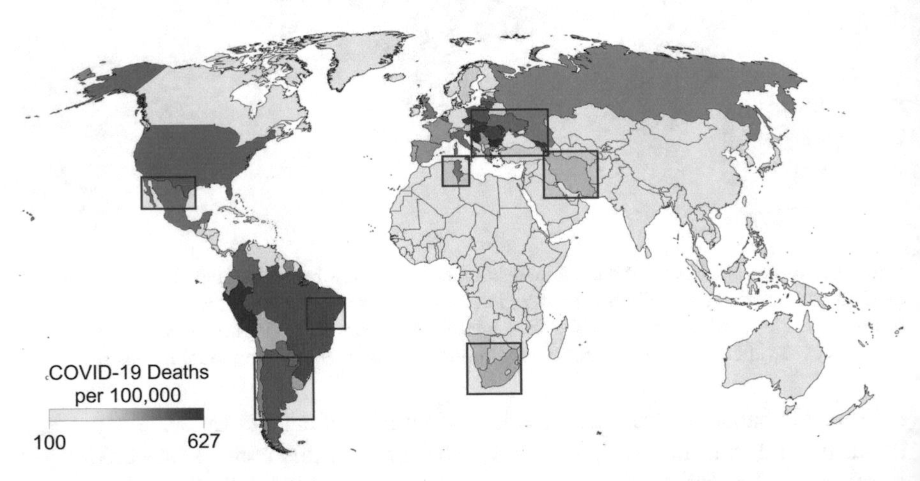

Fig. 20.3 Countries with COVID-19 death rates greater than 100 deaths/100,000 in the population. Mortality data represent cumulative deaths from the onset of the COVID-19 pandemic through January 21, 2022, and were downloaded from the Johns Hopkins University Coronavirus Resource Center (https://coronavirus.jhu.edu/) and visualized in ArcGIS (version 4.1). Blue squares indicate regions identified as "moderate–severe" water stress and "low–moderate" social adaptability by Huggins et al. (2022)

to 2018 from 8297 households in 29 sites across North America, South America, Africa, the Middle East, and Asia indicate that 46% of households lacked the ability to regularly wash their hands or had to acquire water from outside the household, which results in a lack of physical (social) distancing (Stoler et al., 2021). Moreover, 71% of these households experienced water-related problems which can enhance transmission and impacts of COVID-19, including a lack of clean water for drinking, taking medication, and bathing (Stoler et al., 2021). While this lack of access to clean water was a global health emergency prior to the appearance of COVID-19, the pandemic intensified the threat to human health and well-being associated with these preexisting environmental challenges.

In a study of 25 sub-Saharan Africa countries (excluding South Africa) during the COVID-19 pandemic, approximately 46% of households did not have in-house access to water or a sanitation facility (Ekumah et al., 2020). In five of these countries, less than 2% of the households had access to these resources (Ekumah et al., 2020). The lack of in-house clean water, including soap for handwashing, hinders the ability to directly limit the spread of COVID-19 (Jiwani & Antiporta, 2020; Amuakwa-Mensah et al., 2021). Moreover, the lack of in-home water and sanitation facilities forces residents to travel for these services, increases their contact with others, and subsequently increases the likelihood that residents will be exposed and serve as vectors for COVID-19. Almost three million people live in water-stressed areas in sub-Saharan Africa, many of whom would view COVID-19 prevention recommendations (e.g., frequent and sustained handwashing) as a luxury (Anim & Ofori-Asenso, 2020). The COVID-19 pandemic in sub-Saharan Africa has provided a clear reminder of the importance of improving water access for vulnerable populations.

Brazil has had the third highest number of reported COVID-19 cases and second highest number of reported COVID-19 deaths among countries. Most of the Amazon River basin is contained within the borders of Brazil and access to water is not a significant concern relative to other regions in the world. However, access to basic sanitation within the Amazon River basin is lacking in many areas. This is of particular concern because COVID-19 can be shed in feces from humans for at least 2 days after the virus is no longer detected in the respiratory tract (Parasa et al., 2020; Repici et al., 2020), with untreated wastewater serving as an important vector for COVID-19 (Lodder & Husman, 2020). There are thousands of communities located near the Amazon River which include approximately 180 native groups and an estimated 306,000 indigenous people (da Costa et al., 2020). This combination of indigenous livelihoods, lack of developed sanitation systems, and a global pandemic represents one of the many facets of the challenges presented by COVID-19 and the need for place-based approaches for the development of effective water resource infrastructure.

In India, which has had the second highest number of reported COVID-19 cases and third highest number of reported COVID-19 deaths among countries, reduced human activities associated with the pandemic have contributed to water quality improvements in rural regions, yet urban areas have not experienced the same improvements (Balamurugan et al., 2021). The increase in water demand, presumably related to increased attention to personal hygiene (e.g., handwashing), has resulted in up to 25% increases in water consumption in some areas (Balamurugan et al., 2021). Accordingly, the frequency of handwashing with soap appears to have increased in rural areas, although 13% of households in some regions have reported barriers to handwashing (Bauza et al., 2021). However, many cities in India lack an uninterrupted water supply (Hughes et al., 2017), with these constraints more intense during summer because of decreased water storage (Balamurugan et al., 2021). While diverting water from the industrial and information technology sectors as workers adopt a work-from-home work approach may improve conditions, limited tracking of water use among sectors makes water management a challenge. The issue of limited water management data is not restricted to India and points to the need to increase infrastructure monitoring, even in developed countries, to better address water issues.

The impacts of COVID-19, while global in nature, are particularly challenging and dangerous for those that lack the support of robust infrastructure, which can occur in developing regions as well as in transient communities in developed countries (Parikh et al., 2020). The previous examples from Africa, Brazil, and India represent a subset of the variety of ways that access to clean water and sanitation interact with social systems and the constraints imposed by the COVID-19 pandemic (see McDonald et al., 2020, for additional examples). In all cases, access to water, sanitation, and hygiene (WASH) facilities is critical to the mitigation of COVID-19 impacts. The development of WASH infrastructure is related to at least nine of the 13 United Nations Sustainable Development Goals (Parikh et al., 2020), indicating the intertwined nature of water with all aspects of sustainable development and social resilience, as seen during the COVID-19 pandemic.

COVID-19, Water Quality, and Freshwater Ecosystems

The relationships between water availability and direct outcomes from the COVID-19 pandemic (i.e., disease transmission, mortality) have rightly been the primary focus of attention during the past 2 years. However, the COVID-19 pandemic has also provided a unique opportunity to understand how human activities influence freshwater ecosystems. As COVID-19 spread across the globe in 2020, many political regions (cities, states, countries) imposed restrictions on the general public including limiting in-person work and school attendance, public gatherings, and complete lockdowns of social activities. While the goal of these restrictions was to restrict the spread of COVID-19, the altered human activities also, in some cases, greatly modified how humans interacted with freshwater resources. These alterations in human activities have provided the opportunity for a natural global experiment where humans' relationships with the environment and the resilience of environmental systems to human activities can be assessed across large geographic regions.

Human activities influence natural freshwater systems on all parts of the planet. When natural landscapes are transformed to agriculture or urban areas, surface runoff washes excess nutrients, sediment, and contaminants into rivers and lakes, often degrading water quality and aquatic ecosystems. During 2020, social and economic activities were significantly curtailed in many parts of the world because of the COVID-19 pandemic. Industry, transportation, and human activity decreased dramatically during this time. This shutdown presented a unique opportunity to observe the resilience of freshwater systems to decreased perturbations from humans and a natural experiment that will likely not be fully understood until the pandemic has passed.

Although the major impacts of COVID-19 have only been felt since 2020, researchers are already investigating how changes in human activities associated with the pandemic have influenced natural freshwater systems. In Vembanad Lake, the longest lake in India, suspended particulate matter, which can indicate habitat degradation, decreased by up to 36% during the lockdown in the first quarter of 2020 (Yunus et al., 2020). Liu et al. (2022) examined 21 water quality variables from 65 stations on rivers in eastern China, also during early 2020. While water quality improvements were seen across the sites in China, responses of various environmental variables (e.g., nitrogen, dissolved oxygen, pH) were different across geographic space and time (Liu et al., 2022), suggesting the importance of local conditions on water resources. In the main branch of the Ganges River in eastern India, water quality improvements resulting in water reaching a potable level were evident at only one site among seven (Muduli et al., 2021). However, another study at multiple sites at the mouth of the Ganges River indicated dramatic increases in several water quality variables (Roy et al., 2021), suggesting the cumulative downstream effects of reduced human activities farther up in the watershed.

While improved surface water conditions associated with decreased human activities is a positive outcome of the COVID-19 pandemic, these types of responses are not consistent among regions and can provide insights into factors regulating regional disparities in freshwater resource quality. Wetz et al. (2022) examined Pacific, Atlantic, and Gulf of Mexico coastal systems in the United States, investigating whether reduced tourism would impact water quality. In most cases, decreased human activity did not result in significantly improved water quality compared to long-term pre-pandemic trends, although localized and ephemeral decreases in fecal indicator bacteria (enterococci) were evident in Gulf of Mexico coastal sites (Wetz et al., 2022). These results, when compared to less developed regions in India and regions in China with more stringent lockdowns, suggest the importance of robust water infrastructure and the degree of influence of human activities on water quality. In a more nuanced investigation, water pollution and human health risk decreased by up to 39% in the Zarjoub River in northern Iran during the lockdown (Haghnazar et al., 2022). Within these changes, the contribution of municipal wastewater to water pollution increased from 23% to 50%, while industrial effluents and solid waste contributions were reduced from 64% to 45%. Agricultural effluent and vehicular pollution were also reduced from 13% to 5% during the lockdown period compared to the post-lockdown period. These changes in water quality can have direct impacts on humans, but also can result in cascading effects on freshwater ecosystems and the ecologically and economically important biodiversity occupying these systems.

COVID-19, Freshwater Biodiversity, and Fisheries

Water quality is tightly related to freshwater ecosystem functioning and productivity, which provide the foundation supporting freshwater biodiversity. Improvements in water quality associated with altered human activities during the COVID-19 pandemic may provide positive benefits to freshwater species resulting in increased population sizes (Stokes et al., 2020; Cooke et al., 2021). Documenting the extent and duration of these potential responses will likely require several years because of the amount of effort required to effectively characterize overall species diversity across space and time in freshwater habitats. However, more immediate investigation of the relationships between social systems and freshwater fisheries is supported by significant infrastructure and baseline knowledge and has already provided a foundation for understanding interactions between the COVID-19 pandemic, social systems, and freshwater species.

Fisheries are a common and important way that people interact with freshwater biodiversity, with fisheries playing an important role in addressing several of the Sustainable Development Goals of the United Nations 2030 Sustainable Development Agenda (United Nations, 2020). Improvements in water quality and reductions in fishing effort and harvest associated with decreased human activities during COVID-19 lockdowns are suggested to positively impact fishery production (Cooke et al., 2021). Moreover, temporary reductions in irrigation and industrial water use may have benefited fisheries, particularly in temperate regions during early 2020 when fish reproduction is common. Detrimental impacts associated with COVID-19 are also possible (Stokes et al., 2020; Cooke et al., 2021). Increased and unregulated fishing pressure in areas where fisheries provide subsistence protein may result in decreases in fish abundance as well as elevated extinction risk to threatened species (Pinder et al., 2020), which are often of large size or abundant and have significant functional importance in freshwater ecosystems.

As with many aspects of the COVID-19 pandemic, the interactions between altered human activities and fisheries likely vary based on regional social and environmental conditions. According to a global survey of fisheries managers, increases in local fishing pressure are expected in southeastern Asia and eastern Africa, which are presumably due to high unemployment rates resulting in increases in the need for local fisheries to provide subsistence protein (Stokes et al., 2020). In Australia, decreased fishing pressure is expected to result from fishing restrictions. Lockdowns and restrictions on indoor gatherings have also resulted in a shift to outdoor recreational activities, particularly in North America, where retail sales of fishing equipment and licenses have increased dramatically (e.g., Howarth et al., 2021). This increased focus on outdoor recreational activities likely contributes to perceived increases in fishing pressure (Stokes et al., 2020), yet whether this pressure results in negative impacts on fisheries is unclear, as dependence of North American residents on fisheries as a form of subsistence may vary based on local social and economic conditions.

Although COVID-19 lockdowns and restricted social activities are relatively recent, evidence of the impacts of the pandemic on fisheries and consumption are starting to appear. In Kenya, residents around Lake Victoria are experiencing high fish prices and consuming less fish, likely because of reduced fishing and trading activities due to concerns about being infected with COVID-19 (Fiorella et al., 2021). During early 2020, as lockdowns and social restrictions were implemented in Kenya, the region experienced heavy rainfall and flooding, possibly associated with ongoing changes in climate. The apparent interaction between pandemic-altered social activity and climate change had complex effects on the local fisheries and fish consumption. Livelihood losses resulted from reduced fishing efforts and limited trading; however, fish abundance improved due to reduced fishing effort and excess water supply (Aura et al., 2020). The catch, supply, and demand for fishes have also decreased in India, with fish harvests reduced by up to 44% during lockdown periods and price increases reaching 40% relative to previous years (Das et al., 2022). Similarly, in Bangladesh, altered social activities related to the pandemic have altered the catch, supply, and trade of fishes and disrupted the fish farming season, which could have longer-term nutritional and economic effects (Sunny et al., 2021).

Conclusions

The importance of access to clean water cannot be underestimated, as availability of this resource is at the core of social stability and human health. While access to clean water was a significant issue in many parts of the world before the appearance of COVID-19, residents of water-stressed regions have been particularly vulnerable during the pandemic because of limited ability to maintain necessary water-related hygiene practices as well as avoid close social interactions while traveling to collect clean water. Beyond direct impacts on humans, the COVID-19 pandemic has resulted in a range of responses from freshwater ecosystems. In many cases, water quality has improved during lockdowns due to reduced social and economic activities. The overall impacts of freshwater biodiversity are not clear, although post-pandemic investigation of this issue may reveal significant outcomes related to altered human activity and improved water quality. The responses of freshwater fishery resources are apparently mixed, with assessment and prediction of these responses likely complicated by complex economic, social, and environmental conditions in particular regions.

The full extent of the impacts of COVID-19 and associated modifications to society are still being realized and likely will not be fully appreciated until long after the pandemic ends. While the health and well-being of humans have rightly been the primary focus during the pandemic, the opportunity to learn from the past 2 years, and coming years, is enormous. Our ability to mobilize and respond to a global pandemic has been tested and we have seen the areas, both expected and unexpected, that can be improved. Access to clean water is of primary importance in the response to COVID-19 and likely any global pandemic. Ongoing changes in climate will add additional stresses to water resources around the world. The challenges faced during the COVID-19 pandemic of providing clean water and the implications of failing to provide clean water provide a framework and justification for further developing sustainable water resources.

Disclaimer The views in this report are solely of the authors and do not represent those of the US government agencies or any of the organizations mentioned. Assumptions made within the analysis are not a reflection of the position of any US government entity.

References

Amuakwa-Mensah, F., Klege, R. A., Adom, P. K., & Kohlin, G. (2021). COVID-19 and handwashing: Implications for water use in sub-Saharan Africa. *Water Resources and Economics, 36*, 100189.

Anim, D. O., & Ofori-Asenso, R. (2020). Water scarcity and COVID-19 in sub-Saharan Africa. *Journal of Infection, 81*, E108–E109.

Aura, C. M., Nyamweya, C. S., Odoli, C. O., Owiti, H., Njiru, J. M., Otuo, P. W., Waithaka, E., & Malala, J. (2020). Consequences of calamities and their management: The case of COVID-19 pandemic and flooding on inland capture fisheries in Kenya. *Journal of Great Lakes Research, 46*, 1767–1775.

Balamurugan, M., Kasiviswanathan, K. S., Ilampooranan, I., & Soundharajan, B. S. (2021). COVID-19 lockdown disruptions on water resources, wastewater, and agriculture in India. *Frontiers in Water, 3*, 603351.

Bauza, V., Sclar, G. D., Bisoyi, A., Majorin, F., Ghugey, A., & Clasen, T. (2021). Water, sanitation, and hygiene practices and challenges during the COVID-19 pandemic: A cross-sectional study in rural Odisha, India. *American Journal of Tropical Medicine and Hygiene, 104*, 2264–2274.

Cooke, S. J., Twardek, W. M., Lynch, A. J., Cowx, I. G., Olden, J. D., Funge-Smith, S., Lorenzen, K., Arlinghaus, R., Chen, Y. S., Weyl, O. L. F., Nyboer, E. A., Pompeu, P. S., Carlson, S. M., Koehn, J. D., Pinder, A. C., Raghavan, R., Phang, S., Koning, A. A., Taylor, W. W., et al. (2021). A global perspective on the influence of the COVID-19 pandemic on freshwater fish biodiversity. *Biological Conservation, 253*, 108932.

da Costa, J. S., Rodrigues, L. D., Da Silva, A. G. C., Neto, R. A., Batista, I. H., De Albuquerque, C. C., De Melo, M. D. G., & Liberato, M. A. R. (2020). Water, sanitation and COVID-19 in the Amazon. *Holos, 36*, e10803.

Das, B. K., Roy, A., Som, S., Chandra, G., Kumari, S., Sarkar, U. K., Bhattacharjya, B. K., Das, A. K., & Pandit, A. (2022). Impact of COVID-19 lockdown on small-scale fishers (SSF) engaged in floodplain wetland fisheries: Evidences from three states in India. *Environmental Science and Pollution Research, 29*, 8452–8463.

Donde, O. O., Atoni, E., Muia, A. W., & Yillia, P. T. (2021). COVID-19 pandemic: Water, sanitation and hygiene (WASH) as a critical control measure remains a major challenge in low-income countries. *Water Research, 191*, 116793.

Ekumah, B., Armah, F. A., Yawson, D. O., Quansah, R., Nyieku, F. E., Owusu, S. A., Odoi, J. O., & Afitiri, A. R. (2020). Disparate on-site access to water, sanitation, and food storage heighten the risk of COVID-19 spread in Sub-Saharan Africa. *Environmental Research, 189*, 109936.

Ficklin, D. L., Robeson, S. M., & Knouft, J. H. (2016). Impacts of recent climate change on trends in baseflow and stormflow in United States watersheds. *Geophysical Research Letters, 43*, 5079–5088.

Ficklin, D. L., Abatzoglou, J. T., Robeson, S. M., Null, S. E., & Knouft, J. H. (2018). Natural and managed watersheds show similar responses to recent climate change. *Proceedings of the National Academy of Sciences, 115*, 8553–8557.

Fiorella, K. J., Bageant, E. R., Mojica, L., Obuya, J. A., Ochieng, J., Olela, P., Otuo, P. W., Onyango, H. O., Aura, C. M., & Okronipa, H. (2021). Small-scale fishing households facing COVID-19: The case of Lake Victoria, Kenya. *Fisheries Research, 237*, 105856.

Haghnazar, H., Cunningham, J. A., Kumar, V., Aghayani, E., & Mehraein, M. (2022). COVID-19 and urban rivers: Effects of lockdown period on surface water pollution and quality- A case study of the Zarjoub River, north of Iran. *Environmental Science and Pollution Research*. in press.

Howarth, A., Jeanson, A. L., Abrams, A. E. I., Beaudoin, C., Mistry, I., Berberi, A., Young, N., Nguyen, V. M., Landsman, S. J., Kadykalo, A. N., Danylchuk, A. J., & Cooke, S. J. (2021). COVID-19 restrictions and recreational fisheries in Ontario, Canada: Preliminary insights from an online angler survey. *Fisheries Research, 240*, 105961.

Huggins, X., Gleeson, T., Kummu, M., Zipper, S. C., Wada, Y., Troy, T. J., & Famiglietti, J. S. (2022). Hotspots for social and ecological impacts from freshwater stress and storage loss. *Nature Communications, 13*, 439.

Hughes, S., Chu, E. K., & Mason, S. G. (2017). *Climate change in cities: Innovations in multi-level governance* (Vol. 378). Springer-Science.

Jiwani, S. S., & Antiporta, D. A. (2020). Inequalities in access to water and soap matter for the COVID-19 response in sub-Saharan Africa. *International Journal for Equity in Health, 19*, 82.

Knouft, J. H., & Ficklin, D. L. (2017). The potential impacts of climate change on biodiversity in flowing freshwater systems. *Annual Review of Ecology, Evolution, and Systematics, 48*, 111–133.

Liu, D., Yang, H., Thompson, J. R., Li, J. L., Loiselle, S., & Duan, H. T. (2022). COVID-19 lockdown improved river water quality in China. *Science of the Total Environment, 802*, 149585.

Lodder, W., & Husman, A. M. R. (2020). SARS-CoV-2 in wastewater: Potential health risk, but also data source. *Lancet Gastroenterology and Hepatology, 5*, 533–534.

McDonald, D. A., Spronk, S. J., & Chavez, D. (Eds.). (2020). *Public water and Covid-19: Dark clouds and silver linings*. Municipal Service Project (Kingston), Transnational Institute (Amsterdam) and Latin American Council of Social Sciences (CLACSO) (Buenos Aires).

Muduli, P. R., Kumar, A., Kanuri, V. V., Mishra, D. R., Acharya, P., Saha, R., Biswas, M. K., Vidyarthi, A. K., & Sudhakar, A. (2021). Water quality assessment of the Ganges River during COVID-19 lockdown. *International journal of Environmental Science and Technology, 18*, 1645–1652.

Neal, N. J. (2020). COVID-19 and water resources management: Reframing our priorities as a water sector. *Water International, 45*, 435–440.

Parasa, S., Desai, M., Chandrasekar, V. T., Patel, H. K., Kennedy, K. F., Roesch, T., Spadaccini, M., Colombo, M., Gabbiadini, R., Artifon, E. L. A., Repici, A., & Sharma, P. (2020). Prevalence of gastrointestinal symptoms and fecal viral shedding in patients with coronavirus disease 2019 a systematic review and meta-analysis. *JAMA Network Open, 3*, 2011335.

Parikh, P., Diep, L., Gupte, J., & Lakhanpaul, M. (2020). COVID-19 challenges and WASH in informal settlements: Integrated action supported by the sustainable development goals. *Cities, 107*, 102871.

Pinder, A. C., Raghavan, R., Britton, J. R., & Cooke, S. J. (2020). COVID-19 and biodiversity: The paradox of cleaner rivers and elevated extinction risk to iconic fish species. *Aquatic Conservation-Marine and Freshwater Ecosystems, 30*, 1061–1062.

Repici, A., Pace, F., Gabbiadini, R., Colombo, M., Hassan, C., & Dinelli, M. (2020). Endoscopy units and the coronavirus disease 2019 outbreak: A multicenter experience from Italy. *Gastroenterology, 159*, 363.

Roy, M. B., Ghosh, M., & Roy, P. K. (2021). Assessment of water quality of River Ganga during COVID-19 lockdown. *Desalination and Water Treatment, 223*, 26–33.

Sivakumar, B. (2021). COVID-19 and water. *Stochastic Environmental Research and Risk Assessment, 35*, 531–534.

Stokes, G. L., Lynch, A. J., Lowe, B. S., Funge-Smith, S., Valbo-Jorgensen, J., & Smidt, S. J. (2020). COVID-19 pandemic impacts on global inland fisheries. *Proceedings of the National Academy of Sciences of the United States of America, 117*, 29419–29421.

Stoler, J., Miller, J. D., Brewis, A., Freeman, M. C., Harris, L. M., Jepson, W., Pearson, A. L., Rosinger, A. Y., Shah, S. H., Staddon, C., Workman, C., Wutich, A., & Young, S. L. (2021). Household water insecurity will complicate the ongoing COVID-19 response: Evidence from 29 sites in 23 low- and middle-income countries. *International Journal of Hygiene and Environmental Health, 234*, 113715.

Sunny, A. R., Mithun, M. H., Prodhan, S. H., Ashrafuzzaman, M., Rahman, S. M. A., Billah, M. M., Hussain, M., Ahmed, K. J., Sazzad, S. A., Alam, M. T., Rashid, A., & Hossain, M. M. (2021). Fisheries in the context of attaining sustainable development goals (SDGs) in Bangladesh: COVID-19 impacts and future prospects. *Sustainability, 13*, 9912.

United Nations. (2020). *The sustainable development goals report*. https://unstats.un.org/sdgs/report/2020/

Wetz, M. S., Powers, N. C., Turner, J. W., & Huang, Y. X. (2022). No widespread signature of the COVID-19 quarantine period on water quality across a spectrum of coastal systems in the United States of America. *Science of the Total Environment, 807*, 150825.

WHO (World Health Organization). (2019). *Progress on household drinking water, sanitation and hygiene 20 0 0–2017. Special focus on inequalities*. United Nations Children's Fund (UNICEF) and World Health Organization.

WHO (World Health Organization). (2020). *Coronavirus disease (COVID-19) advice for the public*. https://www-who-int.ezp.slu.edu/emergencies/diseases/novel-coronavirus-2019/advice-for-public

Yunus, A. P., Masago, Y., & Hijioka, Y. (2020). COVID-19 and surface water quality: Improved lake water quality during the lockdown. *Science of the Total Environment, 731*, 139012.

Chapter 21
Preventing Pandemics: Earth Observations for One Health

Maya V. Mishra

Personal Story
I study ecology & evolution, a field that enables me to get outside and take in the wonder of the natural world. So it wasn't exactly optimal when I, and thousands of other college students, had to start taking classes from a laptop in the living room in March of 2020. I dearly missed the walks in the woods with my classmates and professors, drawing out the delicate web of connections between trees and deer and ourselves. Inundated with pandemic news, I saw plenty of discussion of pangolins and wet markets, but little of our larger relationship with animals and nature, which I view as integral to explaining the emergence of COVID-19. In the evolutionary history of *Homo sapiens*, nature has been our adversary, something to resist and exploit for our own survival. The persistence of this deep-rooted urge has contributed to our attempted conquest of and separation from the wild world. But the pandemic has emphasized that we're just as much a part of nature as we always have been: pull on one thread in the vast tapestry of life, and you might not expect what will unravel. Humans cannot survive in a vacuum, nor would that world be nearly as vibrant as the one we have today. Understanding of and connection with the natural world is vital to not only conservation, but our own health and wellbeing. So take a walk outside; observe the life that persists everywhere you look. It's all connected—and so are we. *Maya Mishra*

Introduction

Somewhere in a cave in southern China, a colony of bats rustles and squeaks. The stone floor is coated in guano, or bat excrement, lending a musky scent to the air. Soon the bats will take flight for the evening, soaring through the sky in search of prey. During their nightly ranging, these bats will consume many insects, keeping

M. V. Mishra (✉)
Department of Ecology and Evolutionary Biology, Alumna, Princeton University, Princeton, NJ, USA
e-mail: mmishra@alumni.princeton.edu

M. Laituri et al. (eds.), *The Geographies of COVID-19*, Global Perspectives on Health Geography, https://doi.org/10.1007/978-3-031-11775-6_21

populations in check and therefore playing an important role in their ecosystems (Kunz et al., 2011). They also provide prey for avian raptors (Wu et al., 2019), and for humans, they possess commercial value for their role in traditional medicine (El-Sayed & Kamel, 2021).

Bats also have a hidden physiological talent: they can play host to a number of diseases, enabling them to spread pathogens to each other and to other creatures without becoming sick themselves (O'Shea et al., 2014). Unfortunately, this natural advantage also threatens human health: diseases can be transmitted to other animals through contact with the bats or their feces and can then make the leap to humans, sometimes to devastating worldwide effects (Wang & Eaton, 2007).

In our third year of the COVID-19 pandemic, many of us have become unexpectedly familiar with zoonotic diseases, those infections originating in animals. But we tend to lose sight of the broader context in which this pandemic emerged. The zoonotic origins of COVID-19 are not unique: about three-quarters of emerging diseases newly recognized by epidemiologists can be traced to animals (Rabozzi et al., 2012), and zoonoses make up 60% of the total number of infectious diseases affecting humans (Meurens et al., 2021).

In the pandemic age, there is no need to overstate the necessity of monitoring disease emergence and investigating the factors that affect this process. Humanity's increasing encroachment into the Earth's wild places has brought people and livestock into a higher degree of contact with wildlife (Rabozzi et al., 2012). Diseases are therefore more likely to be transmitted from wildlife to domestic animals, where they then can make the jump to humans in an event known as a “"spillover” (Rabozzi et al., 2012).

However, we know much less about the specific conditions that influence disease spillovers. Studying disease emergence involves not only human factors, but also those concerning animals and the environment. These three areas are united through the framework of One Health. This term encompasses a way of approaching scientific and legislative issues with an interdisciplinary perspective that recognizes the overlapping priorities of human, animal, and environmental health. Today, this approach has been institutionalized by such organizations as the United States' Centers for Disease Control and Prevention (CDC), which has had a One Health Office since 2009 (One Health, 2021), and has been adopted as a key principle by others, including the World Organization for Animal Health (“One Health,” n.d.).

The One Health framework also clarifies the role of Earth observations (EO). EO, which encompass a wide range of data types on relevant environments and habitats worldwide, can be gathered via remote sensing by spaceborne satellites. The instruments aboard these satellites can collect data within each facet of One Health: on the human and animal side, they can track animal presence, land use change, and deforestation from space, identifying the areas in which humans or livestock might encounter wild animals. In terms of environmental health, satellites can assess ecosystem health through various imaging methods as well as gathering precipitation and climate data that govern where disease-transmitting species can occur. The National Aeronautics and Space Administration (NASA)'s fleet of satellites has been collecting these and similar EO for years.

In this chapter, I discuss uses of EO in a One Health framework to prevent disease spillover and reduce disease burden worldwide. The data enable three broad approaches that align with the three One Health areas: wildlife-based approaches, which center around the wild animals that carry disease; environment-based approaches, which focus on climate and habitat variables that dictate where species may occur; and human-based approaches, which track human impacts on wild areas and assess the risk of spillover. Critically, all of these categories can be integrated with other interdisciplinary forms of data—including social, economic, and political information—to succeed in preventing disease. I conclude by discussing past examples of EO use to predict and prevent disease through a One Health framework and emphasize how preventative measures must address causes, rather than symptoms, of relevant human behaviors in order to avert disease emergence and decrease burden.

Innovative combinations of EO data have the potential to decrease the burden of existing diseases as well as to assess the risk of spillover events in time and space. If we can predict where and when spillovers are likely to occur, we can take steps to prevent them. As the COVID-19 pandemic has emphasized, informed preventative measures—enabled by Earth observations—could save millions of lives from disease and death.

Wildlife-Based Approaches

Zoonotic diseases, by definition, have animal hosts, or reservoir species, in which the pathogens circulate. Aquatic birds are the reservoir for influenza (Parrish et al., n.d.); SARS, and likely COVID-19, has their reservoir in bats (Li et al., 2005). Reservoirs are part of the reason that zoonotic diseases are so difficult to eliminate: even if cases in humans can be significantly reduced—a difficult task in itself—future interactions between humans and the host species could facilitate new spillovers of the pathogen, leading to additional outbreaks.

It is therefore critical to investigate where these reservoir species are located, as well as any movements and behaviors that may bring them into close contact with humans. To this end, EO provide invaluable insight that can be paired with in situ data for more comprehensive wildlife tracking. A plethora of innovative combinations of data are available to inform future efforts to decrease potentially disease-spreading human-wildlife interactions by promoting conservation, benefiting both humans and other species.

The keystone of wildlife-based approaches to preventing disease is spatial tracking data on the reservoir species themselves. In the past, radio trackers required researchers to carry an antenna and monitor the creature of interest on foot or in a vehicle or aircraft, requiring significant time and energy investments (Robertson et al., 2012). Satellites have streamlined this process, allowing scientists to collect larger quantities of accurate location data without spending long hours in the field.

GPS trackers placed on individual creatures rely on orbiting satellites to deliver location data to scientists. While one might envision the classic large collar on an elephant or lion, researchers worldwide are collaborating through the International Cooperation for Animal Research Using Space (Icarus) to tag and track smaller animals using a satellite-based system (About Icarus, n.d.). Icarus' focal species include a number of reservoir taxa, including birds and bats. Animals can be tracked in real time, with the data routed to Movebank, an open-access database (Kranstauber et al., 2011; The Internet of Animals, n.d.). Icarus and other similar initiatives can provide platforms for forecasting movement and migration patterns to reduce human-wildlife contact, therefore also lessening the risk of a spillover event.

Direct tracking data can be supplemented with other spatial data to paint a clearer picture of species presence, abundance, and behavior. Camera trapping data can be used to estimate the abundance of reservoir species individuals in a set area. The Wildlife Insights database represents one such collaboration in the field, containing over 22 million camera trap images analyzed by artificial intelligence, which can provide data on species presence and abundance around the world (Wildlife Insights, n.d.).

Citizen science data can also be used to monitor species locations and distributions, through widely accessible phone applications such as iNaturalist (iNaturalist, n.d.) and eBird (eBird Science, n.d.). Anyone may submit species sightings, which provide large and spatially far-ranging datasets to scientists. While the quality of these data can vary, citizen science can provide an important supplement to other data sources. By tracking which species have been observed by people in different locations, it can also provide insight, if not an exact measure, of the degree of human-wildlife interaction, which could inform estimates of spillover risk.

Knowing where reservoir species are located is critical information to preventing spillover events. However, as suggested by the One Health framework, species distributions must be integrated with human population characteristics in order to accurately assess and mitigate the risk of disease transmission.

Environment-Based Approaches

All species have specific environmental requirements to persist long-term in their habitats—they can only survive in certain temperature ranges, for example, or they might require particular levels of rainfall. As climate change increases temperatures around the world, affecting seasons and weather patterns, the geographic ranges in which species can exist are also changing (Lippi et al., 2019). Mapping key environmental variables allow us to project the future range shifts of key species in order to determine populations at risk. The environmental variables observed by satellites, including temperature, rainfall, and land cover type, are critical to projecting where and when species will exist, in turn contributing to predictions of where disease may emerge. This allows preventative measures to be deployed to get ahead of outbreaks and manage risk.

Thus far, this work has largely been conducted with insect vectors of disease, such as mosquitoes (Lippi et al., 2019), ticks (Slatculescu et al., 2020), and flies (Moore & Messina, 2010). These vectors carry many known diseases, including malaria, Lyme disease, and sleeping sickness, and previously unknown illnesses are emerging at a rapid rate (Bartlow et al., 2019). The World Health Organization estimates that vector-borne diseases cause over 700,000 deaths globally each year, with cases well into the hundreds of millions (Vector-Borne Diseases, n.d.).

Mosquitoes are colloquially known as the world's deadliest animal, causing between 700,000 and 2.5 million deaths every year (Kamerow, 2014). Different species of mosquito carry a number of zoonotic pathogens, the deadliest being malaria, which is carried by *Anopheles* mosquitoes (CDC, 2020): half the world's population lives at risk of the disease (Kamerow, 2014). Other species of mosquito harbor yellow fever, Zika, and West Nile virus, among others (Avŝiĉ-Zupanc, 2013). These diseases' high burden makes tracking mosquito abundance and distribution critical to preventing significant morbidity and mortality.

Because of the temperature and rainfall dependence of mosquitoes' life cycles, the highest burden of mosquito-borne disease cases occurs in warm, tropical areas where the insect can reproduce and cause infection year-round (Le et al., 2019). Mosquito forecasting based on temperature and precipitation variables has become an important tool, especially in areas that experience this increased burden. EO provide consistent and accurate data to drive these models (Chuang et al., 2012), with variables such as vegetation density and water content measured by the Landsat and Sentinel satellites (Tsantalidou et al., 2021). Predictions of mosquito density allow public health institutions to roll out preventative measures, including public awareness campaigns and insecticidal efforts (Giordano et al., 2018).

In a similar manner, tick abundance can also be predicted using EO. Ticks are another significant contributor to the overall vector-borne disease burden worldwide, carrying Lyme disease, tick-borne encephalitis, and other illnesses (Slatculescu et al., 2020). For ticks, evaluating a combination of climate variables as well as habitat suitability in terms of land cover and use is necessary to determining where these species may establish in our changing world (Slatculescu et al., 2020). Past studies have relied on satellites for these data, including Landsat for land cover imagery (Slatculescu et al., 2020) as well as Terra's Moderate Resolution Imaging Spectroradiometer (MODIS) for the normalized difference vegetation index (NDVI), or "greenness," of plant matter (Da Re et al., 2019). Tick species distribution models predict that warming temperatures will significantly increase the abundance of ticks in northern areas, which previously had low burdens of insect presence and consequently tick-borne illness (Sagurova et al., n.d.; Slatculescu et al., 2020). With science providing a foundation for proactive management, these areas can now prepare for heightened tick presence with campaigns to educate the public on tick-safe behavior and make healthcare providers aware of the risk.

Ticks and mosquitoes are two of the most widely studied insect vectors, but the principles of climate and habitat discussed here can also be applied to other species, including the tsetse fly, which carries sleeping sickness (Moore & Messina, 2010), and the kissing bug, which transmits Chagas disease (Klotz et al., 2014). Vector ranges are predicted to shift and expand with our changing climate (Bartlow et al., 2019), and EO data can be employed to train models and predict these shifts under various future climate scenarios. For the communities projected to be at a heightened risk of disease burden, these studies have and will provide a critical head start on disease prevention.

While species distribution modeling for health-related purposes has historically been applied largely to insect species, it can also be used to an extent for mammalian and avian reservoir species. Variations in climate or ecosystem structure can affect habitat use, migration patterns, and other relevant animal behaviors, and these key factors can in turn affect the degree of human contact with species that could spread disease.

Human-Based Approaches

In a final application of the One Health framework, measuring the presence and effects of humans on wildlife habitat is especially critical to determining disease risk. Human activities, including logging, agriculture, and construction, among others, have increasingly encroached on the habitats of many species. These endeavors increase the likelihood of human-animal contact that may lead to a spillover event in two main ways. First, the people conducting these activities spend a large amount of time in largely intact natural habitat, increasing the chance they may encounter wildlife. Additionally, there is evidence across multiple reservoir taxa and study systems to suggest that when habitat is fragmented or destroyed, disease transmission from wildlife increases (Goldberg et al., 2008; Plowright et al., 2011; Suzán et al., 2008). This makes it more likely that people living and working in these areas will not only encounter wildlife, but also that this wildlife will be carrying and transmitting disease.

EO have long been used to visualize habitat degradation, including tracking deforestation with Landsat imagery (Souza et al., 2013) and wildfire effects with instruments aboard the Suomi National Polar-orbiting Partnership (NPP) satellite (Ba et al., 2020). EO can also provide data on human encroachment by measuring urban expansion with MODIS (Mertes et al., 2015) as well as land use change for agricultural intensification, again with Landsat (Kontgis et al., 2015). These data can all inform integrated models of spillover risk that incorporate human, animal, and environmental factors.

A concrete example of the potential for data integration and intervention across the fields of One Health to prevent disease involves logging. Logging activity, aside from destroying habitat, involves many individuals spending a large amount of time in forests, and many loggers hunt bushmeat for supplemental income and food (Poulsen et al., 2009). This latter activity unsustainably decreases wildlife populations (Poulsen et al., 2009) and poses a high risk of spillover (Schilling et al., 2020), as hunters come into close contact with the animal's bodily fluids, facilitating disease transmission. Maps of logging activity combined with species distributions can therefore be used to evaluate the risk of encountering reservoir species and advise loggers accordingly. EO can contribute to both of these parts: as previously discussed, wildlife can be directly tracked or modeled using climate and habitat data, and logging roads and forest degradation can be mapped using LiDAR data (Affek et al., 2017), which can be obtained from the Global Ecosystem Dynamics Investigation (GEDI) instrument orbiting aboard the International Space Station (Rangel Pinagé et al., 2019).

Besides affecting habitat, the indirect effects of humans on wildlife could alter the risk of disease spillover. For instance, artificial nightlight from human electricity use has been shown to alter the activity patterns and behaviors of birds and bats (Sanders et al., 2021), which could change the degree of potential human exposure and contact with these reservoirs. Nightlight can be measured with Suomi NPP's Visible Infrared Imaging Radiometer Suite (VIIRS) and can be integrated with animal tracking data (Ditmer et al., 2021) for a comprehensive look at behavior change and possible risks. The same principles can be applied to other indirect human effects, such as noise pollution.

The human side of the One Health trio is the one over which we have the most control: through innovative scientific investigation, we can track how our behaviors and policies with regards to nature affect our own health. Integrating human data from varied and diverse sources with those on animals and the environment is critical to creating policies and campaigns to prevent disease emergence and spread.

Discussion

Even in the most densely populated cities, humans are deeply intertwined with the natural world in a rich and beautiful web. When we tug on one strand, the reverberations echo in directions we might not expect. Cutting down trees for timber might fulfill an economic need, but will also destroy habitat, driving reservoir species closer to humans and livestock and increasing the risk of disease transmission. The decisions we make in our stewardship of the natural world are critical to not only animal and environmental health, but also human health.

Earth observations can play a key role in this mission—from tracking individuals within reservoir species, to contributing climate data to species distribution models, to measuring human presence and expansion, satellites have become a critical source of information. EO data can be combined across these different fields to better estimate the disease burden and risk of spillover. This will allow us to employ preventative measures and hopefully avoid the next global pandemic.

While the number of these interdisciplinary projects is still small, there are several past examples of disease prevention projects that combine data across the areas of One Health. EO played a crucial role in an avian influenza early warning system, which was developed as a collaborative project between scientists at NASA, the US Department of Agriculture, and the Department of Defense (*Avian Influenza Early Warning System: Using NASA Data to Predict Pandemics*, 2010). Avian influenza has unique spillover potential, as it is spread by highly mobile waterbirds, which can transmit the virus to flocks of domestic fowl (Avian Influenza Early Warning System, 2010). It can then more easily make the jump to human farmers, where it can cause severe illness: death rates between strains range from 2.5% to 50% (Skeik & Jabr, 2008).

The NASA models projected influenza outbreaks in time and space, relying on a comprehensive combination of data (Avian Influenza Early Warning System, 2010). Climate variables, including temperature, rainfall, radiance, and NDVI, were obtained from NASA satellites and integrated with maps of migratory flyways, species prevalence data, and local socioeconomic conditions to create a weekly forecast of illness (Avian Influenza Early Warning System, 2010). This type of interdisciplinary scientific work, which touches on each of the three One Health fields, will be critical in the future to avert possible pandemics. The framework of this system can hopefully be expanded to other zoonotic diseases in the future, such as those carried by bats, another highly mobile species.

Attempts to elucidate the social and environmental conditions surrounding spillover events are similarly critical to preventing future disease outbreaks. This work has been conducted with *Ebolaviruses*, which are thought to have their reservoirs in bats; confirmed spillovers were found to be associated with locations around forest edges and contact with bushmeat, with different strains exhibiting different characteristics (Judson et al., 2016). EO have also been employed in this space, with Landsat imagery supporting an association of Ebola outbreaks with human-intruded tropical forest areas (Tucker et al., 2002). Spillovers of Hendra virus (McFarlane et al., 2011), Nipah virus (McKee et al., 2021), and human monkeypox (Fuller et al., 2011) have also been associated through EO and other data with various environmental, human, and wildlife variables. This type of analysis, if applied more widely to known zoonoses, can be used to create spillover forecasts, warning people of the risk and therefore decreasing the likelihood of disease occurrence.

As the One Health space continues to grow and evolve, new combinations of data will lend us yet more insight on how we can predict and prevent disease. However, the results then need to be applied through policy and public education in order to be effective. To date, studies have emphasized that human-driven climate change and habitat destruction are not only devastating ecosystems, but also putting our own health at risk by increasing the burden of disease and the probability of spillover. Worldwide, we must reframe how we look at the natural world, transitioning from a lens of exploitation to one of conservation and value. This should be reflected in legislation and funding on local, national, and international scales.

The COVID-19 pandemic has also underscored just how interconnected humans are as well. Disease proliferation anywhere is a threat to people everywhere on Earth and, therefore, the ethos of collaboration across borders and disciplines that has arisen during the pandemic should continue to be supported after we emerge. This is facilitated by databases like Movebank and Wildlife Insights, which make data widely available to scientists around the world. Investigations that focus on human-animal interaction will also need to incorporate social, political, and economic data to a higher degree in order to more fully depict the complexities of human society—and this will necessitate changes beyond the scientific.

For instance, simply legislating against hunting bushmeat is not a feasible mechanism to reduce the risk of spillover that accompanies it. Not only have attempts to eliminate the practice without significant control measures been ineffective in the past (Schilling et al., 2020) but this approach harms communities that rely on bushmeat for subsistence (Brashares et al., 2011), rather than targeting the trade driven by higher-wealth areas that accounts for more of the overhunting and thus the spillover risk. Researchers must investigate what drives hunters to ply their trade, and legislators must pick up the mantle of addressing the roots of the issue (which are generally economic) rather than the symptoms. These principles apply to many of the human-animal interactions that may facilitate spillover events, including logging, agriculture, and construction in wild areas.

No one area of science is sufficient to predict and prevent zoonotic disease emergence and burden worldwide, just as no single government or organization can completely eliminate risk through legislation. In order to prevent the next pandemic, we must look beyond the boundaries of disciplines and nations, combining data and collaborating to protect the health of people, animals, and the environment worldwide.

References

About Icarus. (n.d.). Retrieved December 6, 2021, from https://www.icarus.mpg.de/28056/about-icarus

Affek, A. N., Zachwatowicz, M., Sosnowska, A., Gerlée, A., & Kiszka, K. (2017). Impacts of modern mechanised skidding on the natural and cultural heritage of the Polish Carpathian Mountains. *Forest Ecology and Management, 405*, 391–403. https://doi.org/10.1016/j.foreco.2017.09.047

Avian Influenza Early Warning System: Using NASA Data to Predict Pandemics. (2010). NASA Langley Research Center.

Avŝiĉ-Zupanc, T. (2013). Mosquito-borne diseases—A new threat to Europe? *Clinical Microbiology and Infection, 19*(8), 683–684. https://doi.org/10.1111/1469-0691.12215

Ba, R., Song, W., Lovallo, M., Lo, S., & Telesca, L. (2020). Analysis of multifractal and organization/order structure in Suomi-NPP VIIRS normalized difference vegetation index series of wildfire affected and unaffected sites by using the multifractal detrended fluctuation analysis and the Fisher–Shannon analysis. *Entropy, 22*(4), 415. https://doi.org/10.3390/e22040415

Bartlow, A. W., Manore, C., Xu, C., Kaufeld, K. A., Valle, S. D., Ziemann, A., Fairchild, G., & Link to external site, this link will open in a new window, Fair, J. M., & Link to external site, this link will open in a new window. (2019). Forecasting zoonotic infectious disease response to climate change: Mosquito vectors and a changing environment. *Veterinary Sciences, 6*(2), 40. http://dx.doi.org.ezproxy.princeton.edu/10.3390/vetsci6020040

Brashares, J. S., Golden, C. D., Weinbaum, K. Z., Barrett, C. B., & Okello, G. V. (2011). Economic and geographic drivers of wildlife consumption in rural Africa. *Proceedings of the National Academy of Sciences, 108*(34), 13931–13936. https://doi.org/10.1073/pnas.1011526108

Chuang, T.-W., Henebry, G. M., Kimball, J. S., VanRoekel-Patton, D. L., Hildreth, M. B., & Wimberly, M. C. (2012). Satellite microwave remote sensing for environmental modeling of mosquito population dynamics. *Remote Sensing of Environment, 125*, 147–156. https://doi.org/10.1016/j.rse.2012.07.018

Da Re, D., De Clercq, E. M., Tordoni, E., Madder, M., Rousseau, R., & Vanwambeke, S. O. (2019). Looking for ticks from space: Using remotely sensed spectral diversity to assess Amblyomma and Hyalomma tick abundance. *Remote Sensing, 11*(7), 770. https://doi.org/10.3390/rs11070770

Ditmer, M. A., Stoner, D. C., Francis, C. D., Barber, J. R., Forester, J. D., Choate, D. M., Ironside, K. E., Longshore, K. M., Hersey, K. R., Larsen, R. T., McMillan, B. R., Olson, D. D., Andreasen, A. M., Beckmann, J. P., Holton, P. B., Messmer, T. A., & Carter, N. H. (2021). Artificial nightlight alters the predator–prey dynamics of an apex carnivore. *Ecography, 44*(2), 149–161. https://doi.org/10.1111/ecog.05251

eBird Science—EBird. (n.d.). Retrieved December 8, 2021, from https://ebird.org/ebird/science

El-Sayed, A., & Kamel, M. (2021). Coronaviruses in humans and animals: The role of bats in viral evolution. *Environmental Science and Pollution Research, 28*(16), 19589–19600. https://doi.org/10.1007/s11356-021-12553-1

Fuller, T., Thomassen, H. A., Mulembakani, P. M., Johnston, S. C., Lloyd-Smith, J. O., Kisalu, N. K., Lutete, T. K., Blumberg, S., Fair, J. N., Wolfe, N. D., Shongo, R. L., Formenty, P., Meyer, H., Wright, L. L., Muyembe, J.-J., Buermann, W., Saatchi, S. S., Okitolonda, E., Hensley, L., et al. (2011). Using remote sensing to map the risk of human monkeypox virus in the Congo Basin. *EcoHealth, 8*(1), 14–25. https://doi.org/10.1007/s10393-010-0355-5

Giordano, B. V., Turner, K. W., & Hunter, F. F. (2018). Geospatial analysis and seasonal distribution of West Nile virus vectors (Diptera: Culicidae) in Southern Ontario, Canada. *International Journal of Environmental Research and Public Health, 15*(4), 614. https://doi.org/10.3390/ijerph15040614

Goldberg, T. L., Gillespie, T. R., Rwego, I. B., Estoff, E. L., & Chapman, C. A. (2008). Forest fragmentation as cause of bacterial transmission among nonhuman primates, humans, and livestock, Uganda. *Emerging Infectious Diseases, 14*(9), 1375–1382. https://doi.org/10.3201/eid1409.071196

Home | Wildlife Insights. (n.d.). Retrieved December 7, 2021, from https://www.wildlifeinsights.org/

INaturalist. (n.d.). INaturalist. Retrieved December 8, 2021, from https://www.inaturalist.org/

Judson, S. D., Fischer, R., Judson, A., & Munster, V. J. (2016). Ecological contexts of index cases and spillover events of different ebolaviruses. *PLoS Pathogens, 12*(8), e1005780. https://doi.org/10.1371/journal.ppat.1005780

Kamerow, D. (2014). The world's deadliest animal. *BMJ: British Medical Journal*, 348. http://www.jstor.org/stable/26514797

Klotz, S. A., Dorn, P. L., Mosbacher, M., & Schmidt, J. O. (2014). Kissing bugs in the United States: Risk for vector-borne disease in humans. *Environmental Health Insights, 8s2*, EHI.S16003. https://doi.org/10.4137/EHI.S16003

Kontgis, C., Schneider, A., & Ozdogan, M. (2015). Mapping rice paddy extent and intensification in the Vietnamese Mekong River Delta with dense time stacks of Landsat data. *Remote Sensing of Environment, 169*, 255–269. https://doi.org/10.1016/j.rse.2015.08.004

Kranstauber, B., Cameron, A., Weinzerl, R., Fountain, T., Tilak, S., Wikelski, M., & Kays, R. (2011). The Movebank data model for animal tracking. *Environmental Modelling & Software, 26*(6), 834–835. https://doi.org/10.1016/j.envsoft.2010.12.005

Kunz, T. H., Braun de Torrez, E., Bauer, D., Lobova, T., & Fleming, T. H. (2011). Ecosystem services provided by bats. *Annals of the New York Academy of Sciences, 1223*(1), 1–38. https://doi.org/10.1111/j.1749-6632.2011.06004.x

Le, P. V. V., Kumar, P., Ruiz, M. O., Mbogo, C., & Muturi, E. J. (2019). Predicting the direct and indirect impacts of climate change on malaria in coastal Kenya. *PLoS One, 14*(2), e0211258. https://doi.org/10.1371/journal.pone.0211258

Li, W., Shi, Z., Yu, M., Ren, W., Smith, C., Epstein, J. H., Wang, H., Crameri, G., Hu, Z., Zhang, H., Zhang, J., McEachern, J., Field, H., Daszak, P., Eaton, B. T., Zhang, S., & Wang, L.-F. (2005). Bats are natural reservoirs of SARS-like coronaviruses. *Science, 310*(5748), 676–679.

Lippi, C. A., Stewart-Ibarra, A. M., Loor, M. E. F. B., Zambrano, J. E. D., Lopez, N. A. E., Blackburn, J. K., & Ryan, S. J. (2019). Geographic shifts in Aedes aegypti habitat suitability in Ecuador using larval surveillance data and ecological niche modeling: Implications of climate change for public health vector control. *PLoS Neglected Tropical Diseases, 13*(4), e0007322. https://doi.org/10.1371/journal.pntd.0007322

McFarlane, R., Becker, N., & Field, H. (2011). Investigation of the climatic and environmental context of Hendra virus spillover events 1994–2010. *PLoS One, 6*(12), e28374. https://doi.org/10.1371/journal.pone.0028374

McKee, C. D., Islam, A., Luby, S. P., Salje, H., Hudson, P. J., Plowright, R. K., & Gurley, E. S. (2021). The ecology of Nipah virus in Bangladesh: A nexus of land-use change and opportunistic feeding behavior in bats. *Viruses, 13*(2), 169. https://doi.org/10.3390/v13020169

Mertes, C. M., Schneider, A., Sulla-Menashe, D., Tatem, A. J., & Tan, B. (2015). Detecting change in urban areas at continental scales with MODIS data. *Remote Sensing of Environment, 158*, 331–347. https://doi.org/10.1016/j.rse.2014.09.023

Meurens, F., Dunoyer, C., Fourichon, C., Gerdts, V., Haddad, N., Kortekaas, J., Lewandowska, M., Monchatre-Leroy, E., Summerfield, A., Wichgers Schreur, P. J., van der Poel, W. H. M., & Zhu, J. (2021). Animal board invited review: Risks of zoonotic disease emergence at the interface of wildlife and livestock systems. *Animal, 15*(6), 100241. https://doi.org/10.1016/j.animal.2021.100241

Moore, N., & Messina, J. (2010). A landscape and climate data logistic model of tsetse distribution in Kenya. *PLoS One, 5*(7), e11809. https://doi.org/10.1371/journal.pone.0011809

O'Shea, T. J., Cryan, P. M., Cunningham, A. A., Fooks, A. R., Hayman, D. T. S., Luis, A. D., Peel, A. J., Plowright, R. K., & Wood, J. L. N. (2014). Bat flight and zoonotic viruses. *Emerging Infectious Diseases, 20*(5), 741–745. https://doi.org/10.3201/eid2005.130539

One Health. (n.d.). *OIE - World Organisation for animal health*. Retrieved December 8, 2021, from https://www.oie.int/en/what-we-do/global-initiatives/one-health/

One Health | CDC. (2021, November 17). https://www.cdc.gov/onehealth/index.html

Parrish, C. R., Murcia, P. R., & Holmes, E. C. (n.d.). Influenza virus reservoirs and intermediate hosts: Dogs, horses, and new possibilities for influenza virus exposure of humans. *Journal of Virology, 89*(6), 2990–2994. https://doi.org/10.1128/JVI.03146-14

Plowright, R. K., Foley, P., Field, H. E., Dobson, A. P., Foley, J. E., Eby, P., & Daszak, P. (2011). Urban habituation, ecological connectivity and epidemic dampening: The emergence of Hendra virus from flying foxes (Pteropus spp.). *Proceedings of the Royal Society B: Biological Sciences, 278*(1725), 3703–3712. https://doi.org/10.1098/rspb.2011.0522

Poulsen, J. R., Clark, C. J., Mavah, G., & Elkan, P. W. (2009). Bushmeat supply and consumption in a tropical logging concession in Northern Congo. *Conservation Biology, 23*(6), 1597–1608. https://doi.org/10.1111/j.1523-1739.2009.01251.x

Prevention, C.-C. for D. C. and. (2020, July 16). *CDC - Malaria—About Malaria—Biology*. https://www.cdc.gov/malaria/about/biology/index.html

Rabozzi, G., Bonizzi, L., Crespi, E., Somaruga, C., Sokooti, M., Tabibi, R., Vellere, F., Brambilla, G., & Colosio, C. (2012). Emerging zoonoses: The "One Health approach.". *Safety and Health at Work, 3*(1), 77–83. https://doi.org/10.5491/SHAW.2012.3.1.77

Rangel Pinagé, E., Keller, M., Duffy, P., Longo, M., dos-Santos, M. N., & Morton, D. C. (2019). Long-term impacts of selective logging on Amazon forest dynamics from multi-temporal airborne LiDAR. *Remote Sensing, 11*(6), 709. https://doi.org/10.3390/rs11060709

Robertson, B., Holland, J., & Minot, E. (2012). Wildlife tracking technology options and cost considerations. *Wildlife Research, 38*, 653–663. https://doi.org/10.1071/WR10211

Sagurova, I., Ludwig, A., Ogden, N. H., Pelcat, Y., Dueymes, G., & Gachon, P. (n.d.). Predicted northward expansion of the geographic range of the tick vector Amblyomma americanum in North America under future climate conditions. *Environmental Health Perspectives, 127*(10), 107014. https://doi.org/10.1289/EHP5668

Sanders, D., Frago, E., Kehoe, R., Patterson, C., & Gaston, K. J. (2021). A meta-analysis of biological impacts of artificial light at night. *Nature Ecology & Evolution, 5*(1), 74–81. https://doi.org/10.1038/s41559-020-01322-x

Schilling, M. A., Estes, A. B., Eblate, E., Martin, A., Rentsch, D., Katani, R., Joseph, A., Kindoro, F., Lyimo, B., Radzio-Basu, J., Cattadori, I. M., Hudson, P. J., Kapur, V., Buza, J. J., & Gwakisa, P. S. (2020). Molecular species identification of bushmeat recovered from the Serengeti ecosystem in Tanzania. *PLoS One, 15*(9), e0237590. https://doi.org/10.1371/journal.pone.0237590

Skeik, N., & Jabr, F. I. (2008). Influenza viruses and the evolution of avian influenza virus H5N1. *International Journal of Infectious Diseases, 12*(3), 233–238. https://doi.org/10.1016/j.ijid.2007.07.002

Slatculescu, A. M., Clow, K. M., McKay, R., Talbot, B., Logan, J. J., Thickstun, C. R., Jardine, C. M., Ogden, N. H., Knudby, A. J., & Kulkarni, M. A. (2020). Species distribution models for the eastern blacklegged tick, Ixodes scapularis, and the Lyme disease pathogen, Borrelia burgdorferi, in Ontario, Canada. *PLoS One, 15*(9), e0238126. https://doi.org/10.1371/journal.pone.0238126

Souza, J., Siqueira, J. V., Sales, M. H., Fonseca, A. V., Ribeiro, J. G., Numata, I., Cochrane, M. A., Barber, C. P., Roberts, D. A., & Barlow, J. (2013). Ten-year landsat classification of deforestation and forest degradation in the Brazilian Amazon. *Remote Sensing, 5*(11), 5493–5513. https://doi.org/10.3390/rs5115493

Suzán, G., Marcé, E., Giermakowski, J. T., Armién, B., Pascale, J., Mills, J., Ceballos, G., Gómez, A., Aguirre, A. A., Salazar-Bravo, J., Armién, A., Parmenter, R., & Yates, T. (2008). The effect of habitat fragmentation and species diversity loss on hantavirus prevalence in Panama. *Annals of the New York Academy of Sciences, 1149*(1), 80–83. https://doi.org/10.1196/annals.1428.063

The Internet of Animals. (n.d.). Retrieved December 6, 2021, from https://www.icarus.mpg.de/28546/icarus-internet-of-animals

Tsantalidou, A., Parselia, E., Arvanitakis, G., Kyratzi, K., Gewehr, S., Vakali, A., & Kontoes, C. (2021). MAMOTH: An earth observational data-driven model for mosquitoes abundance prediction. *Remote Sensing, 13*(13), 2557. https://doi.org/10.3390/rs13132557

Tucker, C., Wilson, J., Mahoney, R., Anyamba, A., Linthicum, K., & Myers, M. (2002). Climatic and ecological context of the 1994-1996 Ebola outbreaks. *Photogrammetric Engineering and Remote Sensing, 68*.

Vector-borne diseases. (n.d.). Retrieved December 7, 2021, from https://www.who.int/news-room/fact-sheets/detail/vector-borne-diseases

Wang, L.-F., & Eaton, B. T. (2007). Bats, civets and the emergence of SARS. *Wildlife and Emerging Zoonotic Diseases: The Biology, Circumstances and Consequences of Cross-Species Transmission, 315*, 325–344. https://doi.org/10.1007/978-3-540-70962-6_13

Wu, X., Pang, Y., Luo, B., Wang, M., & Feng, J. (2019). Function of distress calls in least horseshoe bats: A field study using playback experiments. *Acta Chiropterologica, 20*(2), 455–464. https://doi.org/10.3161/15081109ACC2018.20.2.015

Chapter 22
Enabling Accelerated Research in Times of Need: The National Science Foundation's Response to COVID-19 in 2020

Ashley M. Pierce and Amanda R. Shores

Personal Story

When the pandemic struck, both Ashley Pierce and Amanda Shores were transitioning into the realm of science policy and funding as AAAS Science & Technology Policy (S&TP) Fellows at the National Science Foundation. Ashley was 6 months into the fellowship and Amanda was going through the fellowship interview process while teaching at Colorado State University when lockdown commenced in the United States. The transition to at-home work was stark; Ashley had just finished a two-day in-person review panel in the NSF building in Alexandria, VA, not realizing it would be the last time she would enter the building for 8 months. Amanda had just sent her two classes off to spring break not realizing it would be the last time she saw her students in person that semester and before she started her fellowship. Instead of the intense, week-long interview process in DC that historically was the rave experience for potential AAAS S&TP fellows, Amanda and the other interviewing fellows learned about potential fellowship opportunities over numerous zoom meetings. At NSF, Ashley was busy collecting and tracking all the data (COVID-19 research inquiries and their outcomes) coming into the Engineering Directorate and coordinating and analyzing all of NSF's COVID-19 RAPID data collection. This required tracking thousands of inquiries for RAPID research that came into NSF, helping respond to and direct inquiries to the correct program at NSF and analyzing the outcomes of these

Ashley M. Pierce Equally Contributed for this Chapter.

A. M. Pierce (✉)
Unaffiliated, Homewood, CA, USA
e-mail: ash.pie4@gmail.com

A. R. Shores (✉)
Ecosystem Science and Sustainability, Colorado State University, Fort Collins, USA
e-mail: amanda.shores@futureearth.org

M. Laituri et al. (eds.), *The Geographies of COVID-19*, Global Perspectives on Health Geography, https://doi.org/10.1007/978-3-031-11775-6_22

inquiries. When Amanda started her fellowship at NSF, she was interested in how funding was used to support research addressing the pandemic and reached out to Ashley as the person who made a critical impact in tracking such projects. Working together, they developed this chapter to present a qualitative analysis of COVID-19 research funded at NSF for FY 2020. While Amanda has still never entered the NSF building, working with Ashley gave her a sense of connection to NSF and the people there. It has been an incredible experience to, in a small way, help respond to a global pandemic and heartening to regularly read about NSF-funded researchers in the news who are helping to better understand COVID-19. *Ashley Pierce and Amanda Shores*

Introduction

This chapter explores the role of the US National Science Foundation (NSF) in supporting accelerated research in times of urgent need, with a look at the recent history of disruptors, including the COVID-19 pandemic. The NSF supports 27% of the United States' total federal budget for basic research and provides the majority of support for social science research (National Science Foundation, 2021b). At the time this chapter was written, the NSF was organized into seven Directorates, which broadly correspond to general scientific disciplines, in addition to the Office of the Director (OD) which manages special funding programs that are new or do not fit in a specific Directorate (the new Technology, Innovation and Partnerships (TIP) Directorate launched in 2022). However, many topics, including COVID-19, cross both disciplines and geographies and, thus, COVID-19 projects are funded by all Directorates and by the OD. Congress appropriated USD 75 million directly to NSF to support research on COVID-19 through the Coronavirus Aid, Relief, and Economic Security (CARES) Act (2020) allowing NSF to support more projects than awarded in a typical fiscal year (National Science Foundation, 2020a). Inquiries to NSF from the research community for funding for COVID-19-related research began in January 2020, before the scientific name of SARS-CoV-2 had been established, and the first NSF-funded research officially began March 1, 2020. To shed light on how the research community and the NSF respond to urgent research needs we (1) evaluate 10 years of NSF Rapid Response Research (RAPID) funding and Dear Colleague Letters (DCLs) that highlight RAPID research opportunities, (2) conduct a broad characterization of COVID-19 research supported across NSF Directorates through RAPID and other funding mechanisms in fiscal year (FY) 2020, and (3) spotlight COVID-19 projects that demonstrate the inter-/transdisciplinary and spatial approaches necessary to understand and alleviate societal challenges related to the pandemic.

History and Evolution of RAPIDs

RAPIDs are a unique NSF funding mechanism that enable the agency to support research with a severe urgency regarding availability of or access to data, facilities, or specialized equipment, particularly during unanticipated events, such as natural or anthropogenic disasters (National Science Foundation, 2021a). The award mechanism preceding RAPIDs was the Small Grants for Exploratory Research (SGER) funding, which was introduced in 1989 and provided support at any time up to USD 200k for up to 2 years (McCullough, 2007). During an interview, Dennis Wenger explained based on his prior experience as Program Director for Infrastructure Systems Management and Extreme Events at NSF, SGERs were first awarded to researchers in 2001 to study the impacts from the September 11th attacks on the World Trade Center in New York City and the US Pentagon in Virginia. However, there was no coordinated call for SGERs on a specific topic until 2005 when the first DCL for rapid research called for SGERs focused on Hurricane Katrina (Wenger, 2021 & Institute of Medicine, 2015).

In 2009, SGERs were reconceptualized into both RAPIDs and EArly-concept grants for Exploratory Research (EAGERs), distinguishing between the need for urgent research during disruptive events and potentially transformative, early-stage research for promising, untested, exploratory ideas. EAGERs support proposals up to USD 300k for 2 years and can be requested at any time. RAPID funding can also be requested at any time and can provide up to USD 200k for up to 1 year. Both EAGER and RAPID proposals may be reviewed internally, though program officers may obtain external review as needed. Both RAPIDs and EAGERs require an initial email explaining the research to the relevant funding program at NSF and undergo an initial internal vetting before researchers are invited to submit a full proposal. This unique process allows for a shorter time (1–5 weeks) between submission of a proposal and fund disbursement compared to 6 months on average for standard research grants at the NSF or 3–20 months depending on the grant mechanism at the National Institutes of Health research grants (National Science Foundation, 2021a; Office of Budget, Finance and Award Management, 2021; and National Institutes of Health, 2020). The fast response rate of the NSF's RAPID mechanism provided an important first response to the pandemic.

Approach

Annual reports on the NSF Merit Review Process were used to examine funding of RAPID proposals from FY 2010 to 2019, where a fiscal year starts October 1 and ends September 30. The Merit Review report for FY 2020 RAPID funding had not been published at the time of writing this chapter and is not included here. The nsf.gov database search engine was used to identify all NSF-funded COVID-19 projects in FY 2020 and to capture all DCLs directing researchers towards RAPID funding from FY 2010 to 2020. COVID-19 awards were included in this analysis if the funded project was focused on understanding the virus and/or the impacts of the pandemic. Projects tangentially related to COVID-19 (e.g., projects for which the award abstract mentions COVID-19 but that do not directly address the virus or impacts from the pandemic) were considered beyond the scope of this analysis. The authors divided unique funded projects into research categories. Projects that demonstrated the social, transdisciplinary, and geospatial nature of addressing the pandemic are highlighted and described based on information gathered through interviews with the researchers, publications, websites, and publicly available abstracts.

Findings

RAPID DCL History

DCLs are one way in which NSF directs researchers to research topics of interest, communicating opportunities in one or multiple Directorates or offices for potential funding. Though DCLs are not used in all circumstances for various reasons, in some critical circumstances, DCLs are used to direct researchers towards RAPID research (here referred to as RAPID DCLs, Fig. 22.1). Publishing a RAPID DCL is an action employed in special situations as a communication tool and has been used a total of 14 times from FY 2010 to FY 2020 (Fig. 22.1). While six of the 14 DCLs called only for RAPID research proposals, the other eight called for a variety of funding mechanisms in addition to RAPIDs, including programmatic proposals, EAGERS, and supplemental funding to ongoing research. In all, there were 11 RAPID DCLs during this time focusing on natural crises, which can be further broken down into categories of disease (five, including two on COVID-19 in FY 2020), earthquakes (three), and hurricanes (three). The other three RAPID DCLs focused on man-made challenges (two focused on oil spills and one on nanoplastics). The first RAPID DCL that focused on disease before the COVID-19 pandemic addressed Ebola in 2015. The 14 DCLs called for research spanning large geographic areas from regions within a country to global. Across these 14 DCLs that include RAPID research, six had participation from all of the NSF Directorates and ten had participation from at least two NSF Directorates. The participation of multiple Directorates likely indicates that topics span multiple disciplines.

FY 2020 was an anomalous year compared to FY 2010–2019 regarding RAPID funding trends (Fig. 22.1). The previous maximum number of RAPID awards was nearly 300 in FY 2010, coinciding with the previous maximum number of RAPID DCLs in a year (three). In FY 2020, 880 RAPID proposals were funded, 805 of which focused on COVID-19. The additional USD 75 million appropriated by Congress in the CARES Act to address COVID-19 allowed the NSF to fund more RAPIDs and EAGERs on this topic than would have been possible otherwise. Prior to this, the most recent instance of money being appropriated for NSF outside of the

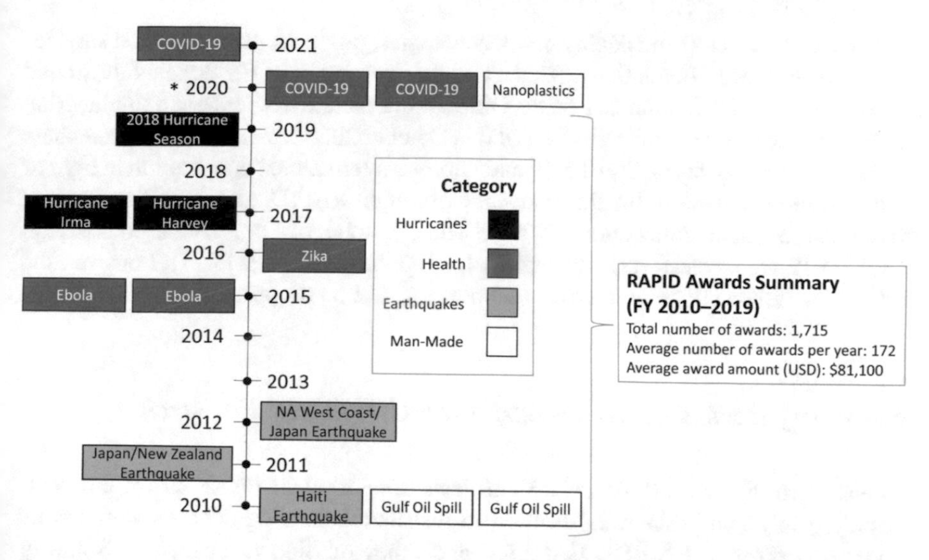

Fig. 22.1 Timeline of NSF DCLs that direct researchers to RAPID (RAPID DCL) research published during FY 2010–2020 including October 1, 2020, from FY 2021. The breakout box displays a summary of the total number of RAPID awards, the mean number of awards funded per year, and the mean award amount for FY 2010–2019. Summary numbers are generated from Merit Review Reports for FY 2010–2019. Multiple DCLs on a specific topic address different aspects of importance
*The DCL issued by the Division of Industrial Innovation and Partnerships (IIP) in the Engineering Directorate did not call for RAPID research and is not included in this figure

normal budget cycle was the 2009 American Recovery and Reinvestment Act (ARRA) following the Great Recession of 2008 (National Science Foundation, 2014).

COVID-19 DCLs

Three DCLs focused on COVID-19 research were released by the NSF in FY 2020, with two being RAPID DCLs. On March 5, 2020, the first COVID-19-related, NSF-wide DCL was published calling for nonmedical, non-clinical-care research proposals to understand the spread of COVID-19, to educate on virus transmission and prevention, and to encourage processes and actions to address COVID-19, including standard and RAPID research opportunities (National Science Foundation, 2020b). The Office of Advanced Cyberinfrastructure (OAC) in the Directorate for Computer and Information Science and Engineering (CISE) released the next COVID-19 DCL calling for projects that addressed COVID-19 challenges through data and/or software infrastructure development through RAPID and supplemental funding to ongoing research (National Science Foundation, 2020c). The third DCL focused on COVID-19 was issued by the Directorate for Engineering's Division of Industrial Innovation and Partnerships (IIP) and focused on non-RAPID seed funding for US-based small businesses to develop and deploy new technologies, products, processes, and services (National Science Foundation, 2020d). Funding continued in FY 2021 as illustrated by a COVID-19 DCL published by the Office of International Science and Engineering (OISE) in the Office of the Director (OD) in October 2020 which directed researchers to explore the impacts of COVID-19 and improve international cooperation in light of the disruptions caused by the pandemic through RAPID and EAGER funding (National Science Foundation, 2020e). Thus, a total of four DCLs focused on COVID-19 were released (including the DCL published in FY 2021); however, the DCL released by IIP did not provide support for RAPID proposals.

Summary Across All NSF-Supported COVID-19 Research

There were 991 COVID-19-related projects supported in FY 2020 by the NSF, resulting in about USD 162.4 million in distributed funding. These awards were supported through RAPIDs, EAGERs, and other funding mechanisms. Some of these awards were collaborative, meaning multiple awards were made to different researchers working together on the same project. After condensing the collaborative awards with their lead project, there were 859 unique projects funded to address the COVID-19 pandemic in FY 2020, the focus of the next section. In FY 2020, NSF-supported COVID-19 research focused on fundamental research to understand the virus and predict spread, develop novel ways to address and mitigate the spread of the virus, and understand the impacts of the pandemic on society, vulnerable communities, and built and natural environments. COVID-19 research projects were categorized by the authors into 11 research topics: (1) biology of the virus and zoonosis; (2) human infection, viral detection, and antiviral designs; (3) transmission, contact tracing, and epidemiological modeling; (4) effects on and tools for healthcare and health outcomes; (5) impacts of human distancing behavior; (6) public policy and decision-making; (7) risk perception and behavior; (8) information access and misinformation; (9) supply chain and food security; (10) education and online instruction; and (11) impacts on ecosystems, environment, and infrastructure. Categories such as biology of the virus tended to be narrower in scope, and other categories such as impacts on ecosystems, environment, and infrastructure and risk perception and behavior were broader (Fig. 22.2). Nearly a quarter (23%) of all funded projects fell within the human infection, viral detection, and antiviral design category, followed by transmission, contact tracing, and epidemiological modeling (16%), human distancing behavior (12%), and effects on and tools for healthcare and health outcomes (11%).

Based on a simple text search for "geospatial" and "gis," research that utilized geospatial approaches and perspectives were found across 9 of the 11 categories, with many of these projects falling within the transmission, contact tracing, and epidemiological modeling category, public policy and decision-making category, and human distancing behavior category. Each NSF Directorate tended to make awards across many of the 11 categories, indicating a cross-disciplinary approach

Category	Project Examples	BIO n=138	CISE n=133	EHR n=51	ENG n=251	GEO n=23	MPS n=57	SBE n=191	OD n=16
Biology of the virus and zoonosis (n=68)	Using CRISPR to understand zoonotic transmission of the virus (Project ID: 2032072) Understanding the structure and binding of the SARS-CoV-2 Spike Protein (Project ID: 2028803) Understanding immunity in bats, natural coronavirus reservoirs (Project ID: 2032157)	59					9		
Human infection, viral detection, and antiviral designs (n=200)	Role of mucus in virus pathogenesis (Project ID: 20330465) Sensors for detecting the virus in the air (Project ID: 2031142) identifying antimicrobial coatings for high touch surfaces (Project ID: 2040273)	31	24		110	2	27	3	3
Transmission, contact tracing, and epidemiological modeling (n=141)	Viral movement, transfer, and persistence of COVID-19 in droplets, aerosols, and surfaces (Project ID: 2027306) Privacy-preserving contact tracing (Project ID: 2027647) Developing indicators of real-time spread of COVID-19 using municipal wastewater samples (Project ID: 2027679)	31	40	2	39	1	13	15	
Effects on and tools for healthcare and health outcomes (n=95)	Contact-free healthcare with video sensing (Project ID: 2030502) Used face mask disinfection to increase access to face masks during supply limits (Project ID: 2031223) Social interactions and health outcomes during the pandemic (Project ID: 2030074)		14		45		6	28	3
Impacts of human distancing behavior (n=105)	Social vulnerabilities and impacts of concurrent weather hazards and COVID-19 (Project ID: 2029918) Health and socioeconomic well-being of minorities in rural meatpacking communities (Project ID: 2033920) Stay-at-home attitudes and impacts on the pandemic (Project ID: 2031095)	4	22	2	10	3		61	2
Public policy and decision making (n=51)	Accelerated prison release dates on crime and COVID-19 transmission (Project ID: 2029890) Effects of crises on public perception of the rule of law (Project ID: 2027653) Ethical implications of policy changes during a pandemic (Project ID: 2032752)	1	6	2	4	3		33	2
Risk perception and behavior (n=44)	Risk perception in response to social distancing (Project ID: 2028412) Behavior, perception, and policies across geographic and economic gradients (Project ID: 2028297) Factors that affect understanding the risks of COVID -19 (Project ID: 2027822)		2	1	10	2	1	28	
Information access and misinformation (n=22)	Information access and practices to reduce the spread of COVID-19 in low-income areas (Project ID: 2027470) How misinformation spreads (Project ID: 2031768) Improving information spread to vulnerable populations (Project ID: 2029880)		5	2	2			12	1
Education and online instruction (n=75)	Addressing Impacts of remote learning and equity in remote STEM teaching (Project ID: 2029642) Faculty adaptability and community engagement when teaching in a crisis (Project ID: 2027471) Impacts of unprecedented shift to online learning on students' cognitive load (Project ID: 2027637)	4	7	42	10	4	1	6	1
Supply chain and food security (n = 21)	Supply chains for medical and personal protective equipment (Project ID: 2027927) Food production, supply, and security (Project ID: 2032065) Online platforms matching supply and demand of medical equipment (Project ID: 2029072)		3		9	1		4	4
Impacts on ecosystems, environment, and infrastructure (n=37)	Impact of ship noise reduction on whale health (Project ID: 2032896) Stream water quality (Project ID: 2031614) Impacts of stay-at-home orders on mobile networks (Project ID: 2027650) Impacts of stay-at-home orders on building plumbing (Project ID: 2027049)	7	10		12	7		1	

Fig. 22.2 Summary of unique COVID-19 projects (n = 859) by research topic (y-axis) and by Directorate and the Office of the Director (OD) at the NSF (x-axis)

from Directorates to address COVID-19 research questions. The Engineering (ENG), Social, Behavioral, and Economic Sciences (SBE), and Computer and Information Science and Engineering (CISE) Directorates spanned the broadest range of COVID-19-related research categories (Fig. 22.2), while Directorates including Biological Sciences (BIO), Mathematics and Physical Sciences (MPS), Education and Human Resources (EHR), and Geosciences (GEO) tended to fund projects that fell in select categories. The research categories that received funding from the most Directorates included the education and online instruction category; the transmission, contact tracing, and epidemiological modeling category; the impacts of human distancing behavior category; and the human infection, viral detection, and antiviral design category.

Research Highlights

COVID-19 projects that demonstrate the inter-/transdisciplinary and spatial approaches necessary to understand and alleviate societal challenges related to the pandemic are highlighted in the following sections.

Highlights from COVID-19 RAPID Response Projects

RAPID funding was the primary funding mechanism for COVID-19 in FY 2020 due to the ephemeral nature of pandemic impacts and response and the need to quickly allocate funds to initiate research. Of the 859 unique projects, 702 were funded as RAPIDs. One such RAPID project (project ID: 2035161) within the supply chain and food security category (Fig. 22.2), supported through the NSF Office of Polar Programs in the Geosciences Directorate, evaluated how COVID-19 has impacted access to food and traditional food knowledge (Duncombe, 2021). This work was conducted by Dr. Stephanie Carroll (Ahtna) and her team at the University of Arizona Native Nations Institute and the University of Colorado. This team is part of a larger Research Coordinating Network (project ID: 1745499) supported by the NSF working on food security obstacles faced by Indigenous communities in the Arctic and the US southwest, which allowed the team to quickly pivot to address the pandemic. The team is coproducing knowledge with Indigenous research partners using participatory, community-driven approaches. The goal of this research is to identify methods for enhancing food-system resilience when these communities are faced with crises and illustrates the importance of having teams with established networks of trust in place before a crisis occurs.

Within the impact on ecosystems, environment, and infrastructure category, research on green-space usage during the pandemic was supported by the Division of Chemical, Bioengineering, Environmental, and Transport Systems in the Engineering Directorate. A project led by Dr. Franco Montalto at Drexel University (project ID: 2027600) supported citizen scientists to use digital technology to observe different behaviors in parks, aiming to understand use of the park by the community, the impacts on vulnerable unhoused populations, and the park's potential as a pathway for transmission. The team also evaluated if residents could be quickly mobilized during crisis situations to aid in data collection.

Within the education and online instruction category, Dr. Laura Rademacher at the University of the Pacific, Stockton, and team led a collective action to share expertise and activities for distanced-based geological learning to address the gap in field-based opportunities (project ID: 2029920). This effort used a variety of geospatial tools as a pathway to provide virtual experiential learning experiences for students.

Falling within the transmission, contact tracing, and epidemiological modeling category, Dr. Tyler Radniecki and team at Oregon State University focused on spatial sampling of SARS-CoV-2 RNA in sewage and stormwater (project ID: 2027679). This information was combined with other geospatial demographic information to determine if, at the community level, wastewater treatment facilities can be used to conduct real-time COVID-19 monitoring.

In the information access and misinformation category, Dr. Elena Zheleva and team from the University of Chicago developed algorithms to detect attitudes regarding stay-at-home orders from social media and integrated this information with geospatial location and socioeconomic data to understand how attitudes across the landscape influenced public health outcomes (project ID: 2031095). These studies demonstrate the broad applicability of inter-/transdisciplinary and geospatial approaches towards addressing the COVID-19 pandemic under a variety of circumstances.

The COVID-19 virus spread and mutated very quickly. New information had to be gathered, integrated, analyzed, and disseminated in near real time. To help facilitate data sharing and collaboration, Dr. Florence Hudson from the Northeast Big Data Innovation Hub hosted by Columbia University, in collaboration with the Midwest Big Data Innovation Hub, the South Big Data Innovation Hub, and the West Big Data Innovation Hub, proposed the COVID Information Commons (CIC, project ID: 2028999). The CIC is intended to be a resource for researchers and decision-makers to leverage findings, invest in, and/or accelerate the most promising research to mitigate the broad societal impacts of the COVID-19 pandemic (COVID Information Commons, 2022). COVID-19 research is compiled with publicly available datasets and organized into categories of research areas and/or geography. The CIC web portal allows project teams to publish or link to their data and present project information in ways that are most relevant and user-friendly for researchers in academia, industry, and government. Though the initial project was a RAPID award for 1 year, the CIC will continue past the initial award period and serve as a model for potential future public health challenges to integrate knowledge sharing and collaboration in times of need.

Highlights from Longer-Term COVID-19 Projects

While RAPID projects allow the research community to quickly develop new data-driven strategies for addressing the pandemic, longer-term projects as well as alternative funding avenues, such as funds for workshops, are needed to help recover and be better prepared in the future. Out of 991 total projects, 157 were funded as EAGERs (60 awards of up to 2 years' duration) or by another funding mechanism (97 awards spanning 1–5 years' duration). A project led by Dr. Julie Chen at the University of Massachusetts, Lowell, was funded to set up transdisciplinary stakeholder networks that can rapidly pivot to produce products at scale without existing supply chains or manufacturing lines (project ID: 2036917). This project, funded by the NSF Future Manufacturing program, focuses on understanding what shared resources and tools and what education and workforce development is needed to quickly pivot manufacturing to address crises. The project will have long-term implications for crisis response, resilience, and preparedness for supply chains into the future.

Another mechanism for developing ideas and building networks is through NSF-funded workshops to bring researchers together around a topic. One such project was funded by the Build and Broaden Program in the Division of Social and Economic Sciences in the Directorate for Social, Behavioral, and Economic Sciences. This award, led by Katia Goldfarb and team, supported a conference to bring together a transdisciplinary group of researchers and community stakeholders to understand the array of long-term impacts faced by the local Hispanic community because of the pandemic (project ID: 2038458). In an interview, Dr. Goldfarb shared that building trust in the community is vital to the pandemic response and that she is concerned that the pandemic has thwarted the Hispanic representation gains made over the last 20 years within STEM fields Goldfarb (2021). She views this conference as a way to broaden participation and strengthen transdisciplinary networks of Hispanic scholars and early career scientists from within the Hispanic community.

Technologies for online evidence-based learning and teaching platforms received newfound direction during the pandemic as a way to effectively teach in a distanced setting. Ashley Rowland from Codon Learning, Inc., received a Small Business and Innovation Research (SBIR) grant from the Division of IIP in the Directorate for Engineering to scale the development of a digital design tool that enables an instructor to quickly create and implement an active learning STEM course (project ID: 2015112). With beta testing originally taking place in six remote biology classrooms, information was collected on the ease of adoption and effectiveness of meeting learning objectives across a wide diversity of students (Rowland, 2021). Investment in this small business will help scale this educational technology that is important for remote learning during pandemics and beyond.

Conclusions

The unique ability for the NSF to quickly distribute research funds for projects focusing on urgent topics allows the NSF to nimbly respond to crises, a process that funneled additional appropriated funds to researchers during the COVID-19 pandemic. Researchers quickly mobilize to request this funding, particularly if their own research is based in, or can quickly pivot to, addressing the current crisis. To further mobilize the research community around urgently needed research topics, the NSF effectively employs DCLs as a communication tool to disseminate funding opportunities. In FY 2020, the use of three DCLs, two of which pointed to RAPID research, was unprecedented in the use of DCLs on a single topic at the NSF, coinciding with the largest amount of RAPID funding distributed from FY 2010 to FY 2020. Certain research categories including education, viral transmission, human infection, and human distancing behavior crossed disciplinary boundaries and were funded across all or most NSF Directorates. The NSF is able to quickly focus funding around urgently needed research to address various national and international disruptive events in an inter- or transdisciplinary manner, with almost half of all RAPID DCL's having participation from all NSF Directorates and the OD.

The impacts of COVID-19 will be felt for many years. What is needed is research that spans decades and monitors ecosystems at the interface of human and natural systems to better understand the effects of this disruption to natural, built, and social systems over the long term. This decadal research will be necessary to understand future stressors and where, why, and when spillover of disease or natural disaster may occur next. This research may require cross-agency, interagency, and international collaboration to cover the many complex facets of a global crisis. The Advisory Committee for Environmental Research and Education (AC-ERE) released a report in 2021 on Environmental and Human Health that suggested that environmental and public health research should be integrated in transdisciplinary research and that institutions should foster research teams necessary for performing such research which may require partnerships with other agencies (Advisory Committee for Environmental Research and Education, 2021).

From human-wildlife interactions increasing zoonotic disease transfer to more frequent, higher-intensity, natural disasters caused by climate change, rapidly deploying research funds to respond to these disruptive events is becoming increasingly necessary for human security and resilience. Our analysis demonstrates that these global challenges involve complex interacting systems, the solution to which requires research integration with society, the economy, and policy across spatial and temporal scales. In a time of rapid change, breaking down barriers to conducting urgently needed, transdisciplinary research is required to support humanity's resilience to crises.

Acknowledgments We would like to thank Dr. Paul Morris, Cecile Gonzalez, Dr. Lina Patino, Dr. Melinda Laituri, Dr. Louise Howe, Dr. Dennis Wenger, Dr. Katia Goldfarb, Dr. Stephanie Russo Carroll (Ahtna), Dr. Noor Johnson, Dr. Daniel Ferguson, Mary Beth Jager (Citizen Potawatomi), and Ashley Roland for their contributions to this paper.

References

Advisory Committee for Environmental Research and Education. (2021). Environmental and human health: Research priorities. *National Science Foundation*. Retrieved from https://www.nsf.gov/ere/ereweb/reports/AC-ERE-Environmental-and-Human-Health-Report_June7-508.pdf

Congress of the United States of America. Coronavirus Aid, Relief, and Economic Security Act, H.R. 748, 117th Cong. (2020). https://www.congress.gov/116/bills/hr748/BILLS-116hr748enr.pdf

COVID Information Commons. (2022). *About the COVID information commons*. Retrieved from https://covidinfocommons.datascience.columbia.edu/content/about-cic

Duncombe, P. (2021). *Network connects indigenous knowledges in the arctic and U.S. southwest*. Retrieved from https://eos.org/articles/network-connects-indigenous-knowledges-in-the-arctic-and-u-s-southwest

Goldfarb, K. P. (2021, February 12). Personal Communication [Personal Interview].

Institute of Medicine. (2015). *Enabling rapid and sustainable public health research during disasters: Summary of a joint workshop*. The National Academies Press.

McCullough, J. (2007). Study of small grants for exploratory research. NSF MOBIS Contract no. GS10F0554N. Requisition DACS-06D186. SRI International.

National Institutes of Health. (2020). *Comparing popular research project grants—R01, R03, or R21*. National Institute of Allergy and Infectious Diseases (NIAID). https://www.niaid.nih.gov/grants-contracts/research-project-grants

National Science Foundation. (2014). *Recovery Act*. https://www.nsf.gov/recovery/

National Science Foundation. (2020a). *Review of the National Science Foundation CARES Act Spending Plan Report No. OIG 20-6-001*. https://www.nsf.gov/oig/_pdf/20-6-001_CARES_Act_Spending_Plan.pdf

National Science Foundation. (2020b). *Dear Colleague Letter on the Coronavirus Disease 2019 (COVID-19)*. https://www.nsf.gov/pubs/2020/nsf20052/nsf20052.jsp

National Science Foundation. (2020c). *Dear Colleague Letter: Provisioning Advanced Cyberinfrastructure to Further Research on the Coronavirus Disease 2019 (COVID-19).* https://www.nsf.gov/pubs/2020/nsf20055/nsf20055.jsp

National Science Foundation. (2020d). *Dear Colleague Letter: Request for SBIR/STTR Phase I Proposals Addressing COVID-19.* https://www.nsf.gov/pubs/2020/nsf20065/nsf20065.jsp

National Science Foundation. (2020e). *Dear Colleague Letter: Future of International Research Collaboration Post COVID-19.* https://www.nsf.gov/pubs/2020/nsf20132/nsf20132.jsp

National Science Foundation. (2021a). Proposal & Award Policies & Procedures Guide (PAPPG). *The National Science Foundation.* https://www.nsf.gov/pubs/policydocs/pappg22_1/index.jsp

National Science Foundation. (2021b). About the National Science Foundation. *The National Science Foundation.* https://www.nsf.gov/about

Office of Budget, Finance and Award Management. (2021). *National Science Foundation.* https://dellweb.bfa.nsf.gov/awdfr3/default.asp

Rowland, A. (2021, July 8). Personal Communication [Personal Interview].

Wenger, D. (2021, February 9). Personal Communication [Personal Interview].

Chapter 23
Conclusion: The Consequences of COVID-19 – What Is Next?

Melinda Laituri, Robert B. Richardson, and Junghwan Kim

In some countries, the poorest people are nearly four time more likely to die from COVID-19 as the richest. Inequality Kills, Oxfam, 2022

The international spread of emerging infectious diseases…[has] created a world where regular outbreak and spillover events are the new norms. Gruetzmacher et al., 2020

The preparation of this volume is bracketed by two events: first, the global pandemic that continues to this day, and second, the emerging conflict in Ukraine where Russia has invaded a sovereign nation. This juxtaposition of global and regional events demonstrates the need to understand the geography of places and consider how geospatial approaches can inform policy and solutions. To address the pandemic, the national governments have put into practice place-based procedures and policies to protect and save lives; we now enter a phase of global uncertainty where conflict upends those policies and lives are lost. The coincidence of the pandemic and conflict compounds the global health emergency particularly in fragile states exposed to conflict and violence where provisioning basic services (such as food, water, and shelter) is essential (Richardson, Chap. 3). As the Ukraine invasion unfolds, we observe in real time the impact of conflict and the emergence of the Ukrainian diaspora and consider the hidden toll the virus will take in concert with violent loss of life. This conflict will hinder and complicate public health responses and this is only one example of a nation-state, Russia advancing a specific agenda

M. Laituri (✉)
Colorado State University, Fort Collins, CO, USA
e-mail: melinda.laituri@colostate.edu

R. B. Richardson
Michigan State University, East Lansing, MI, USA

J. Kim
Virginia Tech, Blacksburg, VA, USA

M. Laituri et al. (eds.), *The Geographies of COVID-19*, Global Perspectives on Health Geography, https://doi.org/10.1007/978-3-031-11775-6_23

with the continued pandemic as a backdrop. Armed conflict results in people on the move and this means the virus will also be on the move. The UN reports that 24 million of the 98 million confirmed to have contracted the virus live in countries facing humanitarian or refugee crises (UN Press Release, 2021), and this number is sure to increase as the conflict rages and expands.

The intersection of COVID-19 and conflict should not be lost as we enter the next stage of the pandemic. In March 2020, the UN called for a global ceasefire to "create opportunities for life-saving aid, open windows for diplomacy and bring hope to people suffering in conflict zones who are particularly vulnerable to the pandemic" (Guterres, 2020). Sheely et al. (2021) of MercyCorps explore this coincidence through interviews with people in Colombia, Nigeria, and Afghanistan representing communities in over 40 countries. Their findings reveal common factors that include distrust in government and institutions, increasing economic hardship, and eroding social cohesion. The World Bank works in 19 of the most fragile places around the world to strengthen health systems and mitigate the pandemic's risks by improving community engagement, addressing the root causes of conflict, and building trust to provisioning of basic services. An underexamined result of pandemic policies is border closures and impacts on movement where international flows of migrant remittances made sending money home difficult due to unemployment and prohibiting worker mobility to work overseas (Economist Intelligence Unit, 2021). Brown and Blanc (2020) of the Carnegie Endowment for International Peace examine 12 conflicts from around the world where fragile states contend with the multiple challenges of the continuing pandemic compounded by sanctions (Syria and Iran), empowerment of non-state actors (Yemen and Somalia), and increasing numbers of conflict-displaced people (Afghanistan and Libya). The Armed Conflict Location & Event Data Project (ACLED) created the COVID-19 Disorder Tracker analyzing changes in demonstration activity, state repression, mob attacks, and overall rates of armed conflict (Kishi, 2021). The tracker and infographic "spotlights" emphasize the consequential geography of conflict and the pandemic.

The consequential geography of the pandemic is further revealed in the indirect effects of the pandemic on violence. Mena et al. (Chap. 12) described the cascading effects of the government lockdown in Quito, Ecuador, across different socioeconomic sectors due to the pandemic. The unintended consequences of protecting health came at the expense of family members under lockdown where the home became a place of danger. The closure of borders in Central America resulted in disruption of criminal economies leading to increasing violence as competition increased between gangs and cartels to carve out new territories of influence (Kishi, 2021). As stay-at-home restrictions created changes in daily life, patterns of criminal activity changed. Nivette et al. (2021) examined 27 cities in 23 different countries and found that generally crime decreased during stay-at-home restrictions, but crime patterns shifted from public-space crime (i.e., theft and assault) to domestic violence in the home and cybercrime in virtual space.

A recurring theme throughout this book is the stark exposure of inequitable places. Our stories echo the results from Oxfam's *Inequality Kills* report where "economic violence" is defined by the structural policy choices made for the richest and most powerful people causing direct harm to the poorest of people, women and girls, and racialized groups (Ahmen et al., 2022). Our case studies include examples from around the world (Fig. 23.1) that reveal how cultural backgrounds and socio-economic challenges are different for distinct populations.

Huang et al. (Chap. 8) and Guhlincoozzi and Wallace (Chap. 9) describe long-standing inequity issues in the United States where socioeconomic disparities disproportionately affect vulnerable populations, specifically African Americans and Latines. Yashadhana et al. (Chap. 15) examined the spatiality of public health restrictions and access to services for Aboriginal and Torres Strait Islanders and the need for culturally relevant approaches to health management. Kebirungi and Mwenyango (Chap. 16) describe the economic and social impacts on the Boda-Boda motorists of Uganda. The World Justice Project (2020) documents the status of Sustainable Development Goal (SDG) 16 – "to promote peaceful and inclusive societies for sustainable development, provide access to *justice for all*, (emphasis added) and build effective, accountable, and inclusive institutions at all levels" (UN Department of Economic and Social Affairs, 2022). The policy brief identifies the "justice gap" that is explicitly linked to place – the geography of justice – where one must have a legal identity to access resources, emergency social protection benefits, and health services.

Several of our case studies focus on cities where transportation networks (Valjarevic, Chap. 17) and the urban economy (Koti, Chap. 19; Karki et al., Chap. 10) were impacted by pandemic policies. Martinez and Short (2021) term this the "pandemic city" and note that the history of cities is closely linked to the history of disease. The establishment of public spaces, sanitation infrastructure, access to clean water, and garbage removal reconfigured cities to be healthier places. However, in the twenty-first century, the city requires a reimaging of urban space for a number

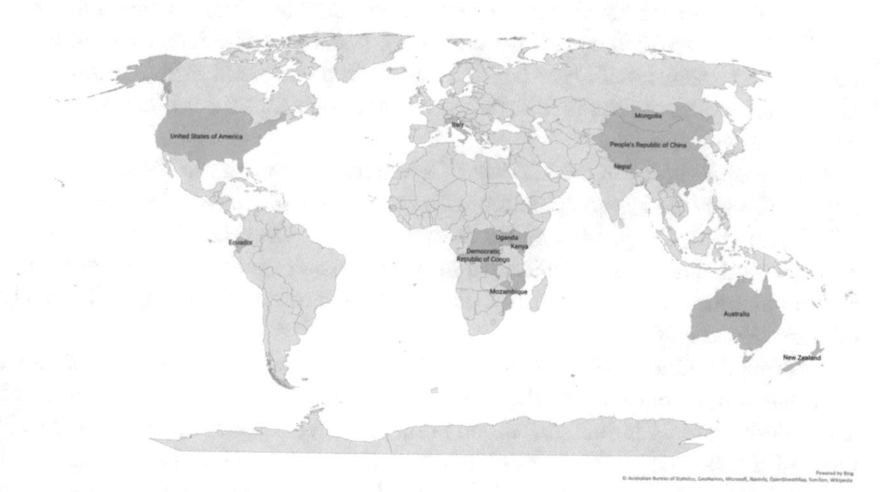

Fig. 23.1 Map of COVID-19 case study countries

of reasons. First, urban expansion has precipitated new connectivities due to land use changes that destroy natural habitats and establishing pathways for disease transmission (Mishra, Chap. 21). Zoonotic diseases have direct trajectories to humans through spillover events from wild animals and spillback effects to both domestic and wild animals (Whalen, 2021; Reaser et al., 2021; Shah, 2022). Second, the twenty-first century is the urban century; by 2050 70% of the world's population will live in cities (Resilient Cities Catalyst, 2022). Population density is linked to disease transmission and cities with large informal settlements will be challenged to provide adequate services (Kim et al., 2022) as well as provide facility-based data for city planning and efforts to meet the Sustainable Development Goals (Mbonglou and John, Chap. 11). Third, cities reflect the global economy. Globalization, a reflection of the interconnectedness of places, has been challenged as borders close and supply chains weaken, exacerbating conditions of inequality. The impact on the informal economy is a key driver for engagement in the local economy in low- and middle-income counties which particularly impacted yet revealed a level of adaptation and resilience for the future (Koti, Chap. 19). Fourth, cities demonstrated how they are places of resiliency. The great pause or anthropause revealed how spaces could change; examples of improved water and air quality were of particular note (Knouft, Chap. 20). Time and space intersected where the temporal design of spaces included timing of road closures on weekends for pedestrian use, expanding the operating hours of public spaces, and increasing the use of roads for pedestrian and bike traffic (Hendron, 2021).

What Is Next?

We revisit the concept of a consequential geography of the pandemic (Laituri, Chap. 2) to envision public health services built upon the One Health approach (Mishra, Chap. 21). We assess the need for data – the backbone of science – to inform the policies and practices of One Health for modeling and monitoring the future. And recommend collaboration in science, policy, and society as the critical framework to facilitate next steps.

One Health builds on the concept of syndemic – referring to the coincidence of multiple factors during an epidemic with socioeconomic and environmental conditions (Campbell, Chap. 14; Courtin & Vineis, 2021). One Health is a collaborative, multisectoral, and transdisciplinary approach – local to global – to achieve optimal health outcomes for both human society and the natural environment (Centers for Disease Control, 2022).

Informing a One Health approach means that we will need data. New data were collected at rapid rates during the pandemic and posted through websites with interactive dashboards. Chuluungaatar and Sumiyasuren (Chap. 7) describe the development of a data dashboard to provide access to health services in underserved and informal areas of Ulaanbaatar and the need for government support to create such resources. However, many of these websites are ephemeral – no longer being updated, broken links, and abandoned. We have learned from the pandemic the need for accurate and near real-time data despite embedded biases and lag times in data generation. We will need to build integrated datasets that reflect the pandemic demographics. Data were collected at the appropriate scale, inclusive of socioeconomic factors (age, ethnicity, gender, vulnerability), sensitive to ensuring privacy and ethical data use (Seidl, Chap. 5; Albanese and Senesi, Chap. 18), inclusion of the unique data associated with long-COVID (conditions that will have long-term impacts on health services everywhere), and tracking of vaccination status (dosage, boosters).

The landscape of data generation has changed with integrated data providing citizens with new applications. Using smart phones, contact tracing, and exposure notification will impact behaviors. The COVID risk calculator can be used similarly as the weather app where one gauges the risk of going out in a storm – one can determine the risk of being exposed to disease, for example, deciding where to go and when based on location, size of the event, indoor or outdoor venue, and community COVID rates (Kruger, 2022). The development of such tools will need to ensure that people's rights, health, and safety are protected and at the same time enable health equity (Landau, 2021).

Building such datasets needs conditions that support collaboration and cooperation. The speed at which the vaccine was developed demonstrates the success of collaborative initiatives where industry, science, and government worked together to create an unprecedented vaccine rollout (Gil, 2021). However, access to the vaccine remains skewed towards richer nations and vaccine production is limited due to profit motives and intellectual property rights. Efforts to address vaccine equity include COVAX sponsored by the World Health Organization to provide no-cost vaccines to low- and middle-income countries, or the need to practice vaccine diplomacy – where countries that have developed vaccines make them available to countries that lack them (Groopman, 2021). Other collaborative activities are suggested that include the Science Readiness Reserves – partnerships on global and multidisciplinary scale between industry, academia, and government. Transformative science is needed to enable these efforts and the US National Science Foundation has been at the forefront in creating innovative spaces (Pierce and Shores, Chap. 22).

The book is composed of a series of case studies examining geography and geospatial approaches to understanding the pandemic. Using methodologies grounded in science, statistics, and modeling and gleaning insights from the emerging COVID-19 literature, we present a global view of the pandemic from local perspectives (Cline and Laituri, Chap. 6). A common theme throughout the case studies is the role of inequality and how spatial analysis and the resulting visualizations (maps, graphs, and infographics) can demonstrate the landscape of a consequential geography through a lens of spatial justice (Richardson, Chap. 3; Kim et al., Chap. 4). Our case studies use multiple statistical methods (Li et al., Chap. 13; Mena et al., Chap. 12), modeling (Huang et al., Chap. 8), and data collection techniques (surveys and apps – Karki et al., Chap. 10; Mbonglou and John, Chap. 11). Spatial results include multiple data dashboards, portals (Mongolia), and analyses (which provide a rich context for examining the pandemic using a geospatial lens). Most specifically, the case studies cover maps and demonstrate geographic concepts that are critical for managing the pandemic – adjacency, connectivity, proximity, scale, diffusion, space, and place (Price, Chap. 1).

References

Ahmen, N., Marriott, A., Dabi, N., Lowthers, M., Lawson, M., & Jugehera, L. (2022, January). Inequality kills: The unparalleled action needed to combat unprecedented inequality in the wake of COVID-19. OXFAM Briefing Paper. https://doi.org/10.21201/2022.8465. https://www.oxfam.org/en/research/inequality-kills

Brown, F., & Blanc, J. (2020, April 14). *Coronavirus in conflict zones: A sobering landscape*. Carnegie Endowment for International Peace. https://carnegieendowment.org/2020/04/14/coronavirus-in-conflict-zones-sobering-landscape-pub-81518

Centers for Disease Control. (2022). *One health*. https://www.cdc.gov/onehealth/basics/index.html

Courtin, E., & Vineis, P. (2021, September 9). COVID-19 as a syndemic. *Frontiers in Public Health*. https://doi.org/10.3389/fpubh.2021.763830

Economist Intelligence Unit. (2021). *COVID-19 and migrant remittances: A hidden crisis looming?* https://www.eiu.com/n/campaigns/covid-migrant-remittances/

Gil, D. (2021). *COVID-19 a year later: What have we learned?* https://www.ibm.com/blogs/research/2021/01/covid-19-a-year-later/

Groopman, J. (2021, April 5). Beyond the vaccine. *The New Yorker*. pp. 55–59.

Gruetzmacher, K., Karesh, W., Amuasi, J., Arshad, A., Farlow, A., Gabrysch, S., Jetzkowitz, J., Lieberman, S., Palmer, C., Winkler, A., & Walzer, C. (2020). The Berlin principles on one health – Bridging global health and conservation. *Science of the Total Environment, 764*(2021), 142919.

Guterres, A. (2020). *United Nations: Ceasefire*. https://www.un.org/en/globalceasefire. Accessed 2 Mar 2022.

Hendron, S. (2021, July 18) A free tool to improve cities. *The New York Times*. Opinion, p.10.

Kim, J., Hagen, E., Muindi, Z., Mbonglou, G., & Laituri, M. (2022). An examination of water, sanitation, and hygiene (WASH) accessibility and opportunity in urban informal settlements during the COVID-19 pandemic: Evidence from Nairobi, Kenya. *Science of the Total Environment, 823*(9), 153398. https://doi.org/10.1016/j.scitotenv.2022.153398

Kishi, R. (2021, April). A year of COVID-19: The pandemic's impact on global conflict and demonstration trends. *The Armed Conflict & Event Data Project (ACLED)*. https://acleddata.com/2021/04/01/a-year-of-covid-19-the-pandemics-impact-on-global-conflict-and-demonstration-trends/

Kruger, A. (2022, January 2). Before heading out, calculate your risk. *The New York Times*, p. 6.

Landau, S. (2021). Digital exposure tools: Design for privacy, efficacy, and equity. *Science, 373*(6560), 1202–1204.

Martinez, L., & Short, J. (2021). The Pandemic City: Urban issues in the time of COVID-19. *Sustainability, 13*, 3295. https://doi.org/10.3390/su1363295

Nivette, A., Zahnow, R., Aguilar, R., et al. (2021). A global analysis of the impact of COVID-19 stay-at-home restrictions on crime. *Nature Human Behavior, 5*(July), 868–877. https://doi.org/10.1038/s41562-021-01139z

Reaser, J., Witt, A., Tabor, G., Hudson, P., & Plowright, R. (2021). Ecological countermeasures for preventing zoonotic disease outbreaks: When ecological restoration is a human health imperative. *Restoration Ecology* – Strategic Issues Article. https://doi.org/10.1111/rec.13357.

Resilient Cities Catalyst. (2022). *The urban century*. https://www.rcc.city/the-urban-century

Shah, S. (2022, January 23). The breeding ground. *The New York Times Magazine*, pp. 18–23.

Sheely, R., Aron, G., Lichtenheld, A., Kelly, D., Hakiman, K., & Fane, M. (2021, June). *A Clash of Contagions: The impact of COVID-19 on conflict in Nigeria, Colombia, and Afghanistan*. https://www.mercycorps.org/research-resources/clash-of-contagions-impact-covid-19-conflict
UN Department of Economic and Social Affairs. (2022). *Goal 16*. https://sdgs.un.org/goals/goal16. Accessed 2 Mar 2022.
UN Press Release. (2021, January 25). *Risk of instability, tension growing, amid glaring inequalities in global COIVD-19 recovery, to United Nations officials warn security council*. https://www.un.org/press/en/2021/sc14422.doc.htm#:~:text=The%20sweeping%20and%20devastating%20effects,impact%20of%20the%20coronavirus%20outbreak
Whalen, E. (2021, September 6–13) The man who saw it coming. *The Nation*, pp. 14–19.
World Justice Project. (2020, October). *The COVID-19 pandemic and the global justice gap*. https://worldjusticeproject.org/our-work/publications/policy-briefs/covid-19-pandemic-and-global-justice-gap

Correction to: Geography of the Pandemic

Melinda Laituri

Correction to:
Chapter 2 in: M. Laituri et al. (eds.), *The Geographies of COVID-19*, Global Perspectives on Health Geography, https://doi.org/10.1007/978-3-031-11775-6_2

The original version of the chapter "Geography of the Pandemic" was inadvertently published with a few errors. These changes have now been included and the chapter and the book have been updated with the changes.

The updated original version of this chapter can be found at
https://doi.org/10.1007/978-3-031-11775-6_2

M. Laituri et al. (eds.), *The Geographies of COVID-19*, Global Perspectives on Health Geography, https://doi.org/10.1007/978-3-031-11775-6_24

Glossary: The Lexicon of the Pandemic Through Geographic and Geospatial Terms

Contact tracing Contact tracing is the process of attempting to identify people who have recently been in contact with someone diagnosed with an infectious disease, especially in order to treat or quarantine them.

Coronavirus Coronaviruses are a family (*Coronaviridae*) of large single-stranded RNA viruses that can cause illnesses such as the common cold, severe acute respiratory syndrome (SARS), and Middle East respiratory syndrome (MERS). These viruses can infect people and some animals.

COVID-19 Coronavirus disease (COVID-19) is a highly contagious respiratory disease caused by the SARS-CoV-2 virus. The most common signs and symptoms of COVID-19 are fever, cough, and trouble breathing. Fatigue, muscle pain, chills, headache, sore throat, runny nose, nausea or vomiting, diarrhea, and a loss of taste or smell may also occur. Some people may not have any symptoms but are still able to spread the virus. Most people with COVID-19 recover without needing special treatment. But other people are at higher risk of serious illness, including older adults and people with serious medical problems, such as heart, lung, or kidney disease, diabetes, cancer, or a weak immune system. Serious illness may include life-threatening pneumonia and organ failure.

COVID-19 pandemic The COVID-19 disease was first identified in the Chinese city of Wuhan in December 2019. A lockdown in Wuhan and other cities in the surrounding Hubei Province failed to contain the outbreak, and it quickly spread to other parts of the People's Republic of China and around the world. The World Health Organization (WHO) declared the SARS-CoV-2 virus and the related COVID-19 disease a Public Health Emergency of International Concern in January 2020 and a pandemic in March 2020.

Community spread People infected by the virus in an area based on local conditions.

Data dashboards An online tool or platform used to share, display, analyze, and download data.

M. Laituri et al. (eds.), *The Geographies of COVID-19*, Global Perspectives on Health Geography, https://doi.org/10.1007/978-3-031-11775-6

Epidemic An epidemic is an outbreak of disease that spreads quickly and affects a disproportionately large number of individuals within a population, community, or region at the same time. An epidemic is often characterized by an unexpected increase in the number of disease cases in a specific geographical area.

First-order impacts The first-order impacts of the COVID-19 pandemic refer to the direct effects of the spread of the disease, such as the number of cases, hospitalizations, and deaths in a given region.

Flatten the curve A phrase from public health strategy aimed at slowing the spread of the virus through protective measures where fewer people get sick over time. The curve is a visual representation of the number of infected people over time.

Geographic Information Systems (GIS) Platform of related tools to store, manage, analyze, visualize, and share geospatial data.

Geospatial Geospatial is a term relating to or denoting data that is associated with a particular location. The term refers to data consisting of, derived from, or relating to data that is linked to specific geographical locations.

mRNA vaccine Modified RNA (mRNA) vaccines use modified mRNA formulated in lipid nanoparticles, or LNPs, that protect the mRNA and shuttle it into cells that create proteins from mRNA.

Pandemic A pandemic is an outbreak of a disease that occurs over a wide geographic area (such as multiple countries or continents) and typically affects a considerable proportion of the population. A pandemic is often characterized by sudden rapid spread, growth, or development.

Personal protective equipment (PPE) Masks, gloves, and medical coverings to reduce exposure to hazardous conditions.

Remote learning Teaching and learning through the use of technology and virtual platforms (i.e., video conferencing, discussion boards).

Remote sensing Detecting and monitoring physical characteristics of a location by measuring reflection or radiation from a distance (i.e., satellite imagery).

RNA Ribonucleic acid (RNA) is a polymeric molecule essential in various biological roles, including decoding genetic information of viruses such as SARS-CoV-2.

RNA virus RNA virus is a virus which has RNA (ribonucleic acid) as its genetic material. The nucleic acid is usually single-stranded RNA (ssRNA), but it may be double-stranded RNA (dsRNA). Notable human diseases caused by RNA viruses include the common cold, influenza, SARS, MERS, COVID-19, dengue virus, hepatitis C, hepatitis E, West Nile fever, Ebola virus disease, rabies, polio, mumps, and measles.

SARS-CoV-2 The severe acute respiratory syndrome coronavirus 2 (SARS-CoV-2) is the virus that causes a respiratory disease called coronavirus disease 2019 (COVID-19). SARS-CoV-2 is a member of a large family of viruses called coronaviruses. SARS-CoV-2 was first known to infect people in 2019. The virus can spread from an infected person's mouth or nose in small liquid particles when they cough, sneeze, speak, sing, or breathe. These particles range from larger respiratory droplets to smaller aerosols. SARS-CoV-2 is the seventh coronavirus known to infect humans.

Second-order impacts The second-order impacts of the COVID-19 pandemic refer to the secondary or indirect effects of the spread of the disease on various sectors of human life, such as economic, social, and environmental impacts.

Social distancing To prevent spread of disease by maintaining physical distance in places where people gather.

Spillover Microbes from animals spread to humans

Super-spreader event Gatherings of large numbers of people where a single infected person can induce a large outbreak.

Telecommuting/remote work Working from home or remote work environments to carry out job duties and responsibilities.

World Health Organization The World Health Organization (WHO) is the United Nations agency responsible for international public health. The WHO connects nations, partners, and people to promote health, keep the world safe, and serve the vulnerable. The WHO has played a leading role in several public health achievements, most notably the eradication of smallpox, the near-eradication of polio, and the development of an Ebola vaccine.

Zoonotic disease Infectious disease transmitted between species from animals to humans.

Index

M. Laituri et al. (eds.), *The Geographies of COVID-19*, Global Perspectives on Health Geography, https://doi.org/10.1007/978-3-031-11775-6

Printed in the United States
by Baker & Taylor Publisher Services